Lecture Notes in Artificial Intelligence   4850

Edited by J. G. Carbonell and J. Siekmann

Subseries of Lecture Notes in Computer Science

W0036758

Max Lungarella   Fumiya Iida
Josh Bongard   Rolf Pfeifer (Eds.)

50 Years
of Artificial
Intelligence

Essays Dedicated to the 50th Anniversary
of Artificial Intelligence

 Springer

Series Editors

Jaime G. Carbonell, Carnegie Mellon University, Pittsburgh, PA, USA
Jörg Siekmann, University of Saarland, Saarbrücken, Germany

Volume Editors

Max Lungarella
Rolf Pfeifer
University of Zurich
Artificial Intelligence Laboratory
Andreasstrasse 15, 8050 Zurich, Switzerland
E-mail: {lunga,pfeifer}@ifi.uzh.ch

Fumiya Iida
Massachusetts Institute of Technology
Robot Locomotion Group Computer Science
and Artificial Intelligence Laboratory
32 Vassar Street, Cambridge, MA 02139, USA
E-mail: iida@csail.mit.edu

Josh Bongard
University of Vermont
Department of Computer Science
329 Votey Hall, Burlington, VT 05405, USA
E-mail: j.bongard@uvm.edu

The illustration appearing on the cover of this book is the work of Daniel Rozenberg
(DADARA).

Library of Congress Control Number: 2007941079

CR Subject Classification (1998): I.2, H.3-5, H.2.8, F.2.2, I.6

LNCS Sublibrary: SL 7 – Artificial Intelligence

ISSN   0302-9743
ISBN-10 3-540-77295-2 Springer Berlin Heidelberg New York
ISBN-13 978-3-540-77295-8 Springer Berlin Heidelberg New York

Springer is a part of Springer Science+Business Media

springer.com

© Springer-Verlag Berlin Heidelberg 2007
Printed in Germany

Typesetting: Camera-ready by author, data conversion by Scientific Publishing Services, Chennai, India
Printed on acid-free paper   SPIN: 12204854   06/3180   5 4 3 2 1 0

Preface

Half a century ago, at the now famous 1956 Dartmouth Conference, the "fathers" of Artificial Intelligence (AI) – among them John McCarthy, Marvin Minsky, Allen Newell, Claude Shannon, Herbert Simon, Oliver Selfridge, and Ray Solomonoff – convened under the premise "that every aspect of learning or any other feature of intelligence can in principle be so precisely described that a machine can be made to simulate it." Fifty years have passed, and AI has turned into an important field whose influence on our daily lives can hardly be overestimated. Many specialized AI systems exist that are at work in our cars, in our laptop computers, and in our personal and commercial technologies. There is no doubt that the impact of AI on our lives in the future will become even more general and ubiquitous.

In this book we provide a representative collection of papers written by the leading researchers in the field of Artificial Intelligence. All of the authors of papers in this volume attended the 50th Anniversary Summit of AI (http://www.ai50.org), held at the Centro Stefano Franscini, Monte Verità, Ascona, Switzerland, July 9–14, 2006. The objective of the summit was fourfold: (1) to celebrate the 50th anniversary of AI as a discipline; (2) to look back and assess the field of AI (what has been done, and where we are); (3) to bring together people with different backgrounds (to enhance interaction between groups and foster future collaborations); and (4) to attract young and talented researchers to generate additional momentum in this exciting field. The AI summit combined discussions from a historical standpoint; scientific exchange on the state of the art; speculations about the future; business, political and educational perspectives; contributions by researchers from different but related areas; presentations of the latest research by top scientists in the field; as well as many informal discussions among the participants and visitors. In this volume, we have tried to maintain the breadth of topics presented and discussed at the summit by including chapters focusing on subjects ranging from the history and prospects of AI, to speech recognition and processing, linguistics, bionics, and consciousness.

We would like to thank all the participants of the summit for helping to make it a successful event, the authors for their contributions to this volume, and the reviewers. We would also like to express our gratitude to the Centro Stefano Franscini, Neuronics AG, Swisscom Innovations, Matek, and Migros Kulturprozent for their support.

September 2007

Max Lungarella
Fumiya Iida
Josh C. Bongard
Rolf Pfeifer

Table of Contents

Morphology and Dynamics

Neurorobotics

Machine Intelligence, Cognition, and Natural Language Processing

Human-Like Intelligence: Motivation, Emotions, and Consciousness

Robot Platforms

Art and AI

AI in the 21st Century – With Historical Reflections

Max Lungarella[1], Fumiya Iida[2], Josh C. Bongard[3], and Rolf Pfeifer[1]

[1] Dept. of Informatics, University of Zurich, Switzerland
[2] Computer Science and Artificial Intelligence Lab, MIT, USA
[3] Dept. of Computer Science, University of Vermont, USA
lunga@ifi.uzh.ch, iida@csail.mit.edu, josh.bongard@uvm.edu,
pfeifer@ifi.uzh.ch

Abstract. The discipline of Artificial Intelligence (AI) was born in the summer of 1956 at Dartmouth College in Hanover, New Hampshire. Half of a century has passed, and AI has turned into an important field whose influence on our daily lives can hardly be overestimated. The original view of intelligence as a computer program – a set of algorithms to process symbols – has led to many useful applications now found in internet search engines, voice recognition software, cars, home appliances, and consumer electronics, but it has not yet contributed significantly to our understanding of natural forms of intelligence. Since the 1980s, AI has expanded into a broader study of the interaction between the body, brain, and environment, and how intelligence emerges from such interaction. This advent of embodiment has provided an entirely new way of thinking that goes well beyond artificial intelligence proper, to include the study of intelligent action in agents other than organisms or robots. For example, it supplies powerful metaphors for viewing corporations, groups of agents, and networked embedded devices as intelligent and adaptive systems acting in highly uncertain and unpredictable environments. In addition to giving us a novel outlook on information technology in general, this broader view of AI also offers unexpected perspectives into how to think about ourselves and the world around us. In this chapter, we briefly review the turbulent history of AI research, point to some of its current trends, and to challenges that the AI of the 21st century will have to face.

1 Introduction

For a long time, humans have been romanced by the idea of understanding how thinking works, or how to construct intelligent machines and replicate the intelligent behavior displayed by many natural systems. Traditional Jewish mysticism, for instance, includes tales of the Golem, a thinking automaton made from the sticky clay of the bank of the river Moldau. In the 17th century, philosopher Gottfried Wilhelm von Leibniz outlined plans for a thinking machine by conceiving an artificial universal language composed of symbols, which could stand for objects or concepts, and logical rules for their manipulation. A little more than half a century ago (when Norbert Wiener was devising Cybernetics [1] and Gray Walter was building robotic tortoises [2], the English mathematician Alan Turing proposed a much-discussed imitation game used as a yardstick for assessing if a machine is intelligent or not, which since then has been known as the Turing Test for artificial intelligence [3, 4].

M. Lungarella et al. (Eds.): 50 Years of AI, Festschrift, LNAI 4850, pp. 1–8, 2007.

The advent of the general-purpose computer in the fifties of last century did substantially alter the dreams and ambitions of people and made AI an even more tangible possibility. So, in 1956, the fathers of "modern" AI, Marvin Minsky, John McCarthy, Allen Newell, Nathaniel Rochester, Claude Shannon, and Herbert Simon came together for a summer school at Dartmouth College (Hanover, New Hampshire) under the premise "that every aspect of learning or any other feature of intelligence can in principle be so precisely described that a machine can be made to simulate it" [5]. That date can be considered the birth of AI because it is only thereafter that numerous research groups around the world began to engage in the construction of artificial systems with the professed goal of emulating, equaling, or even surpassing human mental and physical abilities (a different view on the history of AI can be found in [6]). The many attempts to synthesize intelligence or higher cognition (by formalizing knowledge and crystallizing cognitive principles obtained from the study of human beings) have resulted in many specialized AI systems that nowadays are at work in specific problem domains, such as knowledge discovery and data mining (for example, for the identification of customer profiles), fuzzy logic, probabilistic reasoning, and artificial neural networks (as used in intelligent control for robots, home appliances and consumer electronics), evolutionary optimization (for the design of antennae, circuits, and turbines), as well as statistical and machine learning (often employed in genomics and proteomics).

Although, as these examples show, AI has proven successful in many cases, it is clear that the original goals set out by the first generation of AI visionaries have not yet been reached. The holy grail of AI, the creation of general (human-like) machine intelligence (think of HAL from Stanley Kubrick's *2001* or *Bladerunner*'s Roy Batty), has remained elusive. Natural intelligence is far from being understood and artificial forms of intelligence are still so much more primitive than natural ones. Seemingly simple tasks like object manipulation and recognition – tasks easily solved by a 3-year-old – have not yet been realized artificially. A look at the current research landscape reveals how little we know about how biological brains achieve their remarkable functionalities, how these functionalities develop in the child, or how they have arisen in the course of evolution. Also, we do not understand the cultural and social processes that have helped shape human intelligence. Because basic theories of natural intelligence are lacking and – despite impressive advances – the required technologies for building sophisticated artificial systems are yet not available, it is not surprising that the capabilities of current robots fall far short of even very simple animals. The inevitable conclusion then is that something important must have been missed or still needs to be discovered [7]. There is still no overall theory, no framework, explaining what thinking is and how to build intelligent machines.

Although we do not have any clear cut answer as to what aspects of natural intelligence are key for creating artificial intelligence, we will point at some avenues that might be worth pursuing in realizing the dreams of our ancestors.

2 Trends

The field of artificial intelligence has changed dramatically over the last 50 years. Back in the 1950s, the brain was conceptualized as some kind of powerful computer, and intelligence was thought to be the result of a symbol-crunching computer program located somewhere in our brain (this perspective is also known as the "brain as a computer metaphor" or the "information processing view"; [8, 9]). Since then, the by now "classical", computational, or cognitivistic approach has grown into a large discipline with many facets, and has achieved many successes in applied computer science and engineering. Think, for instance, of the victory of IBM's chess-playing supercomputer (Deep Blue) on reigning world chess champion, Garry Kasparov, on May 1997. While this early view of intelligence is probably adequate in formal or computational domains like chess playing or logical reasoning, in the past 20 years it has become increasingly evident that conceptualizing intelligence as a mere computational process cannot explain natural, adaptive forms of intelligence. The latter kind of intelligence requires a complete physical organism interacting with the real world: in other words, intelligence requires a body [10-15].

The new approach to understanding intelligence has led to a paradigm shift which emphasizes the physical and information-theoretical implications of embodied adaptive behavior, as opposed to the disembodied view of the computational framework [16-19]. The focus in this new paradigm is on systems acting in the real, physical and social world [20]. The implications of this change in perspective are far-reaching and can hardly be overestimated. With the fundamental paradigm shift from a computational to an embodied perspective, the kinds of research areas, theoretical and engineering issues, and the disciplines involved in AI have also changed substantially. The research effort in the field, for instance, has shifted towards understanding the lower level mechanisms and processes underlying intelligent behavior, as well as realizing higher forms of intelligence by first starting with simpler ones [21]. Cognition and action are viewed as the result of emergence and development rather than something that can be directly built (i.e. programmed) into a robot (such research draws heavily on insights from child development; e.g. [17, 22-24]. Automated design methods using ideas borrowed from biological evolution and ontogeny have also provided novel insights into the general nature of intelligence [25].

Physical agents in the real world, whether biological or artificial, are highly complex, and their investigation requires the cooperation of researchers from many different fields. In terms of research disciplines contributing to AI, we observe that in the classical approach computer science, psychology, philosophy, and linguistics played major roles, whereas in the embodied approach, computer science and philosophy [26] are now complemented by robotics, biomechanics [17, 27], material science and biology [28] and neuroscience [29]. The shift from high-level considerations (as raised in psychology and linguistics) to lower level sensory-motor processes – with the neurosciences now offering explanations at both the sensory-motor and cognitive levels of intelligence – is evident [30, 31]. There have also been changes concerning the notions used for describing the research area: a significant number of researchers dealing with embodiment no longer refer to themselves as working in artificial intelligence but rather in robotics, engineering of adaptive systems, artificial life, adaptive locomotion, bio-inspired systems, or neuroinformatics. But more than that, not only have researchers in artificial intelligence migrated into neighboring fields, but researchers in these neighboring fields have also started to contribute to artificial intelligence [29, 32]. This brings us to the issues of cross-disciplinarity and cross-fertilization.

3 Cross-Disciplinarity

Although most researchers in AI today are still working within the boundaries of the field as it was defined 50 years ago, the trend seems to be an expansion of AI into a broader study of the interaction between body, brain, and environment, and how intelligence emerges from such interaction [18, 33]. It has become evident that many fields (linguistics, cognitive sciences, robotics, material science, neuroscience, and morphogenesis) are highly relevant in order to advance the state of the art. The community is interested in locomotion, manipulation, and, in general, how an agent can act successfully in a dynamic and complex world. Concepts such as embodiment, autonomy, situatedness, interactive emergence, development, semiotic dynamics, and social interaction have come to the forefront and have spawned a spectrum of new and interdisciplinary approaches to the study of natural and artificial intelligence. There are important implications of the embodied view of intelligence for science, engineering, economics, and society at large. Here, we only mention but a few.

One of the important insights is that intelligence is not a "box" sitting inside the brain, but is distributed throughout the organism and requires the organism to interact with and explore its local environment. With respect to engineering, it has been demonstrated that systems which are not centrally controlled or do not feature a strict hierarchical structure tend to be more robust and adaptive against external perturbations [34]. It is important to realize that the underlying concepts might be directly mapped onto robot design (e.g. for de-mining operations, waste cleanup in hazardous environments, environmental exploration, service in hospitals and homes, and so on) and onto the design of embedded systems (i.e. systems equipped with sensors and motors which interact autonomously and continuously with their surrounding environment). The application of concepts from embodied AI in embedded systems could have profound economical and technological implications, because such systems are extremely widespread: they are used, for instance, in fuel injection devices, water purification plants, air conditioning systems, remote monitoring and control systems, as well as in many systems designed for human computer interaction. Finally, companies, societies and political bodies with an organizational structure based on local interactions and self-organization have been shown to react more robustly to unpredictable market forces than hierarchically organized ones (a good example of this is the financial difficulties currently experienced by many major airlines, compared to the proliferation and success of smaller, regional airlines).

4 Challenges and Outlook

Niels Bohr once famously quipped: "It's hard to predict, especially the future". Similarly, it is difficult to make predictions, especially about the future of artificial intelligence. However, after 50 years of explorations into artificially intelligent systems (and many false predictions), we have learned enough about the nature of intelligence to venture plausible guesses as to where AI could go as a field and which challenges it might face in the coming decades. In this section, we list some of these challenges and point to chapters in this volume that discuss them.

It seems to be generally accepted nowadays that the behavior of any system physically embedded within a particular environment is not the mere outcome of an internal control structure (e.g. the central nervous system) but is also affected by the system's morphology and the its material properties . Increasing evidence suggests that a direct link exists between embodiment and information: coupled sensory-motor activity and body morphology induce statistical regularities in sensory input and within the control architecture and therefore enhance internal information processing [35]. Mechanisms need to be created for the embodied and embedded systems to collect relevant learning material on their own [36] and for learning to take place in an "ecological context" (i.e. with respect to the environment). These and related issues are addressed in the chapters by Pfeifer et al. [18] and Polani et al. [19]. Potential avenues of how to exploit the interaction between control structure and physical properties (morphology and material characteristics) are described in the chapters by Iida et al. [16] and Behkam and Sitti [18].

Another challenge will be to devise a systematic theory of intelligence [14]. As argued by Nehaniv et al. [17] and Polani et al. [19], a promising approach may have to place special emphasis on information theory as a descriptive and predictive framework linking morphology, perception, action, and neural control. It will also be important to understand how raw (uninterpreted) information from unknown sensors can be used by a developing embodied agent with no prior knowledge of its motor capabilities to bootstrap cognition through a process of autonomous self-structuring in response to a history of self-motivated interaction with a rich environment [18, 19, 37]. This is closely connected to the issue of self-motivation and curiosity which are discussed in the chapters by Kaplan and Oudeyer [22] and Lipson [25]. Research on general capacities such as creativity, curiosity, motivations, action selection, and prediction (i.e. the ability to foresee consequence of actions) will be necessary because, ideally, no tasks should be pre-specified to a robot, but only an internal "abstract" reward function, some core knowledge, or a set of basic motivational (or emotional) "drives." Robots need to be endowed with the capacity to explore which are the activities that are maximally fitted to their current capabilities. Robots controlled by some core motives will hopefully be able to autonomously function in their environments not only to fulfill predefined tasks but also to search for situations where learning happens efficiently.

In the same vein, it will be necessary to address the issue of how robots (and embodied agents in general) can give meaning to symbols and construct meaningful language systems (semiotic systems). A fascinating new insight – explored under the label of "semiotic dynamics" – is that such semiotic systems and the associated information structure may not be static entities, but are continuously invented and negotiated by groups of people or agents which use them for communication and information organization [9]. This quite naturally leads to investigation of the collective dynamics of groups of socially intelligent robots and systems [20].

A further challenge will be how to integrate artificial intelligence and other fields. For instance, Potter [29] argues that neuroscience and artificial intelligence could profit from each other and invites researchers in both fields to venture across the divide (e.g. tools developed in the context of neuroinformatics could be used to design and manage neural databases). Similarly, Steels [9] predicts that many of the insights gained in fields such as nonlinear dynamical systems theory or evolutionary biology may have a direct bearing on the foundations of artificial intelligence as well as on how communication systems and other forms of interaction self-organize. A closer (symbiotic) interaction between artificial and natural systems will also benefit rehabilitation engineering as shown by recent work on human-machine interaction [38, 39]. As envisioned by Fattori et al. [31], robotics and neurophysiology could meet in a research field where bioelectrical signals obtained by single or multi-electrode recordings can be used to drive a robotic device [40]. As a last challenge, one could envision situations where AI could be used in the context of computer-assisted human-human interaction technologies [41].

5 Epilogue

The reconsideration of brain and body as a fundamental unit, physically and informationally, as well as the emergence of a new quantitative framework that links the natural and artificial domains, has begun to produce new insights into the nature of intelligent systems. While much additional work is surely needed to arrive at or even approach a general theory of intelligence, the beginnings of a new synthesis are on the horizon. Perhaps, finally, we will come closer to understanding and building human-like intelligence. The superior adaptability of embodied, distributed systems has been acknowledged for a long time, but there is now theoretical evidence from artificial intelligence research and corroboration from computer simulations supporting this point. In summary, the ideas emerging from the modern, embodied view of artificial intelligence provide novel ways of approaching technological, social, and economic problems in the rapidly changing world of the 21st century.

References

1. Wiener, N.: Cybernetics or Control and Communication in the Animal and the Machine. MIT Press, Cambridge, MA (1948)
2. Walter, G.W.: An imitation of life. Scientific American 182(5), 42–45 (1950)
3. Turing, A.M.: Computing machinery and intelligence. Mind 59(236), 433–460 (1950)
4. Legg, S., Hutter, M.: Tests of machine intelligence. LNCS, vol. 4850. Springer, Heidelberg (2007)
5. McCorduck, P.: Machines Who Think, 2nd edn. A.K. Peters Ltd., Natick, MA (2003)
6. Schmidhuber, J.: 2006: Celebrating 75 years of AI – history and outlook: the next 25 years. LNCS, vol. 4850. Springer, Heidelberg (2007)
7. Brooks, R.A.: The relationship between matter and life. Nature 409, 409–411 (2001)
8. Nilsson, N.: The physical symbol system hypothesis: status and prospects. LNCS, vol. 4850. Springer, Heidelberg (2007)
9. Steels, L.: Fifty years of AI: from symbols to embodiment – and back. LNCS, vol. 4850. Springer, Heidelberg (2007)
10. Brooks, R.A.: New approaches to robotics. Science 253, 1227–1232 (1991)
11. Chiel, H., Beer, R.: The brain has a body: adaptive behavior emerges from interactions of nervous system, body, and environment. Trends in Neurosciences 20, 553–557 (1997)
12. Clark, A.: Being There – Putting Brain, Body, and World Together Again. MIT Press, Cambridge, MA (1997)
13. Pfeifer, R., Scheier, C.: Understanding Intelligence. MIT Press, Cambridge, MA (1999)
14. Pfeifer, R., Bongard, J.C.: How the Body Shapes the Way we Think – A New View of Intelligence. MIT Press, Cambridge, MA (2007)

15. Varela, F.J., Thompson, E., Rosch, E.: The Embodied Mind. MIT Press, Cambridge, MA (1991)
16. Iida, F., Pfeifer, R., Seyfarth, A.: AI in locomotion: challenges and perspectives of underactuated robots. LNCS, vol. 4850. Springer, Heidelberg (2007)
17. Nehaniv, C.L., Mirza, N.A., Olsson, L.: Development via information self-structuring of sensorimotor experience and interaction. LNCS, vol. 4850. Springer, Heidelberg (2007)
18. Pfeifer, R., Lungarella, M., Sporns, O., Kuniyoshi, Y.: On the information theoretic implications of embodiment – principles and methods. LNCS, vol. 4850. Springer, Heidelberg (2007)
19. Polani, D., Sporns, O., Lungarella, M.: How information and embodiment shape intelligent information processing. LNCS, vol. 4850. Springer, Heidelberg (2007)
20. Dautenhahn, K.: A paradigm shift in artificial intelligence: why social intelligence matters in the design and development of robots with human-like intelligence. LNCS, vol. 4850. Springer, Heidelberg (2007)
21. Huelse, M., Wischmann, S., Manoonpong, P., von Twickel, A., Pasemann, F.: Dynamical systems in the sensorimotor loop: on the interrelation between internal and external mechanisms of evolved robot behavior. LNCS, vol. 4850. Springer, Heidelberg (2007)
22. Kaplan, F., Oudeyer, P.-Y.: Intrinsically motivated machines. LNCS, vol. 4850. Springer, Heidelberg (2007)
23. Lungarella, M., Metta, G., Pfeifer, R., Sandini, G.: Developmental robotics: a survey. Connection Science 15(4), 151–190 (2003)
24. Lungarella, M.: Developmental robotics, Scholarpedia, p. 173175 (2007)
25. Lipson, H.: Curious and creative machines. LNCS, vol. 4850. Springer, Heidelberg (2007)
26. Froese, T.: On the role of AI in the ongoing paradigm shift within the cognitive sciences. LNCS, vol. 4850. Springer, Heidelberg (2007)
27. Boblan, I., Bannasch, R., Schulz, A., Schwenk, H.: A human-like robot torso ZAR5 with fluidic muscles: toward a common platform for embodied AI. LNCS, vol. 4850. Springer, Heidelberg (2007)
28. Behkam, B., Sitti, M.: Bacteria integrated swimming robots. LNCS, vol. 4850. Springer, Heidelberg (2007)
29. Potter, S.: What can AI get from neuroscience? LNCS, vol. 4850. Springer, Heidelberg (2007)
30. Arbib, M., Metta, G., van der Smagt, P.: Neurorobotics: from vision to action. In: Siciliano, B., Khatib, O. (eds.) Springer Handbook of Robotics – ch. 63, Springer, Berlin
31. Fattori, P., Breveglieri, R., Nicoletta, M., Maniadakis, M., Galletti, C.: Brain area V6A: a cognitive model for an embodied artificial intelligence. LNCS, vol. 4850. Springer, Heidelberg (2007)
32. O'Regan, K.J.: How to build consciousness into a robot: the sensorimotor approach. LNCS, vol. 4850. Springer, Heidelberg (2007)
33. Vernon, D., Furlong, D.: Philosophical foundations of enactive AI. LNCS, vol. 4850. Springer, Heidelberg (2007)
34. Kitano, H.: Systems biology: a brief overview. Science 295(2002), 1662–1664 (2002)
35. Lungarella, M., Sporns, O.: PLoS Comp. Biol., e144 (2006)
36. Bongard, J.C., Zykov, V., Lipson, H.: Resilient machines through continuous self-modeling. Science 314, 1118–1121 (2006)
37. Bonsignorio, F.P.: Preliminary considerations for a quantitative theory of networked embodied intelligence. LNCS, vol. 4850. Springer, Heidelberg (2007)
38. Hernandez-Arieta, A., Kato, R., Yu, W., Yokoi, H.: The man-machine interaction: FMRI study for an EMG prosthetic hand with biofeedback. LNCS, vol. 4850. Springer, Heidelberg (2007)

39. Clark, A.: Re-inventing ourselves: the plasticity of embodiment, sensing, and mind. J. of Medicine and Philosophy 32, 263–282 (2007)
40. Lebedev, M.A., Nicolelis, M.A.L.: Brain-machine interfaces: past, present, and future. Trends in Neurosciences 29, 536–546 (2006)
41. Weibel, A., Bernardin, K., Woelfel, M.: Computer-supported human-human multilingual communication. LNCS, vol. 4850. Springer, Heidelberg (2007)

The Physical Symbol System Hypothesis: Status and Prospects

Nils J. Nilsson

Stanford Artificial Intelligence Laboratory
Stanford University
nilsson@cs.stanford.edu
http://ai.stanford.edu/~nilsson

Abstract. I analyze some of the attacks against the Physical Symbol System Hypothesis—attacks based on the presumed need for symbol-grounding and non-symbolic processing for intelligent behavior and on the supposed non-computational and "mindless" aspects of brains.

The physical symbol system hypothesis (PSSH), first formulated by Newell and Simon in their Turing Award paper,[1] states that "a physical symbol system [such as a digital computer, for example] has the necessary and sufficient means for intelligent action." The hypothesis implies that computers, when we provide them with the appropriate symbol-processing programs, will be capable of intelligent action. It also implies, as Newell and Simon wrote, that "the symbolic behavior of man arises because he has the characteristics of a physical symbol system."

Newell and Simon admitted that

> The hypothesis could indeed be false. Intelligent behavior is not so easy to produce that any system will exhibit it willy-nilly. Indeed, there are people whose analyses lead them to conclude either on philosophical or on scientific grounds that the hypothesis is false. Scientifically, one can attack or defend it only by bringing forth empirical evidence about the natural world.

Indeed, many people have attacked the PSSH. Their arguments cluster around four main themes. One theme focuses on the presumption that computers can only manipulate meaningless symbols. Intelligence, some people claim, requires more than formal symbol manipulation; it requires some kind of connection to the environment through perception and action in order to "ground" the symbols and thereby give them meaning. Such connectedness is to be achieved through what some of its proponents call "embodiment." Intelligence requires a physical body that senses and acts and has experiences.

---

[1] Allen Newell and Herbert A. Simon, "Computer Science as Empirical Inquiry: Symbols and Search," *Communications of the ACM.* vol. 19, No. 3, pp. 113-126, March, 1976. Available online at:
http://www.rci.rutgers.edu/~cfs/472_html/AI_SEARCH/PSS/PSSH1.html

M. Lungarella et al. (Eds.): 50 Years of AI, Festschrift, LNAI 4850, pp. 9–17, 2007.

Some even claim that to have "*human*-level" intelligence, a machine must have a human-like body. For example, Hubert Dreyfus argues that:[2]

> ... to get a device (or devices) with human-like intelligence would require them to have a human-like being in the world, which would require them to have bodies more or less like ours, and social acculturation (i.e. a society) more or less like ours.

In order to avoid arguments about what kind of body (if any) might be required, I think discussions about this theme would be less confusing if, instead of being about bodies, they were about the need for "grounding" symbols in whatever environment the intelligence is to function. Such an environment might be either the actual physical world or simulated, artificial worlds containing other agents.

Another theme focuses on the presumption that much that underlies intelligent action, especially perception, involves non-symbolic (that is, analog signal) processing. Of course, any physical process can be simulated to any desired degree of accuracy on a symbol-manipulating computer, but an account of such a simulation in terms of symbols, instead of signals, can be unmanageably cumbersome.

The third theme, related to the second, comes from those who claim that "computation," as it is ordinarily understood, does not provide an appropriate model for intelligence. Some have even said that it is time "to do away with the computational metaphor that has been haunting AI for 50 years: the brain is not a computer!"[3] Intelligent behavior requires "brain-style" (not computational) mechanisms.

A fourth theme is based on the observation that much that appears to be intelligent behavior is really "mindless." Insects (especially colonies of insects) and even plants get along quite well in complex environments. Their adaptability and efficacious responses to challenging situations display a kind of intelligence even though they manipulate no symbols. Jordan Pollack extends this claim even to human intelligence. He has written "Most of what our minds are doing involves mindless chemical activity ..."[4]

In light of these attacks, where does the PSSH stand today? Manifestly, we have not yet mechanized human-level intelligence. Is this shortcoming the fault of relying on the PSSH and the approaches to AI that it encourages? Might we need to include, along with symbol manipulation, non-symbolic processing modules in order to produce intelligent behavior? Of course, it could just be that

---

[2] Quote taken from http://en.wikipedia.org/wiki/Hubert_Dreyfus. Dreyfus's point of view about all this is explained in: Hubert L. Dreyfus, "Why Heideggerian AI Failed And How Fixing It Would Require Making It More Heideggerian," a paper written in connection with being awarded the APA's Barwise Prize, 2006.

[3] From a description of the "50th Anniversary Summit of Artificial Intelligence" at http://www.ai50.org/

[4] Jordan B. Pollack, "Mindless Intelligence," *IEEE Intelligent Systems*, p. 55, May/June 2006.

mechanizing intelligence is so much more difficult than we ever imagined it to be that it's not surprising that we haven't done it yet regardless of the approaches we have tried.

Let's look first at the claim that the PSSH is based on manipulating formal (and thus meaningless) symbols and is false for that reason. John Searle, for example, has written:[5]

> What [a computer] does is manipulate formal symbols. The fact that the programmer and the interpreter of the computer output use the symbols to stand for objects in the world is totally beyond the scope of the computer. The computer, to repeat, has a syntax but not semantics.

Searle makes this claim as part of his argument that computers (viewed as symbol-processing systems) cannot be said "to understand" because the objects (in the world) that the symbols stand for are beyond their scope.

Rodney Brooks has also criticized the PSSH, and proposes (in supposed contrast) what he calls "nouvelle AI ... based on the physical grounding hypothesis. This hypothesis states that to build a system that is intelligent it is necessary to have its representations grounded in the physical world."[6]

Searle and Brooks both seem to have ignored an important part of the PSSH. According to Newell and Simon:[7]

> A physical symbol system is a machine that produces through time an evolving collection of symbol structures. Such a system exists in a world of objects wider than just these symbolic expressions themselves.

Regarding this "world of objects," a physical symbol system includes (in addition to its means for formal symbol manipulation) the ability to "designate."

Here is Newell and Simon's definition (my italics):

> "An expression [composed of symbols] designates an object if, given the expression, the system can either *affect the object itself* or *behave in ways dependent on the object.*"

We see that the designation aspect of the PSSH explicitly assumes that, whenever necessary, symbols will be grounded in objects in the environment through the perceptual and effector capabilities of a physical symbol system. Attacks on the PSSH based on its alleged disregard for symbol grounding miss this important point.

In any case, in many applications, it isn't clear that symbol grounding is needed. For example, the "knowledge" possessed by expert systems—expressed

---

[5] John R. Searle, "Minds, Brains, and Programs," *Behavioral and Brain Sciences*, 3(3), pp. 417-457, 1980. Available online at:
http://www.bbsonline.org/documents/a/00/00/04/84/bbs00000484-00/
bbs.searle2.html

[6] Rodney A. Brooks, "Elephants Don't Play Chess," *Robotics and Autonomous Systems*, 6, pp. 3-15, 1990. Available online at:
people.csail.mit.edu/brooks/papers/elephants.pdf

[7] Allen Newell and Herbert A. Simon, *op. cit.*

in symbolic form either as belief networks or as rules—has no direct connection to objects in the world, yet "formal symbol manipulation" of this knowledge delivers intelligent and useful conclusions. Admittedly, robots that perceive and act in real environments (as well as other systems that function in artificial, simulated environments) do need direct connections between some of their symbols and objects in their environments. Shakey most certainly had a body with sensors and effectors, but most of its processing was done by a physical symbol system.

Let's turn now to the second theme, namely that intelligent action requires non-symbolic processing. It is often claimed that much (if not most) of human intelligence is based on our ability to make rapid perceptual judgments using pattern recognition. We are not able to introspect about what underlies our abilities to recognize speech sounds, familiar faces and "body language," situations on a chess board, and other aspects of our environment that we "size-up" and act upon seemingly automatically. Because we cannot introspect about them, it is difficult to devise symbol-based rules for programming these tasks. Instead, we often use a variety of dynamical, statistical, and neural-network methods that are best explained as processing analog rather than discrete symbolic data.

Statistical and neural-network methods are quite familiar to AI researchers. The subject of dynamical systems, however, might not be. In an article in *The MIT Encyclopedia of Cognitive Science*, Tim van Gelder writes:[8]

> A dynamical system for current purposes is a set of quantitative variables changing continually, concurrently, and interdependently over quantitative time in accordance with dynamical laws described by some set of equations. Hand in hand with this first commitment goes the belief that dynamics provides the right tools for understanding cognitive processes.
> . . .
> A central insight of dynamical systems theory is that behavior can be understood geometrically, that is, as a matter of position and change of position in a space of possible overall states of the system. The behavior can then be described in terms of attractors, transients, stability, coupling, bifurcations, chaos, and so forth—features largely invisible from a classical perspective.

I grant the need for non-symbolic processes in some intelligent systems, but I think they supplement rather than replace symbol systems. I know of no examples of reasoning, understanding language, or generating complex plans that are best understood as being performed by systems using exclusively non-symbolic processes.[9] Mostly this supplementation occurs for those perceptual and motor

---

[8] T. J. van Gelder, "Dynamic Approaches to Cognition" in R. Wilson, and F. Keil (eds.), *The MIT Encyclopedia of Cognitive Sciences*, pp. 244-246, Cambridge MA: MIT Press, 1999. Available online at:
http://www.arts.unimelb.edu.au/~tgelder/papers/MITDyn.pdf

[9] In his article on dynamical systems, van Gelder writes "Currently, many aspects of cognition—e.g., story comprehension—are well beyond the reach of dynamical treatment."

activities that are in closest contact with the environment. This point has long been acknowledged by AI researchers as evidenced by the inclusion of "signal-to-symbol transformation" processes in several AI systems.[10]

Pandemonium, an early AI architecture proposed by Oliver Selfridge,[11] was non-commital about the symbolic versus non-symbolic distinction. Its hierarchically organized components, which Selfridge called "demons," could be instantiated either as performing non-symbolic or symbolic processes. In combination, his model would be a provocative proposal for a synthesis of those two processing methods.

Now, let's analyze the phrase "the brain is not a computer," which is the main point of the third theme of attacks against the PSSH. People who make this claim often stress distinctions like:

> Computers have perhaps hundreds of processing units whereas brains have trillions.
> Computers perform billions of operations per second whereas brains perform only thousands.
> Computers are subject to crashes whereas brains are fault tolerant.
> Computers use binary signals whereas brains work with analog ones.
> Computers perform serial operations whereas brains are massively parallel.
> Computers are programmed whereas brains learn.
> Etc.

Aside from the fact that many of these distinctions are no longer valid,[12] comparisons depend on what is meant by "the brain" and what is meant by "a computer." If our understanding of the brain is in terms of its component neurons, with their gazillions of axons, dendrites, and synaptic connections, and if our understanding of a computer is in terms of serial, "von Neumann-style" operation—reading, processing, and writing of bits—all accomplished by transistor circuitry, well then of course, the brain is not *that* kind of a computer. So what?

We don't understand "computation" (the metaphor we are being persuaded to abandon) by reference only to a low-level, von Neumann-style description. We can understand it at any one of a number of description levels. For example,

---

[10] P. Nii, E. Feigenbaum, J. Anton, and A. Rockmore, "Signal-to-Symbol Transformation: HASP/SIAP Case Study," *AI Magazine*, vol 3, Spring 1982.

[11] Oliver. G. Selfridge, "Pandemonium: A Paradigm for Learning," in D. V. Blake and A. M. Uttley, editors, *Proceedings of the Symposium on Mechanisation of Thought Processes*, pages 511-529, London: Her Majesty's Stationary Office, 1959.

[12] For example, a paper written in 2003 claimed that "Google's architecture features clusters of more than 15,000 commodity-class PCs with fault-tolerant software." Undoubtedly, Google uses many more networked computers today. See: Luiz André Barroso, Jeffrey Dean, and Urs Hölzle, "Web Search for a Planet: The Google Cluster Architecture," *IEEE Micro*, March-April, 2003. Available online at: http://labs.google.com/papers/googlecluster-ieee.pdf

computation might be understood as a collection of active recursive functions operating on symbolic list structures. Alternatively, it might be understood as parallel-operating "knowledge sources" reading from, transforming, and writing complex symbolic expressions on a "blackboard." Other possible computational models are a collection of symbol-processing Pandemonium demons, a "dynamic Bayes network" of symbolically-represented propositions,[13] or a loosely-coupled society of simple computational "agents."[14]

Perhaps our gradually increasing understanding of how the brain operates will lead to other useful computational models, such as the graphical models of the neo-cortex proposed by Hawkins; by Hinton, Osindero, and Teh; by Lee and Mumford; and by Dean.[15] Our ideas about what "computation" can be are ever expanding, so those who want to claim that the brain is not a computer will need to be more precise about just what kind of computer the brain is not.

Engineers have no difficulty using several levels of description and neither will brain scientists. Transistors and synapses are best understood and explained using the vocabularies of physics and chemistry. But database systems, for example, are best understood and programmed using higher-level computational concepts—which, by the way, had to be *invented* for those purposes. Similarly I predict, understanding how brains represent declarative knowledge, understand and generate language, and make and carry out plans will require levels of description higher than that of neural circuitry. And just as engineers already have a continuum of bridges connecting an explanation of how transistors work with an explanation of how computers perform database searches, brain scientists will eventually have bridges connecting their explanations of how neurons work with their yet-to-be perfected explanations of how brains carry out those processes we call intelligent.

There is already some exciting progress on developing symbol-based theories of brain operation and on connecting these theories with neural circuitry. For example, Randall C. O'Reilly, writes that the pre-frontal cortex "is critical for maintaining current context, goals, and other information in an active state that guides ongoing behavior in a coherent, task-relevant manner."[16] He even suggests that neural circuits protect against noise in the same way that computers do, namely by employing binary encoding, and that neural circuits are capable of "limited variable binding."

---

[13] Stuart Russell and Peter Norvig, *Artificial Intelligence: A Modern Approach*, Second Edition, Chapter 15, Upper Saddle River, New Jersey: Pearson Education, Inc., 2003.

[14] Marvin Minsky, *The Society of Mind*, New York: Simon and Schuster, 1985.

[15] Jeff Hawkins with Sandra Blakeslee, *On Intelligence*, New York: Times Books, 2004; G. Hinton, S. Osindero, and Y. Teh, "A Fast Learning Algorithm for Deep Belief Networks," *Neural Computation*, 2006, to appear; T. S. Lee and David Mumford, "Hierarchical Bayesian Inference in the Visual Cortex, *J. Opt. Soc. Am. A*, Vol. 20, No. 7, July 2003; Thomas Dean, "Computational Models of the Neocortex," online article at http://www.cs.brown.edu/people/tld/projects/cortex/.

[16] Randall C. O'Reilly, "Biologically Based Computational Models of High-Level Cognition," *Science*, vol. 314, pp. 91-94, October 6, 2006.

In a paper about certain brain sub-systems, Richard Granger writes: "Together the system produces incrementally constructed and selectively reinforced hierarchical representations consisting of nested sequences of clusters."[17] Granger has also told me that "even in brains, many of us find it appropriate to include symbol-processing levels of description (though I should note that the science is evolving, and there are still those who would disagree)."[18]

In his "neural theory" of how the brain understands language, Jerome Feldman employs such computational level, symbolic constructs as "schema," "feature," and "value." He writes, "There is convincing evidence that people organize their perceptions and actions in terms of features and values."[19] Feldman stresses the importance of connecting computational level descriptions in his theory to "key neural properties, including massive parallelism, robustness, spreading activation, context sensitivity, and adaptation and learning."[20]

No doubt AI research will benefit greatly from what computational neuroscientists and cognitive scientists learn about how the brain works. But I don't think it will involve abandoning the computational metaphor.

Now, what about the idea that intelligence is "mindless"? Several examples of mindless processes are cited by adherents of this view. Here are some cited by Jordan Pollack,[21] who coined the word "ectomental" to describe them: The process of evolution, proceeding by random changes in the genome and selective survival of organisms that result from the genome, produced intelligent humans. (But *producing* an intelligent system is different from *being* an intelligent system.) Reinforcement learning produced a neural network that plays better backgammon than human experts. (Pollack failed to note that the inputs to the neural network were symbolic features of the backgammon board and that the best performance was obtained in combination with limited-lookahead symbolic search.) The animal immune system can discriminate between self and non-self without "a central database listing which compounds are in or out." He concludes by writing that "dynamical processes, driven by accumulated data gathered through iterated and often random-seeming processes, can become more intelligent than a smart adult human, yet continue to operate on principles that don't rely on symbols and logical reasoning." So far, no such "dynamical processes" have produced systems that can prove theorems, make and execute plans, and summarize newspaper stories. And, when and if they ever do produce such systems, they will be best explained, I predict, as using "symbols and logical reasoning." Pollack's statement that "Most of what our minds are doing involves mindless chemical activity ..." is no more helpful than would be

---

[17] Richard Granger, "Essential Circuits of Cognition: The Brain's Basic Operations, Architecture, and Representations," in J. Moor and G. Cybenko (eds.), *AI at 50: The Future of Artificial Intelligence*, to be published. Available online at: http://www.dartmouth.edu/~rhg/pubs/RHGai50.pdf

[18] E-mail communication, October 4, 2006.

[19] Jerome A. Feldman, *From Molecules to Metaphor: A Neural Theory of Language*, p. 140, Cambridge, MA: The MIT Press, 2006.

[20] *Ibid*, p. 142.

[21] Jordan B. Pollack, *op. cit.*

a statement like "Most of what airline reservations systems are doing involves mindless electronic currents."

Rodney Brooks has achieved a great deal of success in using his "nouvelle AI" ideas to program rather simple (one is tempted to say mindless) "creatures." Most of his systems lack complex representations, even though his "physical grounding hypothesis" doesn't explicitly disallow them. Nevertheless, the behaviors of these creatures are quite impressive and are described in his "Elephants Don't Play Chess" paper. But the title of that paper belies the difficulty. They don't, do they? Brooks attempts to deflect such criticism by writing "it is unfair to claim that an elephant has no intelligence worth studying just because it does not play chess." But I don't claim that elephant "intelligence" is not worth studying. I only claim that, whatever it is, it isn't human-level intelligence, and I think more complex representations, symbolically manipulated, will be needed for that.

In summary, I don't think any of the four different kinds of attacks on the PSSH diminishes the importance of symbolic processing for achieving human-level intelligence. The first attack is based on the erroneous claim that the PSSH lacks symbol grounding. By all means, let's have symbol grounding when needed. The second attack is based on the need for non-symbolic processing; let's have that too when needed. The third attack, based on the claim that the brain is not a computer, will vanish when people who study brains increasingly use computational concepts to understand brain function. And the fourth attack, based on the idea that brains are mindless, will vanish when it becomes evident that constructs best understood as being mindless achieve only mindless behavior.

So what does all this have to say about the status of the PSSH? Some might say that the PSSH's claim that a physical symbol system is "sufficient" for intelligent action is weakened by acknowledging that non-symbolic processing might also be necessary. Newell, however, seemed not to be willing to concede that point. In a 1988 book chapter, he wrote:[22]

> ... the concept of symbols that has developed in computer science and AI over the years is not inadequate in some special way to deal with the external world."
>
> ...
>
> For example, such symbols are used as a matter of course by the Navlab autonomous vehicle (a van that drives itself around Schenley Park next to Carnegie-Mellon), which views the road in front of it through TV eyes and sonar ears, and controls the wheels and speed of the vehicle to navigate along the road between the trees ... The symbols that float everywhere through the computational innards of this system refer to the road, grass, and trees in an epistemologically adequate, though sometimes empirically inadequate, fashion. These symbols are the symbols of the physical symbol system hypothesis, pure and simple.

---

[22] Allen Newell "Putting It All Together," Chapter 15, of D. Klahr and K. Kotovsky (eds.), *The Impact of Herbert A. Simon*, Hillsdale, NJ: Erlbaum and Associates, 1988.

I'll leave it at that. For those who would rather think about the perception and action routines of Navlab (and of Shakey and Stanley) in terms of signals rather than symbols, the "sufficiency" part of the PSSH is clearly wrong. But the "necessity" part remains uncontested, I think.

What about the future prospects for physical symbol systems in AI? Brooks's "Elephant" paper makes a proposal:

> Traditional [that is, symbolic] AI has tried to demonstrate sophisticated reasoning in rather impoverished domains. The hope is that the ideas used will generalize to robust behavior in more complex domains.
>
> Nouvelle AI tries to demonstrate less sophisticated tasks operating robustly in noisy complex domains. The hope is that the ideas used will generalize to more sophisticated tasks.
>
> Thus the two approaches appear somewhat complementary. It is worth addressing the question of whether more power may be gotten by combining the two approaches.

Here is my prediction about the future of physical symbol systems in AI: They will take on a partner (as Brooks proposes). AI systems that achieve human-level intelligence will involve a combination of symbolic and non-symbolic processing—all implemented on computers, probably networks of computers. Which parts are regarded as symbolic and which parts non-symbolic will depend on choices of the most parsimonious vocabulary and the most useful programming constructs—which, after all, are intimately linked. We will find it most convenient to describe some parts with equations involving continuous and discrete numbers. And, those parts will correspondingly be programmed using operations on continuous and discrete numbers. We will find it most convenient to describe the higher level operations in terms of non-numeric symbols. And, those parts will correspondingly be programmed using symbol-processing operations.

In the not-too-distant future, I hope, the controversies discussed in this paper will be regarded as tempests in a teapot.

Acknowledgments

I would like to thank Jerome Feldman, Richard Granger, Max Lungarella, and Yoav Shoham for their helpful suggestions.

Fifty Years of AI: From Symbols to Embodiment - and Back

Luc Steels

Sony Computer Science Laboratory - Paris
VUB AI Lab, Vrije Universiteit Brussel

Abstract. There are many stories to tell about the first fifty years of AI. One story is about AI as one of the big forces of innovation in information technology. It is now forgotten that initially computers were just viewed as calculating machines. AI has moved that boundary, by projecting visions on what might be possible, and by building technologies to realise them. Another story is about the applications of AI. Knowledge systems were still a rarity in the late seventies but are now everywhere, delivered through the web. Knowledge systems routinely deal with financial and legal problem solving, diagnosis and maintenance of power plants and transportation networks, symbolic mathematics, scheduling, etc. The innovative aspects of search engines like Google are almost entirely based on the information extraction, data mining, semantic networks and machine learning techniques pioneered in AI. Popular games like SimCity are straightforward applications of multi-agent systems. Sophisticated language processing capacities are now routinely embedded in text processing systems like Microsoft's Word. Tens of millions of people use AI technology every day, often without knowing it or without wondering how these information systems can do all these things. In this essay I will focus however on another story: AI as a contributor to the scientific study of mind.

Keywords: history of AI, heuristic search, knowledge systems, behavior-based robotics, semiotic dynamics.

1 Introduction

There has been a tendency to compare the performance of today's machines with human intelligence, and this can only lead to disappointment. But biology is not regarded as a failure because it has not been able to genetically engineer human-like creatures from scratch. That is not the goal of biology, and neither should the construction of human-like robots or programs that pass the Turing test be seen as the goal of AI. I argue that the scientific goal of AI is not to build a machine that somehow behaves like an intelligent system (as Turing proposed originally) but to come up with an explanation how intelligence by physically embodied autonomous systems is possible and could have originated in living systems, like us. AI takes a design stance, investigating in principle by what mechanisms intelligence is achieved, just like aircraft engineers try to figure out by what mechanisms physical systems are

M. Lungarella et al. (Eds.): 50 Years of AI, Festschrift, LNAI 4850, pp. 18–28, 2007.
© Springer-Verlag Berlin Heidelberg 2007

able to fly. Whether the solutions we find are comparable or equal to those used by natural living systems like us is another issue about which AI itself has nothing to say. But one thing is sure, without an adequate operational theory of the mechanisms needed for intelligence, any discussion of human intelligence will always remain at the level of story telling and hand waiving, and such a theory cannot be conceived nor adequately tested only by empirical observation.

There has been a tendency to dismiss the AI achievements of the past, usually by those who are incapable to understand how they work and why they are important. For example, I think it is rather incredible that computer programs today play at grandmaster level. Not that long ago philosophers like Dreyfus were arguing that "computers would never be able to play chess" (Dreyfus, 1972). Now that everyone can see for themselves that they can (and Dreyfus lost the first game he ever played to a computer), critics argue that we can learn nothing about problem-solving or cognition from these systems. I believe these comments come from people who have never bothered to look at what is really behind game-playing programs and who have no notion of what the real issues are that problem solvers, whether natural or artificial, have to face.

I personally stumbled into AI in the early seventies through the door of (computational) linguistics, mainly inspired by the work of Terry Winograd (1972). I was interested in understanding the processes that could explain how language sentences are produced and interpreted but when building my first computer programs to test my hypotheses (using a computer which had less power than today's mobile telephones!), it quickly became clear that my initial intuitions were totally wrong and naive, and so were the intuitions of the psychologists and linguists I encountered, simply because none of us were capable to think in computational terms except as a vague metaphor. I subsequently changed my research methodology completely and never had any regrets about it. It is indeed in the methodology that lies the uniqueness of AI. There is no unique AI theory, nor AI view on intelligence. A multitude of views have been explored, each time making contact and in turn influencing other scientific fields in a big way. I do not think that this multitude will shrink to a single view any time soon and maybe this is not necessary. The richness of AI lies in its openness towards other disciplines and the power of the experimental approach for operationalising and testing mechanistic theories of intelligence.

In this essay, I try to reconstruct, very schematically and from a very personal view of course, some of the key ideas and milestones in the non-linear AI history of the past decades. New movements seem to have come in waves, lasting about seven years before being absorbed in mainstream thinking. Then I give my opinion on what I find the most promising trends in current AI activities and which direction we might want to take in the future.

2 Prologue (1956): The Information Processing View

AI started from a single powerful idea, clearly understood and expressed at the Dartmouth founding conference (McCarthy, et.al., 1955): The brain can be viewed as an information processor. A computer is an example of an information processor but the notion of information processing is much broader. We are looking at a physical system that has a large number of states which systematically correspond to information states, processes that transform these states systematically into others while preserving or implementing some (mathematical) function, and additional information states (programs) which regulate what processes should become active at what moments in time. From a biological point of view, neurons are cells that grow and live within larger networks, and biologists research the question how they can function and sustain themselves, similar to the way they study liver cells or the immune system. The information-processing view introduces a new perspective on the brain, neurons or networks of neurons are viewed as devices that can hold and transform information and so the question becomes: What kind of information is stored and what kind of transformations are carried out over this information. It is interesting that the switch from the biochemical, electrical point of view to the information processing point of view took place at the same time as other areas of biology, and particularly genetics and molecular and developmental biology, also started to emphasise information, viewing the genome as a program that is interpreted by the machinery of the cell. This view has become dominant today (Maynard Smith, 2000).

AI does not address the question of neural information processing directly but takes a design stance. It asks the question: What kind of information processing do living brains need in order to show the astonishing intelligent behavior that we actually observe in living organisms? For example, what information structures and processes are required to fetch from a stream of visual images the face of a person in a fraction of a second, what kind of information processing is necessary to parse a natural language sentence with all its ambiguities and complex syntactic structures and to interpret it in terms of a perceived scene, how is a plan formulated to achieve a complex collaborative action and how is this plan executed, monitored and adjusted? To investigate what kind of processes are in principle needed for intelligence makes sense for two reasons. First of all it is hard to imagine that we will be able to understand the functioning of enormously complex nervous systems with billions of elements and connections if we do not have tools to understand what they might be doing and if we can only do gross-scale measurements like fMRI or PET scanning. It is like trying to understand the structure and operation of computers that are performing millions of instructions per second and are linked with millions of other computers in giant networks like the Internet, without knowing what they might be doing and having only gross statistical measures at our disposal. Second, neural and psychological observations - brain imagining, brain disorders, effects from aging, etc. - can tell us that certain parts of the brain are involved in particular cognitive tasks, but they cannot tell us why certain types of processing are needed. To find that out, we must examine alternative mechanisms (even if they do not occur in nature) and study their impact on performance.

Information processing is a point of view and not an intrinsic property of a system. A computer can be built from anything, including TinkerToys, as Danny Hillis and Brian Silverman showed with their Tic-Tac-Toe playing computer at the MIT AI lab in the nineteen eighties, or with chemistry. So it is perfectly justified to use computers to study the behavior of natural information processing systems. It is also very straightforward to test whether a theory is valid or not. If someone has an idea what kind of information processing is needed for a task, an artificial system can be built that carries it out and anyone can observe whether it has the required performance.

The information processing view was a true revolution in the study of mind and it started to propagate rapidly after the first 1956 AI conference in Dartmouth, with initially a major impact on psychology (particularly through Jerome Bruner, George Miller, Alan Newell, and Herbert Simon) and linguistics. Once the information processing view is adopted, the next step is to do the hard work and investigate in great detail concrete examples. Initially it was thought that there might be some very general, simple principles, like Newton's equations, that could explain all of intelligence. Newell and Simon proposed a General Problem Solver, Rosenblatt proposed Perceptrons, etc., but gradually, as more and more tasks were being tackled, it became clear that intelligence requires massive amounts of highly complex sophisticated information and very adapted information processing. Quite early, Minsky concluded that the brain must be a "huge, branchy, quick-and-dirty kludge", as opposed to a cleanly designed general purpose machine. This is of course not unusual in living systems. After all the human body is also a huge quick-and-dirty kludge with its myriad of interlocking biochemical processes and ad hoc solutions that have a high probability of breaking down. Indeed, if you come to think of it it is amazing that our bodies work so well as they do.

Most AI outsiders (including many philosophers, neuroscientists and psychologists) still cannot believe that so much complexity is required to achieve intelligence. They keep hoping that there is some golden short-cut that makes the hard work of today's vision systems, which use large numbers of algorithms and massive computation to filter, segment, and aggregate the incoming visual streams and match them against sophisticated top-down expectations stimuli, superfluous. Or they believe that language processing cannot be complicated because our brains do it effortlessly at great speeds. This echoes the trouble that biologists had in the 19th century to accept that all of life was ultimately implemented in chemistry. They initially could not believe that you needed vastly complex macromolecules (or even that such molecules could exist in nature), that you had highly complex metabolic cycles, and that a lot of these cycles were self-organised and steered by other molecules. It took hundred years for molecular biology to be accepted as the physical foundation for life and perhaps we just need a bit more time before the complexity of information processing is fully accepted and everybody is focused on what it could be rather than arguing whether the information processing view is appropriate in the study of intelligence.

3 Act I (1960s): Heuristic Search and Knowledge Representation

One reason why a mechanism for "general intelligence", similar to gravity or magnetism, may for ever remain elusive, is that almost any non-trivial problem contains a hidden combinatorial explosion. This was in fact the main lesson from the early game-playing programs. Often it is not possible to decide straight away how to explore a problem and so many avenues need to be searched. For example, almost every word in a sentence has multiple meanings and can be syntactically categorised in a multitude of ways. Hence a parse that fits with the complete sentence can only be found by elaborating different search paths, some possibly being a dead end. The growth of alternative search paths in language processing can be so high that the number of possible paths explodes. It is no longer possible to consider all of them. Such a combinatorial explosion shows up in visual perception, motor planning, reasoning, expert problem solving, memory access, etc., in fact in every domain where intelligence comes into play. It is possible to combat combinatorial explosions with more computing power, but one of the key insights of AI in its first decade (the 60s) is that sooner or later the big numbers catch up on you and you run into a wall. So search can only be dealt with by the application of knowledge, specifically domain and task specific rules of thumb (heuristics) that quickly cut down search spaces and guide the problem solver towards reasonable solutions as fast as possible. The main goal of AI in the sixties hence became the development of techniques for organising the search process, expressing heuristics in evaluation functions or rules, and trying to learn heuristics as part of problem solving. For a while, the field of AI was even called Heuristic Search.

Towards the end of the 60s it became clear that not only the heuristics but also the way that information about the problem domain is represented plays a critical role, both to avoid search and to deal with the infinite variation and noise of the real world. So this lead to the next breakthrough idea: A change in representation can mean the difference between finding a solution or getting swamped in irrelevant details. A host of new research questions was born and explored through an amazing variety of frameworks for knowledge representation and knowledge processing. Many ideas from logic could be pulled into AI and AI gave new challenges to logic, such as the problem of non-monotonic reasoning. Also the kinds of concepts that might be needed in expert problem solving or language understanding were researched and tested by building systems that used them. Initially there was some hope to find the ultimate conceptual primitives (Roger Schank's set of fourteen primitive universal concepts is one example (Schank, 1975)), but, failing that, large projects were started, such as Doug Lenat's CYC project, to do a massive analysis of human concepts and put them in a machine-usable form (Lenat, 1995).

Already in the 70s, the ideas and technologies for handling heuristic search and knowledge representation proved powerful enough to build real applications, as seen in the first wave of expert systems like MYCIN, DENDRAL, or PROSPECTOR. They were designed to mimic expert problem solving in domains like medicine or engineering, and lead to a deeper analysis of problem solving in terms of knowledge level models (Steels and McDermott, 1993). Huge conferences, venture capital, spin-off companies, industrial exhibitions entered the scene and transformed AI forever. From the early 80s, applications became an integral part of AI. The emphasis on applications justified the research effort and proved that AI theory was on the right track, but it also meant that fundamental research slowed down. The initial emphasis on understanding intelligence gave way to a more pragmatic research agenda and a quest for useful applications. This trend is still going on today.

4 Act II (1980s): Neural Networks

The earliest efforts in AI (including the discussion at the Dartmouth conference) already had a kind of dual character. On the one hand, it was clearly necessary to come to grips with "symbolic" processing, and AI laboratories made major contributions to make computers and programming language sufficiently powerful to build and investigate hugely complex symbol systems. On the other hand, sensori-motor control and pattern recognition, needed to relate symbolic structures to the real world, seem to require "subsymbolic" processing. The gap between the symbolic world and the physical world had to be bridged somewhere. Early neural networks (like Rosenblatt's Perceptron) had already shown that some of the processing needed for this aspects of intelligence is better based on the propagation of continuous signals in networks with varying weights and thresholds instead of the transformation of symbolic expressions. However that does not mean that the information processing view is abandoned. Neural networks are just as much 'computational' mechanisms as heuristic search or unification (a core step in logical inference systems).

So there have always been two sides of AI, both developing in parallel. Research and early successes in symbolic processing dominated perhaps the first two decades of AI research and application, but the balance shifted in the 80s, with a renewed interest in neural networks and other forms of biologically inspired computation such as genetic algorithms. Significant advances were made throughout the 80s by studying more complex networks with "hidden" layers, or networks in which signals flow back in the network becoming recurrent, thus making it possible to deal with temporal structures (Elman, 1991). These advances lead to a steady stream of new applications in signal processing and pattern recognition, new technologies to make neural processing fast enough, and a huge impact on neuroscience, which finally began to adopt the information processing view in a serious way.

The field of computational neuroscience is now well established and it is studying what information processing is actually carried out by natural brains as well as offering new ideas to AI about what mechanisms might be needed for intelligence. The renaissance of neural network research does not make earlier work on "symbolic" AI irrelevant. Symbolic techniques remain the most adapted for studying the conceptually oriented aspects of intelligence, as in language processing or expert problem solving, and so far no adequate neural models have been proposed for language understanding, planning, or other areas in which symbolic AI excels. On the other hand, neural network techniques have proven their worth in the grounding of categories in the world or in the smooth interfacing of behavior with sensory stimuli. Intelligence is a big elephant and there is often a tendency by researchers to take the part they happen to focus on as the total. Obviously we need to avoid this trap.

5 Act III (1990s): Embodiment and Multi-agent Systems

But whatever the application successes of symbolic AI and neural networks, there was a growing feeling towards the end of the 80s that some fundamental things were missing, and two important new movements began to emerge. The first one rediscovered the body and the environment as major causal forces in the shaping of intelligent behavior (Steels and Brooks, 1994). Instead of doing very complex calculations for motor control, it was argued that it might just as well be possible to exploit the physical properties of materials and the agent-environment interaction to get smooth real-time behavior. Instead of trying to build complex 'symbolic' world models, which are difficult to extract reliably from real-world signals anyway, it might be possible to set up simple reactive behaviors and exploit the resulting "emergent" behavior. Ideas like this became the main dogmas of the "behavior-based approach to AI", which swept through leading AI laboratories in the early 90s (Pfeifer and Scheier, 2004).

Suddenly researchers started to build animal-like robots again, reminiscent of the cybernetics research that pre-dated the birth of AI in the 50s. Intense interactions started with biology, particularly ethology and evolutionary and developmental biology, and AI researchers were instrumental in helping to found the new field of Artificial Life (Langton, 1989). This wave of activity lead again to the development of new software and hardware tools, new fundamental insights, and the first generation of rather astonishing animal-like robots. The amazing performance in the Robocup challenge testifies that the time was ripe to build real world robots from a behavior-based point of view. The late 90s even saw the first humanoids such as the Honda Asimo and the Sony QRIO. But all these robots are lacking any kind of "symbolic" intelligence (required for planning or natural language dialogue for example), which suggests that embodiment and neural-like dynamics in themselves are not enough to achieve cognition.

A second movement re-discovered that intelligence seldom arises in isolation. Animals and humans live in groups in which common knowledge and communication systems emerge through collective activities. Problems are solved by cooperation with others. The 90s saw the birth of the multi-agent approach to AI, which focused on how intelligence could be distributed over groups of co-operating entities and how the intelligence of individuals could be the outcome of situated interactions with others. Once again, a multitude of theoretical frameworks, programming paradigms, and formal tools sprung up to deal with the many difficult issues related to multi-agent systems and applications found a niche in software agents for the rapidly expanding Internet and for computer games. AI now established intense interactions with sociologists and anthropologists. It contributed to these fields by providing sophisticated agent-based modeling tools and got in turn inspired by integrating a social view on intelligence (Wooldridge, 2002).

6 Act IV (2000s): Semiotic Dynamics

So what is going on in AI right now? A lot of things of course and it will only become clear in retrospect what development has given the deepest long-term impact. It is obvious for example that there is at the moment a very strong trend towards statistical processing, which has shown enormous application potential in such areas as natural language processing (which is now almost entirely statistically based), web-related information retrieval, and robotics (Thrun, Burgard, and Fox, 2005). This trend is successful thanks to the availability of huge data sources and new techniques in machine learning and statistical inference. However my personal subjective choice goes to another line of current research which I believe holds great promise.

We have seen the pendulum swinging in the sixties and seventies towards knowledge representation and cognitive intelligence, counteracted by the pendulum swinging back in the other direction towards dynamics and embodiment in the eighties and nineties. Although symbolic applications of AI (as now heavily used in search engines) continue, basic AI research clearly moved away almost entirely from conceptual thinking and language. So I think this needs to be corrected again. I personally believe that the most exciting question at the moment is to see how grounded symbolic systems can emerge in communicative interactions between embodied agents. Whereas early symbolic AI no doubt overemphasised the importance of symbols, more recent embodied AI simply ignores it, throwing away the baby with the bathwater. A more balanced view is feasable and necessary. Symbolic intelligence is the hallmark of human intelligence (some biologists have called us the "Symbolic Species" (Deacon, 1998)) and we must try to understand how intelligence that creates and builds further on symbols is possible.

In the earliest AI research (and indeed in cognitive science in general), there has always been the tacit assumption that categories and concepts for structuring the world, as required in problem solving or language communication, are static, universally shared, and hence definable a priori. This lead to the search for conceptual primitives, large-scale ontologies, attempts to capture common-sense knowledge once and for all as illustrated by Lenat's CYC project or the semantic web. More recent, statistically based machine learning techniques similarly assume that there is a (static) conceptual or linguistic system out there which can be acquired, and then used without further adaptation. However observations of natural dialogue and human development show that human language and conceptualisation is constantly on the move and gets invented and aligned on the fly. Conceptualisation is often strongly shaped and reshaped by language. Conceptualisations are invented and imposed on the world based on individual histories of situated interactions, and they become shared by a dynamic negotiation process in joint collaborative tasks, including communication (Pickering and Garrod, 2004). From this point of view, ontologies and language are seen as complex adaptive systems in constant flux. Language nor the meanings expressed by language are based on a static set of conventions that can be induced statistically from language data.

Today this viewpoint is being explored under the label of semiotic dynamics (Steels, 2006). Semiotic dynamics studies how ontologies and symbol systems may emerge in a group of agents and by what mechanisms they may continue to evolve and complexify. The study of semiotic dynamics uses similar tools and techniques as other research in the social sciences concerned with opinion dynamics, collective economical decision making, etc. (Axelrod, 2006). There have already been some initial, very firm results. Groups of robots have been shown capable of self-organising a communication system with natural language like properties, including grammatical structure (Steels, 2003). These multi-agent experiments integrate insights from neural networks and embodied AI to achieve the grounding of language in sensori-motor interaction, but they also rely on sophisticated symbolic processing techniques that were developed in AI in the 60s and 70s. The current evolution of the web towards social tagging and collective knowledge development makes this research even more relevant and tangible, because when we collect data from collective human semiotic behavior that arises in these collective systems we see the same sort of semiotic dynamics as in artificial systems. Semiotic dynamics has made a deep contact with statistical physics and complex systems science. These fields have the tools to investigate the self-organising behaviors of large systems, for example by proving that a certain set of rules will lead to a globally shared lexicon or by showing how systems scale with the size of the population.

By researching how symbol systems can be self-constructed by embodied agents, AI is tackling objections raised by philosophers like Searle (in his Chinese Room parable) or biologists like Edelman. It is true that in most AI systems of the past symbols and the conceptualisation of reality they imply were constructed and coded by designers or inferred from human use. But these scholars concluded (wrongly) that it would never be possible for physically embodied information processing systems to establish and handle symbols autonomously, whereas it is now clear that they can.

7 Epilogue: Peeking into the Future

AI is not that different from other sciences in that it is pushed forward by new technologies that enable more powerful experiments, by new challenges coming from novel applications, or by new advances in other sciences, including mathematics, which can lead to new mechanisms or new predictive tools for understanding and hence exploiting existing mechanisms. Let us see how these three aspects could influence AI in the coming decades.

+ The push from technology: In the 80s there was a high hope that new computer architectures (such as the Connection Machine (Hillis, 1986)) could lead to large advances in AI. This has not happened because standard architectures and mass-produced machines became so cheap and progressively so powerful that they overtook more exotic computational ideas. However there are strong indications that new generations of parallel computers are going to become available soon. They will make experiments possible that are too slow today or can only be done by a very limited number of people. If we want to exploit the promise of massively parallel computing, we will need to come up with novel computing paradigms, exploiting metaphors from biology or chemistry rather than from logic and mathematics. Research in membrane computing, molecular computing, amorphous computing, etc. is showing the way. Not only computer technology but also mechanical engineering and materials science are evolving rapidly and this will make it possible to build completely new kinds of artificial systems, possibly on a very small scale. These technologies are already being explored in embodied AI research and they will surely push the boundaries of autonomous systems, as well as clarify the role of information processing in achieving intelligence.

+ The push from applications: We can observe today a global trend towards collective phenomena in the information processing world. The exponential growth of websites, blogs, wikis, peer-to-peer sharing systems, folksonomies, wikipedias, etc., makes it clear that our densely interconnected world is leading to a whole new way of knowledge production and knowledge communication. Centralised knowledge production and control is swept away by powerful waves of collective tagging, collective encyclopedias, collective news media. Dense interconnectivity and the knowledge dynamics it supports can only increase, when more and more physical devices get substantial computing power and become networked so that everyone can access and change information from anywhere. Many information technologies are participating in the invention of new tools for supporting these remarkable developments, and AI can do so as well, even though some of the older ideas will no longer work, such as the reliance on the individual knowledge of an "expert" which used to be the basis for expert systems, or the idea that it is possible to define logically a universal ontology which can be imposed in a top-down fashion on the web.

A second application area which I believe is of extreme importance and to which future AI should contribute with all its force concerns the retooling of our economies so that they become sustainable. Unless we act today, ecological catastrophes are unavoidable in the not so distant future. AI is highly relevant to achieve this goal but the work has hardly started. AI can contribute with methods and techniques for measuring, interpreting, and assessing pollution and stress to our environments, for tracing and tracking the use of natural resources to organise complete healthy production and consumption cycles, for tracking energy use and managing it in a dynamical adaptive way, and for predicting and visualising the outcomes of current ecological trends so that people become more aware of the urgency of action. Biologically-inspired research results from embodied agents research are in my opinion highly relevant to orchestrate the adaptive management and optimisation of energy resources in micro-grids with local distributed production and peer-to-peer energy exchange.

+ Push from other fields: Future steps forward in theory formation are difficult to predict. Evolutionary network theory (Strogatz, 2001), which only developed in the last decade, has proven to be highly relevant for developing the theory of semiotic dynamics, and much remains to be discovered in non-linear dynamical systems theory, that may have a direct bearing on the foundations of embodied intelligence, neural networks, and agent-environment interaction, as well as on the understanding of the collective dynamics of large groups of agents which self-organise communication systems or other forms of interaction. The conceptual frameworks that have been used so far in neuroscience (basically networks of a few simplistic neurons) are much too simple to start tackling the enormous complexity of real brains so there is a growing demand from neuroscience for the kind of more sophisticated conceptual frameworks that have been common in AI. But I predict that most of the potentially useful interaction in the future will come from interactions with evolutionary biology. Biology has a great tradition of conceptual thinking and its most brilliant thinkers have now a firm grasp of the information processing paradigm (Maynard Smith, 2000). What we need to do is shift our attention away from looking at the end product of intelligence and trying to engineer that towards an understanding how intelligence dynamically arises both within a single individual and within our species. Research on the question of the origins of language is one example in this direction.

AI research, due its strong ties to engineering, has always been very pragmatic, directed by societal and industrial needs. This is in principle a good thing and keeps the field relevant. But I personally feel that we must put a much greater effort into basic research again and into distilling the lessons and principles learned from this pragmatic bottom-up AI research to translate them both in a systematic theory and into a format accessible to other sciences and the informed layman, similar to the way biologist have been able to philosophise and communicate their fundamental research results to a very broad public. A good recent example in this direction is Pfeifer and Bongard (2007). At the same time, AI is an experimental science, because that is the only way that claims and arguments from a design stance can be tested and validated. AI is at its best when doing carefully set up experiments to test the mechanisms that may play a role in intelligence. I believe that these experiments must be set up and communicated like scientific experiments (instead of "applications"), because it is only through that route that we will ever get a systematic theory.

In my view, the coming fifty years of AI look extremely bright. We will continue to make major discoveries and tackle challenges important to society. It is my personal experience that other fields are highly interested in our research results, but only if we stick to the design stance and not pretend to be modeling the human brain or human intelligence. We should get into a dialogue with the fields who have this as their major aim (cognitive science, neuroscience, etc.) but as equal partners bringing our own insights to the table.

References

1. Axelrod, R.: Agent-based modeling as a bridge between disciplines. In: Judd, K.L., Tesfatsion, L. (eds.) Handbook of Computational Economics: Agent-Based Computational Economics, North-Holland (2006)
2. Deacon, T.: The Symbolic Species: The Co-evolution of Language and the Brain. Norton Pub., New York (1998)
3. Dreyfus, H.: What Computers Can't Do: The Limits of Artificial Intelligence. Harpers and Row, New York (1972)
4. Elman, J.: Distributed representations: Simple recurrent networks, and grammatical structure. Machine Learning 7(2-3), 195–226 (1991)
5. Hillis, D.: The Connection Machine. MIT Press, Cambridge, MA (1986)
6. Langton, C. (ed.): Artificial Life. Addison-Wesley, Reading (1989)
7. Lenat, B.: Cyc: A large-scale investment in knowledge infrastructure. Comm. ACM 38(11), 33–38 (1995)
8. Maynard-Smith, J.: The concept of information in biology. Philosophy of Science 67, 177–194 (2000)
9. McCarthy, J., Minsky, M., Rochester, N., Shannon, C.: A proposal for the Dartmouth summer research project on artificial intelligence (1955), http://www.formal.stanford.edu/jmc/history/dartmouth/dartmouth.html
10. Pfeifer, R., Scheier, C.: Understanding Intelligence. MIT Press, Cambridge, MA (1999)
11. Pfeifer, R., Bongard, J.: How the Body Shapes the Way We Think: A New View of Intelligence. MIT Press, Cambridge, MA (2007)
12. Pickering, M.J., Garrod, S.: Toward a mechanistic psychology of dialogue. Behavioral and Brain Sciences 27, 169–225 (2004)
13. Schank, R.: Conceptual Information Processing. Elsevier, Amsterdam (1975)
14. Steels, L., McDermott, J. (eds.): The Knowledge Level in Expert Systems. Conversations and Commentary. Academic Press, Boston (1993)
15. Steels, L., Brooks, R. (eds.): The "artificial life" route to "artificial intelligence". Building Situated Embodied Agents. Lawrence Erlbaum Ass., New Haven (1994)
16. Steels, L.: Evolving grounded communication for robots. Trends in Cognitive Sciences 7(7), 308–312 (2003)
17. Steels, L.: Semiotic dynamics for embodied agents. IEEE Intelligent Systems 5(6), 32–38 (2006)
18. Strogatz, S.: Exploring complex networks. Nature 410, 268–276 (2001)
19. Thrun, S., Burgard, W., Fox, D.: Probabilistic Robotics. MIT Press, Cambridge, MA (2005)
20. Winograd, T.: Understanding Natural Language. Academic Press, London (1972)
21. Wooldridge, M.: An Introduction to Multiagent Systems. John Wiley & Sons, Chichester (2002)

2006: Celebrating 75 Years of AI - History and Outlook: The Next 25 Years

Jürgen Schmidhuber

TU Munich, Boltzmannstr. 3, 85748 Garching bei München, Germany
and IDSIA, Galleria 2, 6928 Manno (Lugano), Switzerland
juergen@idsia.ch - http://www.idsia.ch/~juergen

Abstract. When Kurt Gödel layed the foundations of theoretical computer science in 1931, he also introduced essential concepts of the theory of Artificial Intelligence (AI). Although much of subsequent AI research has focused on heuristics, which still play a major role in many practical AI applications, in the new millennium AI theory has finally become a full-fledged formal science, with important optimality results for embodied agents living in unknown environments, obtained through a combination of theory *à la* Gödel and probability theory. Here we look back at important milestones of AI history, mention essential recent results, and speculate about what we may expect from the next 25 years, emphasizing the significance of the ongoing dramatic hardware speedups, and discussing Gödel-inspired, self-referential, self-improving universal problem solvers.

1 Highlights of AI History—From Gödel to 2006

Gödel and Lilienfeld. In 1931, 75 years ago and just a few years after Julius Lilienfeld patented the transistor, Kurt Gödel layed the foundations of theoretical computer science (CS) with his work on universal formal languages and the limits of proof and computation [5]. He constructed formal systems allowing for self-referential statements that talk about themselves, in particular, about whether they can be derived from a set of given axioms through a computational theorem proving procedure. Gödel went on to construct statements that claim their own unprovability, to demonstrate that traditional math is either flawed in a certain algorithmic sense or contains unprovable but true statements.

Gödel's incompleteness result is widely regarded as the most remarkable achievement of 20th century mathematics, although some mathematicians say it is logic, not math, and others call it the fundamental result of theoretical computer science, a discipline that did not yet officially exist back then but was effectively created through Gödel's work. It had enormous impact not only on computer science but also on philosophy and other fields. In particular, since humans can "see" the truth of Gödel's unprovable statements, some researchers mistakenly thought that his results show that machines and Artificial Intelligences (AIs) will always be inferior to humans. Given the tremendous impact of Gödel's results on AI theory, it does make sense to date AI's beginnings back to his 1931 publication 75 years ago.

Zuse and Turing. In 1936 Alan Turing [37] introduced the *Turing machine* to reformulate Gödel's results and Alonzo Church's extensions thereof. TMs are often more

M. Lungarella et al. (Eds.): 50 Years of AI, Festschrift, LNAI 4850, pp. 29–41, 2007.

convenient than Gödel's integer-based formal systems, and later became a central tool of CS theory. Simultaneously Konrad Zuse built the first working program-controlled computers (1935-1941), using the binary arithmetic and the *bits* of Gottfried Wilhelm von Leibniz (1701) instead of the more cumbersome decimal system used by Charles Babbage, who pioneered the concept of program-controlled computers in the 1840s, and tried to build one, although without success. By 1941, all the main ingredients of 'modern' computer science were in place, a decade after Gödel's paper, a century after Babbage, and roughly three centuries after Wilhelm Schickard, who started the history of automatic computing hardware by constructing the first non-program-controlled computer in 1623.

In the 1940s Zuse went on to devise the first high-level programming language (Plankalkül), which he used to write the first chess program. Back then chess-playing was considered an intelligent activity, hence one might call this chess program the first design of an AI program, although Zuse did not really implement it back then. Soon afterwards, in 1948, Claude Shannon [33] published information theory, recycling several older ideas such as Ludwig Boltzmann's entropy from 19th century statistical mechanics, and the *bit of information* (Leibniz, 1701).

Relays, Tubes, Transistors. Alternative instances of transistors, the concept pioneered and patented by Julius Edgar Lilienfeld (1920s) and Oskar Heil (1935), were built by William Shockley, Walter H. Brattain & John Bardeen (1948: point contact transistor) as well as Herbert F. Mataré & Heinrich Walker (1948, exploiting transconductance effects of germanium diodes observed in the *Luftwaffe* during WW-II). Today most transistors are of the field-effect type *à la* Lilienfeld & Heil. In principle a switch remains a switch no matter whether it is implemented as a relay or a tube or a transistor, but transistors switch faster than relays (Zuse, 1941) and tubes (Colossus, 1943; ENIAC, 1946). This eventually led to significant speedups of computer hardware, which was essential for many subsequent AI applications.

The I in AI. In 1950, some 56 years ago, Turing invented a famous subjective test to decide whether a machine or something else is intelligent. 6 years later, and 25 years after Gödel's paper, John McCarthy finally coined the term "AI". 50 years later, in 2006, this prompted some to celebrate the 50th birthday of AI, but this chapter's title should make clear that its author cannot agree with this view—it is the thing that counts, not its name.

Roots of Probability-Based AI. In the 1960s and 1970s Ray Solomonoff combined theoretical CS and probability theory to establish a general theory of universal inductive inference and predictive AI [35] closely related to the concept of Kolmogorov complexity [14]. His theoretically optimal predictors and their Bayesian learning algorithms only assume that the observable reactions of the environment in response to certain action sequences are sampled from an unknown probability distribution contained in a set M of all enumerable distributions. That is, given an observation sequence we only assume there exists a computer program that can compute the probabilities of the next possible observations. This includes all scientific theories of physics, of course. Since we typically do not know this program, we predict using a weighted sum ξ of *all* distributions in M, where the sum of the weights does not exceed 1. It turns out that this is indeed the best one can possibly do, in a very general sense [11, 35]. Although the universal approach is practically infeasible since M contains infinitely many distributions, it does represent the first sound and general theory of optimal prediction based on experience, identifying the limits of both human and artificial predictors, and providing a yardstick for all prediction machines to come.

AI vs Astrology? Unfortunately, failed prophecies of human-level AI with just a tiny fraction of the brain's computing power discredited some of the AI research in the 1960s and 70s. Many theoretical computer scientists actually regarded much of the field with contempt for its perceived lack of hard theoretical results. ETH Zurich's Turing award winner and creator of the PASCAL programming language, Niklaus Wirth, did not hesitate to link AI to astrology. Practical AI of that era was dominated by rule-based expert systems and Logic Programming. That is, despite Solomonoff's fundamental results, a main focus of that time was on logical, deterministic deduction of facts from previously known facts, as opposed to (probabilistic) induction of hypotheses from experience.

Evolution, Neurons, Ants. Largely unnoticed by mainstream AI gurus of that era, a biology-inspired type of AI emerged in the 1960s when Ingo Rechenberg pioneered the method of artificial evolution to solve complex optimization tasks [22], such as the design of optimal airplane wings or combustion chambers of rocket nozzles. Such methods (and later variants thereof, e.g., Holland [10], 1970s), often gave better results than classical approaches. In the following decades, other types of "subsymbolic" AI also became popular, especially neural networks. Early neural net papers include those of McCulloch & Pitts, 1940s (linking certain simple neural nets to old and well-known, simple mathematical concepts such as linear regression); Minsky & Papert [17] (temporarily discouraging neural network research), Kohonen [12], Amari, 1960s; Werbos [40], 1970s; and many others in the 1980s. Orthogonal approaches included fuzzy logic (Zadeh, 1960s), Rissanen's practical variants [23] of Solomonoff's universal method, "representation-free" AI (Brooks [2]), Artificial Ants (Dorigo & Gambardella [4], 1990s), statistical learning theory (in less general settings than those studied by Solomonoff) & support vector machines (Vapnik [38] and others). As of 2006, this alternative type of AI research is receiving more attention than "Good Old-Fashioned AI" (GOFAI).

Mainstream AI Marries Statistics. A dominant theme of the 1980s and 90s was the marriage of mainstream AI and old concepts from probability theory. Bayes networks, Hidden Markov Models, and numerous other probabilistic models found wide applications ranging from pattern recognition, medical diagnosis, data mining, machine translation, robotics, etc.

Hardware Outshining Software: Humanoids, Robot Cars, Etc. In the 1990s and 2000s, much of the progress in practical AI was due to better hardware, getting roughly 1000 times faster per Euro per decade. In 1995, a fast vision-based robot car by Ernst Dickmanns (whose team built the world's first reliable robot cars in the early 1980s with the help of Mercedes-Benz, e. g., [3]) autonomously drove 1000 miles from Munich to Denmark and back, in traffic at up to 120 mph, automatically passing other cars (a safety driver took over only rarely in critical situations). Japanese labs (Honda, Sony) and Pfeiffer's lab at TU Munich built famous humanoid walking robots. Engineering problems often seemed more challenging than AI-related problems.

Another source of progress was the dramatically improved access to all kinds of data through the WWW, created by Tim Berners-Lee at the European particle collider CERN (Switzerland) in 1990. This greatly facilitated and encouraged all kinds of "intelligent" data mining applications. However, there were few if any obvious fundamental algorithmic breakthroughs; improvements / extensions of already existing algorithms seemed less impressive and less crucial than hardware advances. For example, chess world champion Kasparov was beaten by a fast IBM computer running a fairly standard algorithm. Rather simple but computationally expensive probabilistic methods for speech recognition, statistical machine translation, computer vision, optimization, virtual realities etc. started to become feasible on PCs, mainly because PCs had become 1000 times more powerful within a decade or so.

2006. As noted by Stefan Artmann (personal communication, 2006), today's AI textbooks seem substantially more complex and less unified than those of several decades ago, e. g., [18], since they have to cover so many apparently quite different subjects. There seems to be a need for a new unifying view of intelligence. In the author's opinion this view already exists, as will be discussed below.

2   Subjective Selected Highlights of Present AI

The more recent some event, the harder it is to judge its long-term significance. But this biased author thinks that the most important thing that happened recently in AI is the begin of a transition from a heuristics-dominated science (e.g., [24]) to a real formal science. Let us elaborate on this topic.

2.1   The Two Ways of Making a Dent in AI Research

There are at least two convincing ways of doing AI research: **(1)** construct a (possibly heuristic) machine or algorithm that somehow (it does not really matter how) solves a previously unsolved interesting problem, such as beating the best human player of *Go* (success will outshine any lack of theory). Or **(2)** prove that a particular novel algorithm is optimal for an important class of AI problems.

It is the nature of heuristics (case **(1)**) that they lack staying power, as they may soon get replaced by next year's even better heuristics. Theorems (case **(2)**), however, are for eternity. That's why formal sciences prefer theorems.

For example, probability theory became a formal science centuries ago, and totally formal in 1933 with Kolmogorov's axioms [13], shortly after Gödel's paper [5]. Old but provably optimal techniques of probability theory are still in every day's use, and in fact highly significant for modern AI, while many initially successful heuristic approaches eventually became unfashionable, of interest mainly to the historians of the field.

2.2   No Brain Without a Body / AI Becoming a Formal Science

Heuristic approaches will continue to play an important role in many AI applications, to the extent they empirically outperform competing methods. But like with all young sciences at the transition point between an early intuition-dominated and a later formal era, the importance of mathematical optimality theorems is growing quickly. Progress in the formal era, however, is and will be driven by a different breed of researchers, a fact that is not necessarily universally enjoyed and welcomed by all the earlier pioneers.

Today the importance of embodied, embedded AI is almost universally acknowledged (e. g., [20]), as obvious from frequently overheard remarks such as "let the physics compute" and "no brain without a body." Many present AI researchers focus on real robots living in real physical environments. To some of them the title of this subsection may seem oxymoronic: the extension of AI into the realm of the physical body seems to be a step away from formalism. But the new millennium's formal point of view is actually taking this step into account in a very general way, through the first mathematical theory of universal embedded AI, combining "old" theoretical computer science and "ancient" probability theory to derive optimal behavior for embedded, embodied rational agents living in unknown but learnable environments. More on this below.

2.3   What's the I in AI? What Is Life? Etc.

Before we proceed, let us clarify what we are talking about. Shouldn't researchers on Artificial Intelligence (AI) and Artificial Life (AL) agree on basic questions such as: What is Intelligence? What is Life? Interestingly they don't.

Are Cars Alive? For example, AL researchers often offer definitions of life such as: it must reproduce, evolve, etc. Cars are alive, too, according to most of these definitions. For example, cars evolve and multiply. They need complex environments with car factories to do so, but living animals also need complex environments full of chemicals and other animals to reproduce — the DNA information by itself does not suffice. There is no obvious fundamental difference between an organism whose self-replication information is stored in its DNA, and a car whose self-replication information is stored in a car builder's manual in the glove compartment. To copy itself, the organism needs its mothers womb plus numerous other objects and living beings in its environment (such as trillions of bacteria inside and outside of the mother's body). The car needs iron mines and car part factories and human workers.

What is Intelligence? If we cannot agree on what's life, or, for that matter, love, or consciousness (another fashionable topic), how can there be any hope to define intelligence? Turing's definition (1950, 19 years after Gödel's paper) was totally subjective: intelligent is what convinces me that it is intelligent while I am interacting with it. Fortunately, however, there are more formal and less subjective definitions.

2.4   Formal AI Definitions

Popper said: all life is problem solving [21]. Instead of defining intelligence in Turing's rather vague and subjective way we define intelligence with respect to the abilities of universal optimal problem solvers.

Consider a learning robotic agent with a single life which consists of discrete cycles or time steps $t = 1, 2, \ldots, T$. Its total lifetime T may or may not be known in advance. In what follows,the value of any time-varying variable Q at time t ($1 \leq t \leq T$) will be denoted by $Q(t)$, the ordered sequence of values $Q(1), \ldots, Q(t)$ by $Q(\leq t)$, and the (possibly empty) sequence $Q(1), \ldots, Q(t-1)$ by $Q(< t)$.

At any given t the robot receives a real-valued input vector $x(t)$ from the environment and executes a real-valued action $y(t)$ which may affect future inputs; at times $t < T$ its goal is to maximize future success or *utility*

$$u(t) = E_\mu \left[\sum_{\tau=t+1}^{T} r(\tau) \ \middle| \ h(\leq t) \right], \tag{1}$$

where $r(t)$ is an additional real-valued reward input at time t, $h(t)$ the ordered triple $[x(t), y(t), r(t)]$ (hence $h(\leq t)$ is the known history up to t), and $E_\mu(\cdot \mid \cdot)$ denotes the conditional expectation operator with respect to some possibly unknown distribution μ from a set M of possible distributions. Here M reflects whatever is known about the possibly probabilistic reactions of the environment. For example, M may contain all computable distributions [11, 35]. Note that unlike in most previous work by others [36], there is just one life, no need for predefined repeatable trials, no restriction to Markovian interfaces between sensors and environment, and the utility function implicitly takes into account the expected remaining lifespan $E_\mu(T \mid h(\leq t))$ and thus the possibility to extend it through appropriate actions [29].

Any formal problem or sequence of problems can be encoded in the reward function. For example, the reward functions of many living or robotic beings cause occasional hunger or pain or pleasure signals etc. At time t an optimal AI will make the best possible use of experience $h(\leq t)$ to maximize $u(t)$. But how?

2.5   Universal, Mathematically Optimal, But Incomputable AI

Unbeknownst to many traditional AI researchers, there is indeed an extremely general "best" way of exploiting previous experience. At any time t, the recent theoretically optimal yet practically infeasible reinforcement learning (RL) algorithm AIXI [11] uses Solomonoff's above-mentioned universal prediction scheme to select those action sequences that promise maximal future reward up to some horizon, given the current data $h(\leq t)$. Using a variant of Solomonoff's universal probability mixture ξ, in cycle $t + 1$, AIXI selects as its next action the first action of an action sequence maximizing ξ-predicted reward up to the horizon. Hutter's recent work [11] demonstrated AIXI's optimal use of observations as follows. The Bayes-optimal policy p^ξ based on the mixture ξ is self-optimizing in the sense that its average utility value converges asymptotically for all $\mu \in M$ to the optimal value achieved by the (infeasible) Bayes-optimal policy p^μ which knows μ in advance. The necessary condition that M admits self-optimizing policies is also sufficient.

Of course one cannot claim the old AI is devoid of formal research! The recent approach above, however, goes far beyond previous formally justified but very limited AI-related approaches ranging from linear perceptrons [17] to the A^*-algorithm [18]. It provides, for the first time, a mathematically sound theory of general AI and optimal decision making based on experience, identifying the limits of both human and artificial intelligence, and a yardstick for any future, scaled-down, practically feasible approach to general AI.

2.6 Optimal Curiosity and Creativity

No theory of AI will be convincing if it does not explain curiosity and creativity, which many consider as important ingredients of intelligence. We can provide an explanation in the framework of optimal reward maximizers such as those from the previous subsection.

It is possible to come up with theoretically optimal ways of improving the predictive world model of a curious robotic agent [28], extending earlier ideas on how to implement artificial curiosity [25]: *The rewards of an optimal reinforcement learner are the predictor's improvements* on the observation history so far. They encourage the reinforcement learner to produce action sequences that cause the creation and the learning of new, previously unknown regularities in the sensory input stream. It turns out that art and creativity can be explained as by-products of such intrinsic curiosity rewards: good observer-dependent art deepens the observer's insights about this world or possible worlds, connecting previously disconnected patterns in an initially surprising way that eventually becomes known and boring. While previous attempts at describing what is satisfactory art or music were informal, this work permits the first *technical, formal* approach to understanding the nature of art and creativity [28].

2.7 Computable, Asymptotically Optimal General Problem Solver

Using the Speed Prior [26] one can scale down the universal approach above such that it becomes computable. In what follows we will mention general methods whose optimality criteria explicitly take into account the computational costs of prediction and decision making—compare [15].

The recent asymptotically optimal search algorithm for *all* well-defined problems [11] allocates part of the total search time to searching the space of proofs for provably correct candidate programs with provable upper runtime bounds; at any given time it focuses resources on those programs with the currently best proven time bounds. The method is as fast as the initially unknown fastest problem solver for the given problem class, save for a constant slowdown factor of at most $1 + \epsilon, \epsilon > 0$, and an additive constant that does not depend on the problem instance!

Is this algorithm then the *holy grail* of computer science? Unfortunately not quite, since the additive constant (which disappears in the $O()$-notation of theoretical CS) may be huge, and practical applications may not ignore it. This motivates the next section, which addresses all kinds of formal optimality (not just asymptotic optimality).

2.8 Fully Self-referential, Self-improving Gödel Machine

We may use Gödel's self-reference trick to build a universal general, fully self-referential, self-improving, optimally efficient problem solver [29]. A Gödel Machine is a computer whose original software includes axioms describing the hardware and the original software (this is possible without circularity) plus whatever is known about the (probabilistic) environment plus some formal goal in form of an arbitrary user-defined utility function, e.g., cumulative future expected reward in a sequence of optimization tasks - see equation (1). The original software also includes a proof searcher which uses the axioms (and possibly an online variant of Levin's universal search [15]) to systematically make pairs ("proof", "program") until it finds a proof that a rewrite of the original software through "program" will increase utility. The machine can be designed such that each self-rewrite is necessarily globally optimal in the sense of the utility function, even those rewrites that destroy the proof searcher [29].

2.9   Practical Algorithms for Program Learning

The theoretically optimal universal methods above are optimal in ways that do not (yet) immediately yield practically feasible general problem solvers, due to possibly large initial overhead costs. Which are today's practically most promising extensions of traditional machine learning?

Since virtually all realistic sensory inputs of robots and other cognitive systems are sequential by nature, the future of machine learning and AI in general depends on progress in in sequence processing as opposed to the traditional processing of stationary input patterns. To narrow the gap between learning abilities of humans and machines, we will have to study how to learn general algorithms instead of such reactive mappings. Most traditional methods for learning time series and mappings from sequences to sequences, however, are based on simple time windows: one of the numerous feedforward ML techniques such as feedforward neural nets (NN) [1] or support vector machines [38] is used to map a restricted, fixed time window of sequential input values to desired target values. Of course such approaches are bound to fail if there are temporal dependencies exceeding the time window size. Large time windows, on the other hand, yield unacceptable numbers of free parameters.

Presently studied, rather general sequence learners include certain probabilistic approaches and especially recurrent neural networks (RNNs), e.g., [19]. RNNs have adaptive feedback connections that allow them to learn mappings from input sequences to output sequences. They can implement any sequential, algorithmic behavior implementable on a personal computer. In gradient-based RNNs, however, we can *differentiate our wishes with respect to programs,* to obtain a search direction in algorithm space. RNNs are biologically more plausible and computationally more powerful than other adaptive models such as Hidden Markov Models (HMMs - no continuous internal states), feedforward networks & Support Vector Machines (no internal states at all). For several reasons, however, the first RNNs could not learn to look far back into the past. This problem was overcome by RNNs of the *Long Short-Term Memory* type (LSTM), currently the most powerful and practical supervised RNN architecture for many applications, trainable either by gradient descent [9] or evolutionary methods [32], occasionally profiting from a marriage with probabilistic approaches [8].

Unsupervised RNNs that learn without a teacher to control physical processes or robots frequently use evolutionary algorithms [10, 22] to learn appropriate programs (RNN weight matrices) through trial and error [41]. Recent work brought progress through a focus on reducing search spaces by co-evolving the comparatively small weight vectors of individual recurrent neurons [7]. Such RNNs can learn to create memories of important events, solving numerous RL / optimization tasks unsolvable by traditional RL methods [6, 7]. They are among the most promising methods for practical program learning, and currently being applied to the control of sophisticated robots such as the walking biped of TU Munich [16].

3   The Next 25 Years

Where will AI research stand in 2031, 25 years from now, 100 years after Gödel's ground-breaking paper [5], some 200 years after Babbage's first designs, some 400 years after the first automatic calculator by Schickard (and some 2000 years after the crucifixion of the man whose birth year anchors the Western calendar)?

Trivial predictions are those that just naively extrapolate the current trends, such as: computers will continue to get faster by a factor of roughly 1000 per decade; hence they will be at least a million times faster by 2031. According to frequent estimates, current supercomputers achieve roughly 1 percent of the raw computational power of a human brain, hence those of 2031 will have 10,000 "brain powers"; and even cheap devices will achieve many brain powers. Many tasks that are hard for today's software on present machines will become easy without even fundamentally changing the algorithms. This includes numerous pattern recognition and control tasks arising in factories of many industries, currently still employing humans instead of robots.

Will theoretical advances and practical software keep up with the hardware development? We are convinced they will. As discussed above, the new millennium has already brought fundamental new insights into the problem of constructing theoretically optimal rational agents or universal AIs, even if those do not yet immediately translate into practically feasible methods. On the other hand, on a more practical level, there has been rapid progress in learning algorithms for agents interacting with a dynamic environment, autonomously discovering true sequence-processing, problem-solving programs, as opposed to the reactive mappings from stationary inputs to outputs studied in most of traditional machine learning research. In the author's opinion the above-mentioned theoretical and practical strands are going to converge. In conjunction with the ongoing hardware advances this will yield non-universal but nevertheless rather general artificial problem-solvers whose capabilities will exceed those of most if not all humans in many domains of commercial interest. This may seem like a bold prediction to some, but it is actually a trivial one as there are so many experts who would agree with it.

Nontrivial predictions are those that anticipate truly unexpected, revolutionary breakthroughs. By definition, these are hard to predict. For example, in 1985 only very few scientists and science fiction authors predicted the WWW revolution of the 1990s. The few who did were not influential enough to make a significant part of humanity believe in their predictions and prepare for their coming true. Similarly, after the latest stock market crash one can always find with high probability some "prophet in the desert" who predicted it in advance, but had few if any followers until the crash really occurred.

Truly nontrivial predictions are those that most will not believe until they come true. We will mostly restrict ourselves to trivial predictions like those above and refrain from too much speculation in form of nontrivial ones. However, we may have a look at previous unexpected scientific breakthroughs and try to discern a pattern, a pattern that may not allow us to precisely predict the details of the next revolution but at least its timing.

3.1   A Pattern in the History of Revolutions?

Let us put the AI-oriented developments [27] discussed above in a broader context, and look at the history of major scientific revolutions and essential historic developments (that is, the subjects of the major chapters in history books) since the beginnings of modern man over 40,000 years ago [30, 31]. Amazingly, they seem to match a binary logarithmic scale marking exponentially declining temporal intervals [31], each half the size of the previous one, and measurable in terms of powers of 2 multiplied by a human lifetime (roughly 80 years—throughout recorded history many individuals have reached this age, although the average lifetime often was shorter, mostly due to high children mortality). It looks as if history itself will *converge* in a historic singularity or Omega point Ω around 2040 (the term *historic singularity* is apparently due to Stanislaw Ulam (1950s) and was popularized by Vernor Vinge [39] in the 1990s). To convince yourself of history's convergence, associate an error bar of not much more than 10 percent with each date below:

1. $\Omega - 2^9$ lifetimes: modern humans start colonizing the world from Africa
2. $\Omega - 2^8$ lifetimes: bow and arrow invented; hunting revolution
3. $\Omega - 2^7$ lifetimes: invention of agriculture; first permanent settlements; beginnings of civilization
4. $\Omega - 2^6$ lifetimes: first high civilizations (Sumeria, Egypt), and the most important invention of recorded history, namely, the one that made recorded history possible: writing
5. $\Omega - 2^5$ lifetimes: the ancient Greeks invent democracy and lay the foundations of Western science and art and philosophy, from algorithmic procedures and formal proofs to anatomically perfect sculptures, harmonic music, and organized sports. Old Testament written (basis of Judaism, Christianity, Islam); major Asian religions founded. High civilizations in China, origin of the first calculation tools, and India, origin of alphabets and the zero
6. $\Omega - 2^4$ lifetimes: bookprint (often called the most important invention of the past 2000 years) invented in China. Islamic science and culture start spreading across large parts of the known world (this has sometimes been called the most important event between Antiquity and the age of discoveries)
7. $\Omega - 2^3$ lifetimes: the Mongolian Empire, the largest and most dominant empire ever (possibly including most of humanity and the world economy), stretches across Asia from Korea all the way to Germany. Chinese fleets and later also European vessels start exploring the world. Gun powder and guns invented in China. Rennaissance and Western bookprint (often called the most influential invention of the past 1000 years) and subsequent Reformation in Europe. Begin of the Scientific Revolution
8. $\Omega - 2^2$ lifetimes: Age of enlightenment and rational thought in Europe. Massive progress in the sciences; first flying machines; first steam engines prepare the industrial revolution
9. $\Omega - 2$ lifetimes: Second industrial revolution based on combustion engines, cheap electricity, and modern chemistry. Birth of modern medicine through the germ theory of disease; genetic and evolution theory. European colonialism at its short-lived peak

10. $\Omega - 1$ lifetime: modern post-World War II society and pop culture emerges; super-power stalemate based on nuclear deterrence. The 20th century super-exponential population explosion (from 1.6 billion to 6 billion people, mainly due to the Haber-Bosch process [34]) is at its peak. First spacecraft and commercial computers; DNA structure unveiled

11. $\Omega - 1/2$ lifetime (now): for the first time in history most of the most destructive weapons are dismantled, after the Cold War's peaceful end. 3rd industrial revolution based on personal computers and the World Wide Web. A mathematical theory of universal AI emerges (see sections above) - will this be considered a milestone in the future?

12. $\Omega - 1/4$ lifetime: This point will be reached around 2020. By then many computers will have substantially more raw computing power than human brains.

13. $\Omega - 1/8$ lifetime (100 years after Gödel's paper): will practical variants of Gödel machines start a runaway evolution of continually self-improving superminds way beyond human imagination, causing far more unpredictable revolutions in the final decade before Ω than during all the millennia before?

14. ...

The following disclosure should help the reader to take this list with a grain of salt though. The author, who admits being very interested in witnessing Ω, was born in 1963, and therefore perhaps should not expect to live long past 2040. This may motivate him to uncover certain historic patterns that fit his desires, while ignoring other patterns that do not. Perhaps there even is a general rule for both the individual memory of single humans and the collective memory of entire societies and their history books: constant amounts of memory space get allocated to exponentially larger, adjacent time intervals further and further into the past. Maybe that's why there has never been a shortage of prophets predicting that the end is near - the important events according to one's own view of the past always seem to accelerate exponentially. See [31] for a more thorough discussion of this possibility.

References

[1] Bishop, C.M.: Neural networks for pattern recognition. Oxford University Press, Oxford (1995)

[2] Brooks, R.A.: Intelligence without reason. In: Proceedings of the Twelveth Internationl Joint Conference on Artificial Intelligence, pp. 569–595 (1991)

[3] Dickmanns, E.D., Behringer, R., Dickmanns, D., Hildebrandt, T., Maurer, M., Thomanek, F., Schiehlen, J.: The seeing passenger car 'VaMoRs-P'. In: Proc. Int. Symp. on Intelligent Vehicles 1994, Paris, pp. 68–73 (1994)

[4] Dorigo, M., Di Caro, G., Gambardella, L.M.: Ant algorithms for discrete optimization. Artificial Life 5(2), 137–172 (1999)

[5] Gödel, K.: Über formal unentscheidbare Sätze der Principia Mathematica und verwandter Systeme I. Monatshefte für Mathematik und Physik 38, 173–198 (1931)

[6] Gomez, F., Schmidhuber, J., Miikkulainen, R.: Efficient non-linear control through neuroevolution. In: Fürnkranz, J., Scheffer, T., Spiliopoulou, M. (eds.) ECML 2006. LNCS (LNAI), vol. 4212, Springer, Heidelberg (2006)

[7] Gomez, F.J., Miikkulainen, R.: Active guidance for a finless rocket using neuroevolution. In: Cantú-Paz, E., Foster, J.A., Deb, K., Davis, L., Roy, R., O'Reilly, U.-M., Beyer, H.-G., Kendall, G., Wilson, S.W., Harman, M., Wegener, J., Dasgupta, D., Potter, M.A., Schultz, A., Dowsland, K.A., Jonoska, N., Miller, J., Standish, R.K. (eds.) GECCO 2003. LNCS, vol. 2723, Springer, Heidelberg (2003)

[8] Graves, A., Fernandez, S., Gomez, F., Schmidhuber, J.: Connectionist temporal classification: Labelling unsegmented sequence data with recurrent neural nets. In: ICML 2006. Proceedings of the International Conference on Machine Learning (2006)

[9] Hochreiter, S., Schmidhuber, J.: Long short-term memory. Neural Computation 9(8), 1735–1780 (1997)

[10] Holland, J.H.: Adaptation in Natural and Artificial Systems. University of Michigan Press, Ann Arbor (1975)

[11] Hutter, M.: Universal Artificial Intelligence: Sequential Decisions based on Algorithmic Probability. Springer, Berlin (2004) (On J. Schmidhuber's SNF grant 20-61847)

[12] Kohonen, T.: Self-Organization and Associative Memory, 2nd edn. Springer, Heidelberg (1988)

[13] Kolmogorov, A.N.: Grundbegriffe der Wahrscheinlichkeitsrechnung. Springer, Berlin (1933)

[14] Kolmogorov, A.N.: Three approaches to the quantitative definition of information. Problems of Information Transmission 1, 1–11 (1965)

[15] Levin, L.A.: Universal sequential search problems. Problems of Information Transmission 9(3), 265–266 (1973)

[16] Lohmeier, S., Loeffler, K., Gienger, M., Ulbrich, H., Pfeiffer, F.: Sensor system and trajectory control of a biped robot. In: AMC 2004. Proc. 8th IEEE International Workshop on Advanced Motion Control, Kawasaki, Japan, pp. 393–398. IEEE Computer Society Press, Los Alamitos (2004)

[17] Minsky, M., Papert, S.: Perceptrons. MIT Press, Cambridge, MA (1969)

[18] Nilsson, N.J.: Principles of artificial intelligence. Morgan Kaufmann, San Francisco (1980)

[19] Pearlmutter, B.A.: Gradient calculations for dynamic recurrent neural networks: A survey. IEEE Transactions on Neural Networks 6(5), 1212–1228 (1995)

[20] Pfeifer, R., Scheier, C.: Understanding Intelligence. MIT Press, Cambridge (2001)

[21] Popper, K.R.: All Life Is Problem Solving. Routledge, London (1999)

[22] Rechenberg, I.: Evolutionsstrategie - Optimierung technischer Systeme nach Prinzipien der biologischen Evolution. Dissertation, 1971. Fromman-Holzboog (1973)

[23] Rissanen, J.: Modeling by shortest data description. Automatica 14, 465–471 (1978)

[24] Rosenbloom, P.S., Laird, J.E., Newell, A.: The SOAR Papers. MIT Press, Cambridge (1993)

[25] Schmidhuber, J.: Curious model-building control systems. In: Proceedings of the International Joint Conference on Neural Networks, Singapore, vol. 2, pp. 1458–1463. IEEE press, Los Alamitos (1991)

[26] Schmidhuber, J.: The Speed Prior: a new simplicity measure yielding near-optimal computable predictions. In: Kivinen, J., Sloan, R.H. (eds.) COLT 2002. LNCS (LNAI), vol. 2375, pp. 216–228. Springer, Heidelberg (2002)

[27] Schmidhuber, J.: Artificial Intelligence - history highlights and outlook: AI maturing and becoming a real formal science (2006), http://www.idsia.ch/juergen/ai.html

[28] Schmidhuber, J.: Developmental robotics, optimal artificial curiosity, creativity, music, and the fine arts. Connection Science 18(2), 173–187 (2006)

[29] Schmidhuber, J.: Gödel machines: fully self-referential optimal universal problem solvers. In: Goertzel, B., Pennachin, C. (eds.) Artificial General Intelligence, pp. 201–228. Springer, Heidelberg (2006)

[30] Schmidhuber, J.: Is history converging? Again? (2006),http://www.idsia.ch/ juergen/history.html

[31] Schmidhuber, J.: New millennium AI and the convergence of history. In: Duch, W., Mandziuk, J. (eds.) Challenges to Computational Intelligence, Springer, Heidelberg (2007) Also available as TR IDSIA-04-03, cs.AI/0302012

[32] Schmidhuber, J., Wierstra, D., Gagliolo, M., Gomez, F.: Training recurrent networks by EVOLINO. Neural Computation 19(3), 757–779 (2007)

[33] Shannon, C.E.: A mathematical theory of communication (parts I and II). Bell System Technical Journal XXVII, 379–423 (1948)

[34] Smil, V.: Detonator of the population explosion. Nature 400, 415 (1999)

[35] Solomonoff, R.J.: Complexity-based induction systems. IEEE Transactions on Information Theory IT-24(5), 422–432 (1978)

[36] Sutton, R., Barto, A.: Reinforcement learning: An introduction. MIT Press, Cambridge, MA (1998)

[37] Turing, A.M.: On computable numbers, with an application to the Entscheidungsproblem. Proceedings of the London Mathematical Society, Series 2 41, 230–267 (1936)

[38] Vapnik, V.: The Nature of Statistical Learning Theory. Springer, New York (1995)

[39] Vinge, V.: The coming technological singularity. In: VISION-21 Symposium sponsored by NASA Lewis Research Center, and Whole Earth Review, Winter issue (1993)

[40] Werbos, P.J.: Beyond Regression: New Tools for Prediction and Analysis in the Behavioral Sciences. PhD thesis, Harvard University (1974)

[41] Yao, X.: A review of evolutionary artificial neural networks. International Journal of Intelligent Systems 4, 203–222 (1993)

Evolutionary Humanoid Robotics:
Past, Present and Future

Malachy Eaton

Department of Computer Science and Information Systems
University of Limerick, Ireland
malachy.eaton@ul.ie

Abstract. Evolutionary robotics is a methodology for the creation of autonomous robots using evolutionary principles. Humanoid robotics is concerned
specifically with autonomous robots that are human-like in that they mimic the
body or aspects of the sensory, processing and/or motor functions of humans to
a greater or lesser degree. We investigate how these twin strands of advanced
research in the field of autonomous mobile robotics have progressed over the
last decade or so, and their current recent convergence in the new field of
evolutionary humanoid robotics. We describe our current work in the evolution
of controllers for bipedal locomotion in a simulated humanoid robot using an
accurate physics simulator, and briefly discuss the effects of changes in robot
mobility and of environmental changes. We then describe our current work in
the implementation of these simulated robots using the Bioloid robot platform.
We conclude with a look at possible visions for the future.

Keywords: Artificial evolution, humanoid robotics.

1 Introduction and Motivation

Evolutionary humanoid robotics is a branch of evolutionary robotics dealing with the
application of the laws of genetics and the principle of natural selection to the design
of humanoid robots. For a good introduction to the general field see the book by
Nolfi and Floreano[1]. Evolutionary techniques have been applied to the design of
both robot body and 'brain' for a variety of different wheeled and legged robots[2-6].
In this article we are primarily concerned with the application of evolutionary
techniques to autonomous robots whose morphology and/or control/sensory apparatus
is broadly human-like.

In Brooks' paper on the subject [7] he lists two main motivations for the
construction (or evolution) of humanoid robots. He presents the argument that the
form of human bodies may well be critical to the representations we use for both
language and thought. Thus, if we wish (for whatever reason) to build a human-like
intelligence the robot body must also be human-like. This is a view supported by
Pfeifer and Bongard in their recent book, aptly titled "How the body shapes the way
we think". [8]

The second motivation he suggests relates to the area of human-robot interaction.
If the robot has a human-like form then people should find it easier and more natural

M. Lungarella et al. (Eds.): 50 Years of AI, Festschrift, LNAI 4850, pp. 42–52, 2007.

to interact with it just as if it were human. However it is important not to ignore the so-called "uncanny valley" effect as presented by Mashahiro Mori [9] and further discussed by Mac Dorman [10]. This suggests that there is a positive correlation between the likeness of a robot to a human; with how familiar and hence how easy they are to interact with, from a human perspective. This is as we would expect. However after a certain point small further increases produce a sharp decrease in familiarity (the *"uncanny valley"*) which only then increases again as the robot becomes almost indistinguishable from a human. This effect is seen to increase for a moving robot as opposed to a stationary one. It is thought that this unnerving effect is correlated to an innate fear of mortality and culturally evolved mechanisms for coping with this. This may suggest the desirability of, for the present, striving to produce humanoid robots with useful skills (discussed further below), but without at this stage attempting to imbue them with over human-like features or expressions.

A third possible motivation, not discussed in Brooks' paper is that the humanoid robot may be able to operate with ease in environments and situations where humans operate, such as opening and closing door, climbing up and down stairs, bending down to put washing in a washing machine etc. This will allow the robot to be useful in a whole host of situations in which a non-humanoid robot would be quite powerless. A major application in the future could be in the area of home helps for the elderly. In modern developed countries like Italy and Japan fertility rates are dropping dramatically and they are left with a disproportionate elderly population and a relatively small population of potential carers.

The hope is that by using artificial evolution robots may be evolved which are stable and robust, and which would be difficult to design by conventional techniques alone. However we should bear in mind the caveat put forward by Mataric and Cliff [11]; that the effort expended in designing and configuring the evolutionary algorithm should ideally be considerably less than that required to do a manual design.

For a brief introduction to the current state of the art with regard to Humanoid robotics including the HRP-3, KHR-1 and KHR-2, Sony QRIO and Honda ASIMO and P2 see Akachi et al [12]. See references [13-20] for other articles of specific interest in the design of autonomous robots, and humanoid robots in particular. The increasingly important issue of the benchmarking and evaluation of future autonomous robots, which is an area that will be of increasing relevance and significance as intelligent robots, especially of the humanoid variety, play a greater role in our everyday lives is discussed in references [21] and [22].

For other work in the specific area of the evolution of bipedal locomotion see references [23-30]. Space precludes a detailed discussion of this other work, some of it very interesting, however in these papers walking generally operates either solely on a simulated robot[24-26,28-30], and/or on robots with a restricted number of degrees of freedom (typically 6-10 DOF). Some of these systems also require the incorporation of quite a high degree of domain specific knowledge in the genetic algorithm. In addition there is some very interesting work on bipedal locomotion without the use of complex control algorithms in Cornell University and Delft University among others (passive-dynamic walkers)[31], which have five DOF each, and in the MIT learning biped which has six DOF and uses reinforcement learning to acquire a control policy [31], but as these are not evolutionary based techniques they fall outside the scope of this discussion.

2 Evolution of Bipedal Locomotion

We now concisely describe a specific set of experiments for the evolution of bipedal locomotion in a high-DOF simulated humanoid robot. Bipedal locomotion is a difficult task, which, in the past, was thought to separate us from the higher primates. In the experiments outlined here we use a genetic algorithm to choose the joint values for a simulated humanoid robot with a total of 20 degrees of freedom (elbows, ankles, knees, etc.) for specific time intervals (keyframes) together with maximum joint ranges in order to evolve bipedal locomotion. An existing interpolation function fills in the values between keyframes; once a cycle of 4 keyframes is completed it repeats until the end of the run, or until the robot falls over. The humanoid robot is simulated using the Webots mobile robot simulation package and is broadly modeled on the Sony QRIO humanoid robot [32,33]. In order to get the robot to walk a simple function based on the product of the length of time the robot remains standing by the total distance traveled by the robot was devised. This was later modified to reward walking in a forward (rather than backward) direction and to promote walking in a more upright position, by taking the robots final height into account.

In previous experiments [34] it was found that the range of movement allowed to the joints by the evolutionary algorithm: that is the proportion of the maximum range of movement allowed to the robot for each joint, was an important factor in evolving successful walks. Initial experiments placed no restriction on the range of movement allowed and walks did not evolve unless the robot was restricted to a stooped posture and a symmetrical gait, even then results were not impressive. By restricting possible movement to different fractions of the maximum range walks did evolve, however as this was seen as a critical factor in the evolutionary process it was decided in the current work to include a value specifying the fraction of the total range allowed in the humanoid robots genome.

The genome length is 328 bits comprising 4 bits determining the position of the 20 motors for each of 4 keyframes; 80 strings are used per generation. 8 bits define the fraction of the maximum movement range allowed. The maximum range allowed for a particular genome is the value specified in the corresponding to each motor divided by the number of bits set in this 8 bit field plus 1. 8 bits was chosen as reasonable walking patterns were seen to evolve when the range was restricted by a factor of 4 or thereabouts in previous experiments. The genetic algorithm uses roulette wheel selection with elitism; the top string being guaranteed safe passage to the next generation, together with standard crossover and mutation. Maximum fitness values may rise as well as fall because of the realistic nature of the Webots simulation. Two-point crossover is applied with a probability of 0.5 and the probability of a bit being mutated is 0.04. These values were arrived at after some experimentation.

3 Experimental Results

We ran three trials of the evolutionary algorithm on a population size of 80 controllers for 700 generations of simulated robots, taking approximately 2.5 weeks simulation time on a reasonably fast computer, corresponding to approximately 7 weeks of "real time" experimentation. A fitness value over about 100 corresponds to the robot at least staying standing for some period of time, over 500 corresponds to a walk of some description. The results obtained were interesting; walks developed in all three runs, on average after about 30 generations with fine walking gaits after about 300 generations. This is about half the time on average that walking developed with a fixed joint range. We can see from Fig. 1 that the joint range associated with the individual with maximum fitness fluctuates wildly in early generations; typically low values (high movement ranges) initially predominate as the robot moves in a "thrashing" fashion. Then the movement range becomes restricted for the highest performing individuals as a smaller range of movement increases the likelihood that the robot will at least remain standing for a period, while hopefully moving a little. Then in later generations typically the movement range gradually becomes relaxed again, as a greater range of movement facilitates more rapid walking once the robot has "learnt" how to remain upright.

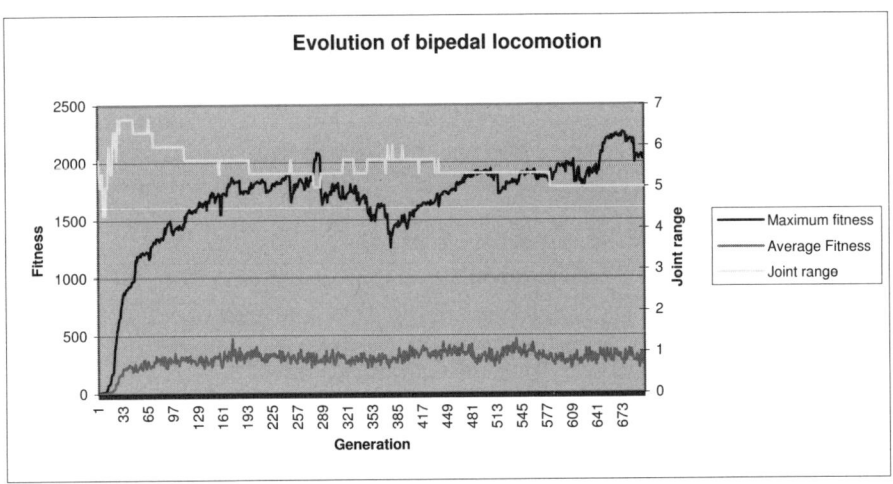

Fig. 1. Fitness and joint range graphs

4   Effect of Restraining Joints and of Environment Modification

We now investigate the effect of restraining motion in part of the robot. We do this by immobilising the robots right knee joint, and both ankle joints. This might correspond to a situation of a person walking with a prosthetic leg. Figure 2 shows the results of this experiment again averaged over 3 runs. The robot learns to walk albeit with a reduced maximum fitness compared to the robot with no constraints. Figure 3 illustrates a typical walk which develops. The right (constrained) leg moves sideways and forwards, coming well off the ground, as the right arm moves backwards in a steadying motion. The left leg follows in a shuffling motion, and the cycle repeats. In order to gain some insight into the evolutionary process we use a slightly modified version of the "*degree of population diversity*" described in by Leung et al. [35]. This measure provides an easy to calculate and useful measure of population diversity: i.e. how alike the different strings in a population. We subtract this value from the genome bit length to produce our inverse degree of population diversity measure (IDOPD). This value will vary from 0 (no similarity in the strings) to a value corresponding to the genome length (all genomes have the same value at every bit location). Diversity is measured on the right hand vertical axis. In addition for these experiments the number of bits encoding the joint range was increased from 8 to 16 giving a total genome length of 336 bits. We have also conducted successful experiments on the robot walking (skating) in low-friction environments, and in conditions of reduced gravity.

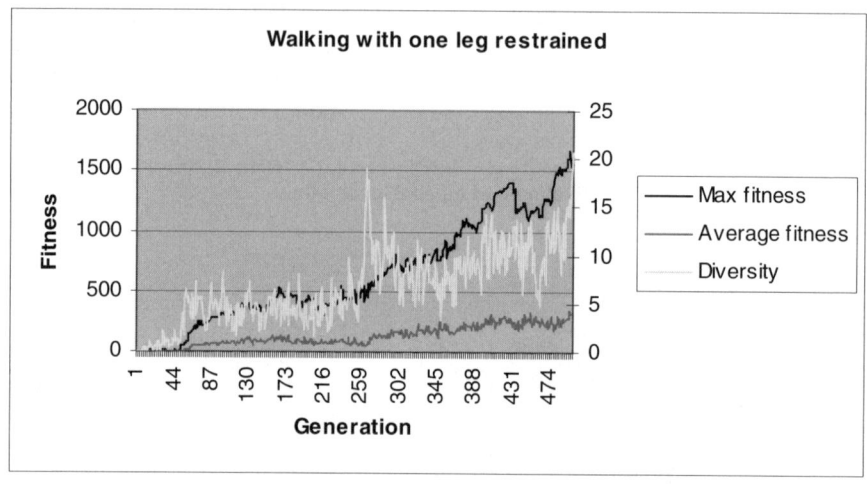

Fig. 2. Walking with right leg immobilised

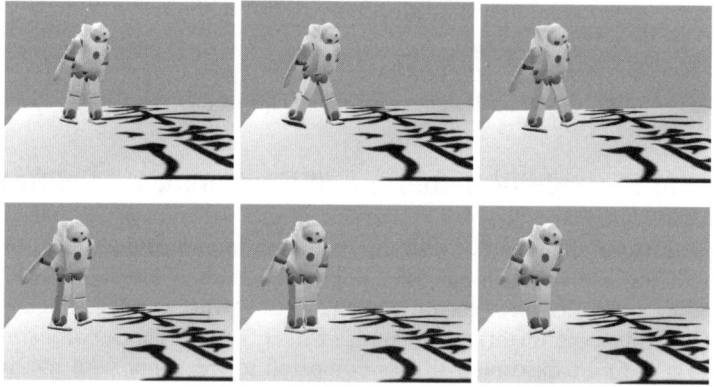

Fig. 3. The evolved sideways walk

5 The Bioloid Robotic Platform

We have been working for some time to implement our simulated robots in the real world using the Bioloid robot platform which is produced by Robotis Inc. Korea. This platform consists of a CPU (the CM-5), a number of senso-motoric actuators (Dynamixel AX12+) and a large number of universal frame construction pieces.

Using this platform it is possible to construct a wide variety of robots, from simple wheeled robots to complex humanoid robots with many degrees of freedom. In addition, because of the ability to construct a range of robots with slightly different morphologies, it lends itself well to evolutionary robotics experiments in which both robot "body and brain" are evolved. To gain initial experience with this kit we initially constructed a "puppy-bot" (Fig. 4) which can walk on four legs, avoid obstacles and perform several cute tricks. With this experience we then constructed the Bioloid humanoid robot which has 18 degrees of freedom in total. A modified version of this humanoid robot was used for Humanoid Team Humboldt in the RoboCup competitions in Bremen 2006. [36]

Two pieces of software are provided with the Bioloid system; the behaviour control programmer, and the motion editor. The behaviour control programmer programs the humanoids' response to different environmental stimuli, while the motion editor describes individual motions based on the keyframe concept described in our work.

Fig. 4. The puppy robot (left) and Bioloid humanoid robot (right)

6   Implementation of Simulated Robots

We have now built an accurate model of the Bioloid humanoid in Webots, and can translate the information in our sequence control file into a format understandable by the Bioloid motion editor. This translation is currently done partly by hand but we are working on fully automating this process. It is now possible to evolve walking, and other behaviours, in Webots using our model, and then transfer the evolved behaviour directly to the Bioloid humanoid robot. The graph below shows the maximum evolved fitness (right hand axis) and joint range and diversity measure (left hand axis) for the simulated Bioloid humanoid, averaged over five runs for 200 generations. Good walking patterns generally developed by generation 100 or so. Note the fitness function is similar to the one used for the PINO-like robot, but as the robot is a lot shorter (approx 35cm), obviously the distance traveled in a successful walk is less, hence the fitness values returned will be lower. Also the joint range values vary as in

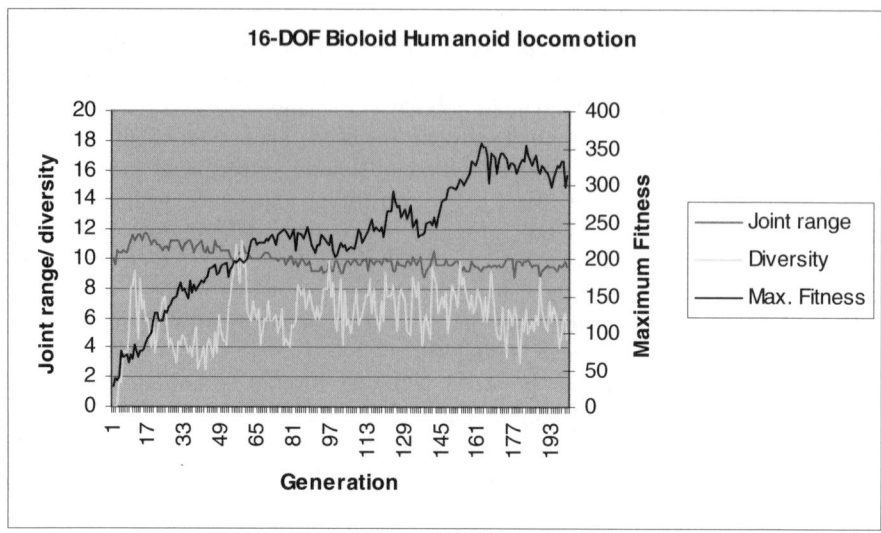

Fig. 5. Maximum fitness, Joint range and diversity graphs for 200 generations of the Webots simulation of the 16-DOF Bioloid Humanoid robot averaged over 5 runs

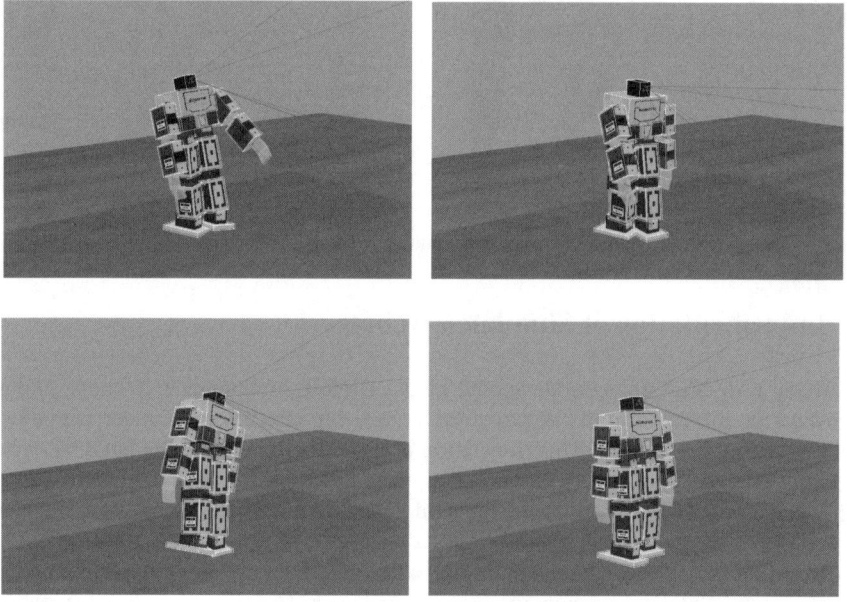

Fig. 6. Simulated snap shots of the Webots simulation of the Bioloid humanoid robot walking

the previous experiment, rising in the early runs before stabilizing out; The difference however is not as marked, as the robot is more inherently stable to begin with, so does not have to restrict its joint ranges as much in the early runs so as to remain standing.

The robot simulated was 16-DOF rather than the 18-DOF of the actual robot; both elbow joints were immobilized. This was done as they operate in an in-out fashion rather than the forward-backward configuration in the original PINO-like robot and this also might interfere with the arm movement. Also four additional 16-bit fields were added to specify the speed of movement for each of the four cycles, however this is not of particular relevance to the current discussion and will be discussed in more detail in a a later article.

When a walk has evolved this can then be transferred to the Bioloid humanoid via the motion editor. An example of an evolved simulated walk is given in Fig. 6 together with the walk as transferred to the Bioloid humanoid in Fig. 7. The fidelity of the transfer is reasonably good, indicating the accuracy of the model, however work remains to be done to fully "cross the reality gap"[37], but these initial results are very promising.

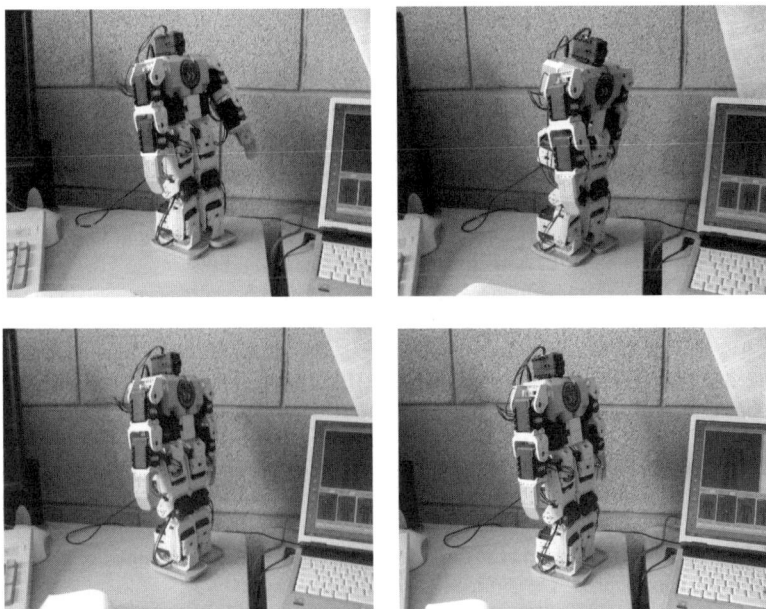

Fig. 7. The actual walking control sequence transferred to the physical robot

7   Discussion and Looking Ahead

In this work we have demonstrated one of the first applications of evolutionary algorithms to the development of complex movement patterns in a many-degree-of-freedom humanoid robot. Perhaps the time has now arrived for a more serious and detailed discussion on the possible ethical ramifications of the evolution of human-like robots. Such robots may be able to take our place in the workforce or in other fields and there may well also be other significant social consequences. Other, more technical, issues arise – while Asimov's three laws of robotics may appear a little dated, it could be important to avoid the appearance in the home or workplace of unexpected evolved "side effects" which may escape a rigorous testing regime. For example one of the walking behaviours evolved in our work involved the robot walking (staggering) in a manner amusingly reminiscent of an intoxicated person. While this gait proved surprisingly effective, not many people would relish the prospect of being shown around a new house for sale by a seemingly drunken robot! In conclusion, if indeed we are now beginning to see the first tentative "steps" towards autonomous humanoid robots, perhaps now is the time to look forward to the harnessing of this technology for the benefit of mankind.

Acknowledgments. Thanks are due to T.J. Davitt for his work on the initial version of the code for this system, and to Diarmuid Scannell for his work on the Webots Bioloid humanoid model. Also much appreciation to Olivier Michel of Cyberbotics for his help and rapid responses to my many queries.

References

1. Nolfi, S., Floreano, D.: Evolutionary Robotics: The Biology, Intelligence, and Technology of Self-Organizing Machines. MIT Press, Cambridge (2000)
2. Holland, O.: From the imitation of life to machine conciousness. In: Kunii, H.S., Jajodia, S., Sølvberg, A. (eds.) ER 2001. LNCS, vol. 2224, pp. 1–37. Springer, Heidelberg (2001)
3. Floreano, D., Urzelai, J.: Evolutionary Robotics: The Next Generation. In: Gomi, T. (ed.) Evolutionary Robotics III, AAI Books, Ontario (Canada) (2000)
4. Brooks, R.: Steps towards living machines. In: Kunii, H.S., Jajodia, S., Sølvberg, A. (eds.) ER 2001. LNCS, vol. 2224, pp. 72–93. Springer, Heidelberg (2001)
5. Harvey, I.: Artificial evolution: a continuing SAGA. In: Kunii, H.S., Jajodia, S., Sølvberg, A. (eds.) ER 2001. LNCS, vol. 2224, pp. 94–109. Springer, Heidelberg (2001)
6. Pollack, J., Lipson, H., Funes, P., et al.: First three generations of evolved robots. In: Kunii, H.S., Jajodia, S., Sølvberg, A. (eds.) ER 2001. LNCS, vol. 2224, pp. 62–71. Springer, Heidelberg (2001)
7. Brooks, R.: Evolutionary robotics: where from and where to. In: Embley, D.W. (ed.) ER 1997. LNCS, vol. 1331, pp. 1–13. Springer, Heidelberg (1997)
8. Pfeifer, R., Bongard, J.: How the body shapes the way we think: a new view of intelligence. MIT press, Cambridge (2007)
9. Mori, M.: Bukimi no tani [the uncanny valley]. Energy 7, 33–35 (1970)
10. MacDorman, K.F.: Mortality salience and the uncanny valley. In: Proceedings of 2005 IEEE-RAS International Conference on Humanoid Robots, IEEE Computer Society Press, Los Alamitos (2005)
11. Mataric, M., Cliff, D.: Challenges in evolving controllers for physical robots. Robotics and autonomous systems 19(1), 67–83 (1996)
12. Akachi, K., Kaneko, K., Kanehira, N., et al.: Development of humanoid robot HRP-3P. In: Proc. of 2005 5th IEEE-RAS International Conference on humanoid robots, pp. 50–55. IEEE Computer Society Press, Los Alamitos (2005)
13. Xie, M., Weng, J., Fontanie, J.: Editorial. International Journal of Humanoid Robotics 1(1) (2004)
14. Brooks, R., Braezel, C., Marjanovic, M., et al.: The Cog project: building a humanoid robot. In: Nehaniv, C.L. (ed.) Computation for Metaphors, Analogy, and Agents. LNCS (LNAI), vol. 1562, Springer, Heidelberg (1999)

15. Pfeifer, R., Scheier, C.: Implications of embodiment for robot learning. In: Proc. 2nd European workshop on advanced mobile robots, pp. 38–43 (October 1997)
16. Brooks, R., Aryananda, L., Edsinger, A., et al.: Sensing and manipulating built-for-human environments. International Journal of Humanoid Robotics 1(1), 1–28 (2004)
17. Full, R.: Using biological inspiration to build artificial life that locomotes. In: Kunii, H.S., Jajodia, S., Sølvberg, A. (eds.) ER 2001. LNCS, vol. 2224, pp. 110–120. Springer, Heidelberg (2001)
18. Ambrose, R., Ambrose, C.: Primate anatomy, kinematics, and principles for humanoid design. International Journal of Humanoid Robotics 1(1), 175–198 (2004)
19. Zykov, V., Bongard, J., Lipson, H.: Evolving Dynamic Gaits on a Physical Robot. In: Deb, K. (ed.) GECCO 2004. LNCS, vol. 3102, Springer, Heidelberg (2004)
20. Bongard, J., Zykov, V., Lipson, H.: Resilient machines through continuous self modeling. Science 314, 1118–1121 (2006)
21. Gat, E.: Towards a principled experimental study of autonomous mobile robots. Autonomous Robots 2, 179–189 (1995)
22. Eaton, M., Collins, J.J., Sheehan, L.: Towards a benchmarking framework for research into bio-inspired hardware-software artefacts. Artificial Life and Robotics 5(1), 40–46 (2001)
23. Boeing, A., Hanham, S., Braunl, T.: Evolving autonomous biped control from simulation to reality. In: Proc. 2nd International Conference on Autonomous Robots and Agents, December 13-15, pp. 440–445 (2004)
24. Bongard, J., Paul, C.: Making evolution an offer it can't refuse: morphology and the extradimensional bypass. In: Kelemen, J., Sosík, P. (eds.) ECAL 2001. LNCS (LNAI), vol. 2159, pp. 401–412. Springer, Heidelberg (2001)
25. Ishiguro, A., Fujii, A., Eggenburger, H.: Neuromodulated control of bipedal locomotion using a polymorphic CPG circuit. Adaptive Behavior 11(1), 7–18 (2003)
26. Endo, K., Maeno, T., Kitano, H.: Co-evolution of morphology and walking pattern of biped humanoid robot using evolutionary computation-consideration of characteristic of the servomotors. In: Proceedings of the 2002 IEEE/RSJ Intl. Conference on Intelligent Robots and Systems, Lausanne, Switzerland, pp. 2678–2683 (October 2002)
27. Zhang, R., Vadakkepat: An evolutionary algorithm for trajectory based gait generation of biped robot. In: Proceedings of the International Conference on Computational Intelligence, Robotics and Autonomous Systems, Singapore (2003)
28. Reil, T., Husbands, P.: Evolution of central pattern generators for bipedal walking in a real-time physics environment. IEEE Trans. On Evolutionary Computation 6(2), 159–168 (2002)
29. Sellers, W.I., Dennis, L.A., Crompton, R.H.: Predicting the metabolic energy costs of bipedalism using evolutionary robotics. The Journal of Experimental Biology 206, 1127–1136 (2003)
30. Miyashita, K., Ok, S., Kazunori, H.: Evolutionary generation of human-like bipedal locomotion. Mechatronics 13, 791–807 (2003)
31. Collins, S., Ruina, A., Tedrake, R., et al.: Efficient bipedal robots based on passive-dynamic walkers. Science 307, 1082–1085 (2005)
32. Michel, O.: Webots: Professional Mobile Robot Simulation. International Journal of Advanced Robotic Systems 1(1), 39–42
33. Mojon, S.: Realization of a Physic Simulation for a Biped Robot. In: Semester project, School of Computer and Communication Sciences EPFL, Switzerland (2003)
34. Eaton, M., Davitt, T.J.: Automatic evolution of bipedal locomotion in a simulated humanoid robot with many degrees of freedom. In: Proceedings of the 11th International Symposium on Artificial Life and Robotics, Japan, January 23-25, 2006, pp. 448–451 (2006)

35. Leung, Y., Gao, Y., Zong-Ben, X.: Degree of population diversity – a perspective on premature convergence in genetic algorithms and its Markov chain analysis. IEEE Transactions on Neural Networks 8(5), 1165–1176 (1997)
36. Hild, M., Jungel, M., Spranger, M.: Humanoid team Humboldt team description 2006. In: Proceedings CD Robocup 2006, Bremen, Germany (2006)
37. Lipson, H., Bongard, J., Zykov, V., Malone, E.: Evolutionary robotics for legged machines: from simulation to physical reality. In: Arai, T., et al. (eds.) Intelligent Autonomous Systems 9(IAS-9), pp. 11–18 (2006)

Philosophical Foundations of AI

David Vernon[1] and Dermot Furlong[2]

[1] Etisalat University College, UAE
[2] Trinity College Dublin, Ireland

Abstract. Artificial Intelligence was born in 1956 as the off-spring of
the newly-created cognitivist paradigm of cognition. As such, it inherited
a strong philosophical legacy of functionalism, dualism, and positivism.
This legacy found its strongest statement some 20 years later in the physical symbol systems hypothesis, a conjecture that deeply influenced the
evolution of AI in subsequent years. Recent history has seen a swing away
from the functionalism of classical AI toward an alternative position that
re-asserts the primacy of embodiment, development, interaction, and,
more recently, emotion in cognitive systems, focussing now more than
ever on enactive models of cognition. Arguably, this swing represents a
true paradigm shift in our thinking. However, the philosophical foundations of these approaches — phenomenology — entail some far-reaching
ontological and epistemological commitments regarding the nature of a
cognitive system, its reality, and the role of its interaction with its environment. The goal of this paper is to draw out the full philosophical
implications of the phenomenological position that underpins the current
paradigm shift towards enactive cognition.

1 Philosophical Preliminaries

Realism is a doctrine which holds that the objects of our perceptions are what
is real and that reality is what is directly perceived; it is through our perceptions that we apprehend the actual real external world. The tradition of modern
realism has an long pedigree, beginning with Ockham and continuing through
Gallileo, Hobbes, Locke, Hume, Moore, and Russell. Gallileo, along with, *e.g.*,
Copernicus, Descartes, and Kepler, heralded the beginning of the scientific age
which placed all empirical measurement and quantification along with rigourous
mathematical (or logical) reasoning as the cornerstones for the construction of
knowledge. This empiricist ethos was strengthed by John Locke, a quintessential realist, who viewed perception as a causal process whereby physical stimuli
act on the sensory apparatus to produce ideas (concepts or representations, in
the modern terminology). Much of today's common understanding of reality is
a legacy of this Lockean frame of mind. In realistic positions, there is the un-
derpinning assumption that reality exists absolutely and, whether rationally by
reason or empirically by sense, we apprehend it and thus come to understand its
form and structure.

Idealism, on the other hand, is a doctrine which posits that reality is ultimately dependent on the mind and has no existence outside of it. If Locke was

M. Lungarella et al. (Eds.): 50 Years of AI, Festschrift, LNAI 4850, pp. 53–62, 2007.

the quintessential realist, then Berkeley was the quintessential idealist. Berkeley developed the philosophy that nothing exists save that which is perceived by a mind. This is neatly summarized by his famous aphorism '*esse est percipi*' — to be is to be perceived. Berkeley's position is that our idea about the world are based on our perceptions of it. In this sense, Berekeley is also taking an empirical position — that our knowledge of the world is gained exclusively from our senses. On the other hand, Berkeley denied the existence of matter: what exists is that which is perceived, and it exists because it is perceived. Reality pervades all perception but corporeal matter has no place in this scheme. This denial of the reality of matter distinguishes Berkeley's empirical idealist notions of perception from the realist, empirical, notion that perception is an abstraction or apprehension of the (material) world *via* a causal process of sensing.

Kant (1724-1804) was also an idealist, but his views differed significantly from those of Berkeley. Kant differentiated between *noumena*, the domain of 'things in themselves' and *phenomena*, or the 'appearances' of things as they are presented to us by our senses. Kant argued that noumena are not accessible to us, and cannot be known directly, whereas the phenomena — the contact we have with these things via our senses and perceptions — are the basis for knowledge. Kant refers to noumena as 'trancendental objects' and his philosophy is sometimes referred to as 'trancendental idealism'. Thus, Kant admits the 'reality' of a domain of objects, the unknowable noumenological domain. On the other hand, he maintains that the objects of our experience are the only knowable objects and it is the mind that shapes and forms these sense data and, hence, for us, these objects are the only objects that really exist and they exist *because* of us and our minds. Reality, then, exists as an unknowable, non-sensible, noumenal domain which gives rise to the phenomenal domain of our senses.[1] The idealist tradition did not stop with Kant and has been added to by, *e.g.*, Schopenhauer, Nietzsche, and Hegel.

There are many variations on these two themes of idealism and realism, perhaps the most well-known of which is *dualism* which holds that reality comprises two distinct 'substances': one physical and one mental. Dualism was first propounded as a philosophical system by Descartes who argued for the existence of *two* domains of reality: one corporeal and one non-corporeal. Both mutually-exclusive domains exist concurrently. It is this mutual exclusivity which has caused dualism most of its problems for, if they are truly mutually exclusive, it is not clear how they can interact. This difficulty has been transposed into modern philosophical debate as the 'mind-body' problem: the paradox that if the body and mind are mutually exclusive entities, then how do they 'communicate'?

In the above, we have attempted the impossible: to summarize five hundred years of philosophical thought in a few paragraphs. Nonetheless, from this cursory look at the history of western philosophy, we can see that the philosophical

---

[1] Although Kant is best known as an idealist, his particular brand of philosophical idealism is sometimes referred to as constructive realism due to the central role played by the observer in shaping phenomenal reality.

positions on reality have been dominated by realism (including dualism). Additionally, the philosophies that have been most closely aligned with the scientific method have also been those of realism. In a sense, this isn't surprising since realism is the more immediately common-sense view: things exist — we perceive them. This world-view has been copper-fastened in the last century by the logical positivists, *e.g.* Schlick and Carnap, who held that reality is exactly that which yields to empirical investigation and anything that is not verifiable by empirical investigation is meaningless.

There were, of course, other developments in philosophical thinking, which begin with Kant's distinction between noumena and phenomena, and which evolved into a type of reconciliation of the idealist and the realist positions. The one that interests us here was developed by Husserl, who held that reality is personally and fundamentally *phenomenological* but is set against an objective spatio-temporal world. However, it was best espoused by Heidegger who denied the dichotomy between the world and 'us' and saw existence or 'being in the world' as our activity in a constitutive domain. Reality does not exist 'outside us'; we are beings in a world, not disjoint from it. From a phenomenological perspective, what we perceive depends on what it is we are. The position taken by phenomenology is subtly, but significantly, different to that taken by either realism or idealism. The position is as follows. We play a role in defining reality, but only insofar as it affects us as individuals (the idealist aspect), that is, insofar as it affects our experience of reality; the reality that we perceive does exist (the realist aspect) but our perception and conception of it is conditioned by our experience. Thus, reality is for us a personal experience, though it derives from a common source and this reality is our experience and is contingent upon the current ontological status of us as entities in that universe. As perceivers, our perceptions of the world are a function of what we are: reality is conditioned by experience and experience is conditioned by the nature of the system and its history of interaction with reality.

The dependence of reality on the ontogenetic state of an individual is the essential characteristic of phenomenology and is often referred to as radical constructivism: we construct our reality as a consequence of our perceptions and experiences. Unfortunately, the term constructivism is also sometimes used to denote an entirely different realist position taken by advocates of the cognitivist approach to artificial intelligence whereby representations of the external world are constructed through perception. Consequently, one must be careful when interpreting the term constructivism to be clear exactly what is meant: the radical constructivism of phenomenology or the representational constructivism of realism.

2   The Birth of AI

The development of cybernetics in 1943 heralded the birth of cognitive science and an attempt to create a formal logical model of cognition and a science of mind [1]. The year 1956 saw the emergence of an approach referred to as *cognitivism* which asserts that cognition involves computations defined over internal representations, in a process whereby information about the world is abstracted by perception, and represented using some appropriate symbolic data-structure, reasoned about, and then used to plan and act in the world.

For cognitivist systems, perception is concerned with the abstraction of faithful spatio-temporal representations of the external world from sensory data. Reasoning itself is symbolic: a procedural process whereby explicit representations of an external world are manipulated to infer likely changes in the configuration of the world that arise from causal actions.

In most cognitivist approaches concerned with the creation of artificial cognitive systems, the symbolic representations are typically the descriptive product of a human designer. This means that they can be directly accessed and understood or interpreted by humans and that semantic knowledge can be embedded directly into and extracted directly from the system. These programmer-dependent representations 'blind' the system [2] and constrain it to an idealized description that is dependent on and a consequence of the programmer's own cognition. Arguably, it is this *a priori* designer- or programmer-dependent knowledge that is embedded in the system that limits the adapability of the cognitive system since this knowledge intrinsically encapsulates the designer's assumptions about the system's environment, it operation, and its space of interaction.

Cognitivism makes the realist assumption that 'the world we perceive is isomorphic with our perceptions of it as a geometric environment' [3]. Today, cognitivist systems will deploy an arsenal of techniques including machine learning, probabilistic modelling, and other techniques in an attempt to deal with the inherently uncertain, time-varying, and incomplete nature of the sensory data that is being used to drive this representational framework. However, ultimately the representational structure is still predicated on the descriptions of the designer.

AI is the direct descendent of cognitivism [4] and represents the empirical side of cognitivist cognitive science. A major milestone in its development occured in 1976 with Newell's and Simon's 'Physical Symbol System' approach [5]. In their paper, two hypotheses are presented:

1. *The Physical Symbol System Hypothesis:* A physical symbol system has the necessary and sufficient means for general intelligent action.
2. *Heuristic Search Hypothesis:* The solutions to problems are represented as symbol structures. A physical-symbol system exercises its intelligence in problem-solving by search, that is, by generating and progressively modifying symbol structures until it produces a solution structure.

The first hypothesis implies that any system that exhibits general intelligence is a physical symbol system *and* any physical symbol system of sufficient size can be configured somehow ('organized further') to exhibit general intelligence.

The second hypothesis amounts to a assertion that symbol systems solve problems by heuristic search, *i.e.* 'successive generation of potential solution structures' in an effective and efficient manner. 'The task of intelligence, then, is to avert the ever-present threat of the exponential explosion of search'.

A physical symbol system is equivalent to an automatic formal system[6]. It is 'a machine that produces through time an evolving collection of symbol structures.' A symbol is a physical pattern that can occur as a component of another type of entity called an expression (or symbol structure): expressions/symbol structures are arrangements of symbols/tokens. As well as the symbol structures, the system also comprises processes that operate on expressions to produce other expressions: 'processes of creation, modification, reproduction, and destruction'. An expression can *designate* an object and thereby the system can either 'affect the object itself or behave in ways depending on the object', or, if the expression designates a process, then the system *interprets* the expression by carrying out the process. In the words of Newell and Simon,

> 'Symbol systems are collections of patterns and processes, the latter being capable of producing, destroying, and modifying the former. The most important properties of patterns is that they can designate objects, processes, or other patterns, and that when they designate processes, they can be interpreted. Interpretation means carrying out the designated process. The two most significant classes of symbol systems with which we are acquainted are human beings and computers.'

What is important about this explanation of a symbol system is that it is more general than the usual portrayal of symbol-manipulation systems where symbols designate only objects, in which case we have a system of processes that produces, destroys, and modifies symbols, and no more. Newell's and Simon's original view is more sophisticated. There are two recursive aspects to it: processes can produce processes, and patterns can designate patterns (which, of course, can be processes). These two recursive loops are closely linked. Not only can the system build ever more abstract representations and reason about those representation, but it can modify itself as a function both of its processing, *qua* current state/structure, and of its representations.

Symbol systems can be instantiated and the behaviour of these instantiated systems depend on the the details of the symbol system, its symbols, operations, and interpretations, and *not* on the particular form of the instantiation.

The *physical symbol system hypothesis* asserts that a physical symbol system has the necessary and sufficient means for general intelligence. From what we have just said about symbol systems, it follows that intelligent systems, either natural or artificial ones, are effectively equivalent because the instantiation is actually inconsequential, at least in principle.

To a very great extent, cognitivist systems are identical to physical symbol systems.

The strong interpretation of the physical symbol system hypothesis is that not only is a physical symbol system sufficient for general intelligence, it is also necessary for intelligence.

It should be clear that cognitivism, and the classical AI of physical symbol systems, are dualist, functionalist, and positivist. They are dualist in the sense that there is a fundamental distinction between the mind (the computational processes) and the body (the computational infrastructure and, where required, the plant that instantiates any physical interaction). The are functionalist in the sense that the actual instantiation and computational infrastructure is inconsequential: any instantiation that supports the symbolic processing is sufficient. They are positivist in the sense that they assert a unique and absolute empirically-accessible external reality that is apprended by the senses and reasoned about by the cognitive processes.

3 Enaction

Cognitivism is not however the only position one can take on cognition. There is a second class of approaches, all based to a lesser or greater extent on principles of emergent self-organization [1,7] and best epitomized by enactive approaches.

The enactive systems research agenda stretches back to the early 1970s in the work of computational biologists Maturana and Varela [8,9,10,11,1,2,12]. In contradistinction to cognitivism, which involves a view of cognition that requires the representation of a given objective pre-determined world [13,1], enaction asserts that cognition is a process whereby the issues that are important for the continued existence of a cognitive entity are brought out or enacted: co-determined by the entity as it interacts with the environment in which it is embedded. Thus, nothing is 'pre-given'. Instead there is an enactive interpretation: a real-time context-based choosing of relevance. Cognition is the process whereby an autonomous system becomes viable and effective in its environment. It does so through a process of self-organization through which the system is continually re-constituting itself in real-time to maintain its operational identity through moderation of mutual system-environment interaction and co-determination [12]. Co-determination implies that the cognitive agent is specified by its environment and at the same time that the cognitive process determines what is real or meaningful for the agent. In a sense, co-determination means that the agent constructs its reality (its world) as a result of its operation in that world. Thus, for emergent approaches, perception is concerned with the acquisition of sensory data in order to enable effective action [12] and is crucially dependent on the richness of the action interface [14]. It is not a process whereby the structure of an absolute external environment is abstracted and represented in a more or less isomorphic manner.

In contrast to the cognitivist approach, many enactive approaches assert that the primary model for cognitive learning is anticipative skill construction rather than knowledge acquisition and that processes that both guide action and improve the capacity to guide action while doing so are taken to be the root capacity for all intelligent systems [15]. While cognitivism entails a self-contained abstract model that is disembodied in principle, the physical instantiation of the systems plays no part in the model of cognition [16,17]. In contrast, enactive approaches are intrinsically embodied and the physical instantiation plays a pivotal constitutive role in cognition.

With enactive systems, one of the key issues is that cognitive processes are temporal processes that 'unfold' in real-time and synchronously with events in their environment. This strong requirement for synchronous development in the context of its environment is significant for two reasons. First, it places a strong limitation on the rate at which the ontogenetic learning of the cognitive system can proceed: it is constrained by the speed of coupling (*i.e.* the interaction) and not by the speed at which internal changes can occur [2]. Second, taken together with the requirement for embodiment, we see that the consequent historical and situated nature of the systems means that one cannot short-circuit the onto-genetic development. Specifically, you can't bootstrap an emergent dynamical system into an advanced state of learned behaviour.

For cognitivism, the role of cognition is to abstract objective structure and meaning through perception and reasoning. For enactive systems, the purpose of cognition is to uncover unspecified regularity and order that can then be con-strued as meaningful because they facilitate the continuing operation and devel-opment of the cognitive system. In adopting this stance, the enactive position challenges the conventional assumption that the world *as the system experiences it* is independent of the cognitive system ('the knower'). The only condition that is required of an enactive system is *effective action*: that it permit the contin-ued integrity of the system involved. It is essentially a very neutral position, assuming only that there is the basis of order in the environment in which the cognitive system is embedded. From this point of view, cognition is exactly the process by which that order or some aspect of it is uncovered (or constructed) by the system. This immediately allows that there are different forms of reality (or relevance) that are dependent directly on the nature of the dynamics making up the cognitive system. This is not a solipsist position of ungrounded subjec-tivism, but neither is it the commonly-held position of unique — representable — realism. It is fundamentally a phenomenological position.

The goal of enactive systems research is the complete treatment of the nature and emergence of autonomous, cognitive, social systems. It is founded on the concept of autopoiesis – literally *self-production* – whereby a system emerges as a coherent systemic entity, distinct from its environment, as a consequence of processes of self-organization.

In the enactive paradigm, linguistic behaviours are at the intersection of onto-genetic and communication behaviours and they facilitate the creation of a com-mon understanding of the shared world that is the environment of the coupled systems. That is, language is the emergent consequence of the structural coupling of a socially-cohesive group of cognitive entities. Equally, knowledge is particular to the system's history of interaction. If that knowledge is shared among a soci-ety of cognitive agents, it is not because of any intrinsic abstract universality, but because of the consensual history of experiences shared between cognitive agents with similar phylogeny and compatible ontogeny. A key postulate of enactive sys-tems is that reasoning, as we commonly conceive it, is the consequence of reflexive[2] use of the linguistic descriptive abilities to the cognitive agent itself [12]. Linguistic capability is in turn developed as a consequence of the consensual co-development of an epistemology in a society of phylogenetically-identical cognitive agents. This

---

[2] Reflexive in the sense of self-referential, not in the sense of a reflex action.

is significant: reasoning in this sense is a descriptive phenomenon and is quite distinct from the self-organizing mechanism (*i.e.* structural coupling and operational closure [12]) by which the system/agent develops its cognitive and linguistic behaviours. Since language (and all inter-agent communication) is a manifestation of high-order cognition, specifically co-determination of consensual understanding amongst phylogenetically-identical and ontogenetically-compatible agents, symbolic or linguistic reasoning is actually the product of higher-order social cognitive systems rather than a generative process of the cognition of an individual agent.

4   Conclusion

The chief point we wish to make in this paper is that the differences between the cognitivist and emergent positions are deep and fundamental, and go far beyond a simple distinction based on symbol manipulation. It isn't principally the symbolic nature of the processing that is at issue in the divide between the cognitivist and the emergent approaches — it is arguable that linguistically-capable enactive systems explicitly use symbols when reasoning. Neither is it the presence or use of a physical body or situated perceptual agents. Cognitivists now readily admit the need for embodiment; in Anderson's words: 'There is reason to suppose that the nature of cognition is strongly determined by the perceptual-motor systems, as the proponents of embodied and situated cognition have argued' [18]. Elsewhere they are compared on the basis of several related characteristics [19] but in this paper, we have contrasted the two paradigms on the basis of their philosophical commitments: the functionalist, dualist, and positivist ground of cognitivist cognition versus the phenomenological agent-specific mutual-specification of enactive cognition.

In the enactive paradigm, the perceptual capacities are a consequence of an historic embodied development and, consequently, are dependent on the richness of the motoric interface of the cognitive agent with its world. That is, the action space defines the perceptual space and thus is fundamentally based in the frame-of-reference of the agent. Consequently, the enactive position is that cognition can only be created in a developmental agent-centred manner, through interaction, learning, and co-development with the environment. It follows that through this ontogenic development, the cognitive system develops its own epistemology, *i.e.* its own system-specific knowledge of its world, knowledge that has meaning exactly because it captures the consistency and invariance that emerges from the dynamic self-organization in the face of environmental coupling. Thus, we can see that, from this perspective, cognition is inseparable from 'bodily action' [20]: without physical embodied exploration, a cognitive system has no basis for development. Despite the current emphasis on embodiment, Ziemke notes that many current approaches in cognitive/adaptive/epigenetic robotics still adhere to the functionalist dualist hardware/software distinction in the sense that the computational model does not in principle require an instantiation [21,22]. Ziemke suggests that this is a real problem because the idea of embodiment in the enactive sense is that the morphology of the system is actually a key component of the systems dynamics. In other words, morphology not only matters, it is a constitutive part of the system's self-organization and structural coupling with the environment and defines its cognition and developmental capacity.

There are many challenges to be overcome in pushing back the boundaries of AI research, particularly in the practice of enactive AI. Foremost among these is the difficult task of identifying the phylogeny and ontogeny of an artificial cognitive system: the requisite cognitive architecture that facilitates both the system's autonomy (*i.e.* its self-organization and structural coupling with the environment) and its capacity for development and self-modification. To allow true ontogenetic development, this cognitive architecture must be embodied in a way that allows the system the freedom to explore and interact and to do so in an adaptive physical form that enables the system to expand its space of possible autonomy-preserving interactions. This in turn creates a need for new physical platforms that offer a rich repertoire of perception-action couplings and a morphology that can be altered as a consequence of the system's own dynamics. In meeting these challenges, we move well beyond attempts to build cognitivist systems that exploit embedded knowledge and which try to see the world the way we designers see it. We even move beyond learning and self-organizing systems that uncover for themselves statistical regularity in their perceptions. Instead, we set our sights on building enactive phenomenologically-grounded systems that construct their own understanding of their world through adaptive embodied exploration and social interaction.

Acknowledgments

The authors would like to acknowledge the helpful suggestions made by the two reviewers of an earlier version of this paper. This work was supported in part by the European Commission, Project IST-004370 RobotCub, under Strategic Objective 2.3.2.4: Cognitive Systems.

References

1. Varela, F.J.: Whence perceptual meaning? A cartography of current ideas. In: Varela, F.J., Dupuy, J.-P. (eds.) Understanding Origins – Contemporary Views on the Origin of Life, Mind and Society. Boston Studies in the Philosophy of Science, pp. 235–263. Kluwer Academic Publishers, Dordrecht (1992)
2. Winograd, T., Flores, F.: Understanding Computers and Cognition – A New Foundation for Design. Addison-Wesley Publishing Company, Inc., Reading, Massachusetts (1986)
3. Shepard, R.N., Hurwitz, S.: Upward direction, mental rotation, and discrimination of left and right turns in maps. Cognition 18, 161–193 (1984)
4. Freeman, W.J., Núñez, R.: Restoring to cognition the forgotten primacy of action, intention and emotion. Journal of Consciousness Studies 6(11-12) ix–xix (1999)
5. Newell, A., Simon, H.A.: Computer science as empirical inquiry: Symbols and search. Communications of the Association for Computing Machinery 19, 113–126 (1976)

6. Haugland, J.: Semantic engines: An introduction to mind design. In: Haugland, J. (ed.) Mind Design: Philosophy, Psychology, Artificial Intelligence, pp. 1–34. Bradford Books, MIT Press, Cambridge, Massachusetts (1982)
7. Clark, A.: Mindware – An Introduction to the Philosophy of Cognitive Science. Oxford University Press, New York (2001)
8. Maturana, H.: Biology of cognition. Research Report BCL 9.0, University of Illinois, Urbana, Illinois (1970)
9. Maturana, H.: The organization of the living: a theory of the living organization. Int. Journal of Man-Machine Studies 7(3), 313–332 (1975)
10. Maturana, H.R., Varela, F.J.: Autopoiesis and Cognition — The Realization of the Living. Boston Studies on the Philosophy of Science. D. Reidel Publishing Company, Dordrecht, Holland (1980)
11. Varela, F.: Principles of Biological Autonomy. Elsevier North Holland, New York (1979)
12. Maturana, H., Varela, F.: The Tree of Knowledge – The Biological Roots of Human Understanding. New Science Library, Boston & London (1987)
13. van Gelder, T., Port, R.F.: It's about time: An overview of the dynamical approach to cognition. In: Port, R.F., van Gelder, T. (eds.) Mind as Motion – Explorations in the Dynamics of Cognition, pp. 1–43. Bradford Books, MIT Press, Cambridge, Massachusetts (1995)
14. Granlund, G.H.: The complexity of vision. Signal Processing 74, 101–126 (1999)
15. Christensen, W.D., Hooker, C.A.: An interactivist-constructivist approach to intelligence: self-directed anticipative learning. Philosophical Psychology 13(1), 5–45 (2000)
16. Vernon, D.: The space of cognitive vision. In: Christensen, H.I., Nagel, H.-H. (eds.) Cognitive Vision Systems: Sampling the Spectrum of Approaches. LNCS, pp. 7–26. Springer, Heidelberg (2006)
17. Vernon, D.: Cognitive vision: The case for embodied perception. Image and Vision Computing, 1–14 (in press, 2007)
18. Anderson, J.R., Bothell, D., Byrne, M.D., Douglass, S., Lebiere, C., Qin, Y.: An integrated theory of the mind. Psychological Review 111(4), 1036–1060 (2004)
19. Vernon, D., Metta, G., Sandini, G.: A survey of artificial cognitive systems: Implications for the autonomous development of mental capabilities in computational agents. IEEE Transaction on Evolutionary Computation 11(1), 1–31 (2006)
20. Thelen, E.: Time-scale dynamics and the development of embodied cognition. In: Port, R.F., van Gelder, T. (eds.) Mind as Motion – Explorations in the Dynamics of Cognition, pp. 69–100. Bradford Books, MIT Press, Cambridge, Massachusetts (1995)
21. Ziemke, T.: Are robots embodied? In: Balkenius, Z., Dautenhahn, K., Breazeal (eds.) Proceedings of the First International Workshop on Epigenetic Robotics — Modeling Cognitive Development in Robotic Systems, Lund University Cognitive Studies, Lund, Sweden. vol. 85, pp. 75–83 (2001)
22. Ziemke, T.: What's that thing called embodiment? In: Alterman, Kirsh (eds.) Proceedings of the 25th Annual Conference of the Cognitive Science Society. Lund University Cognitive Studies, pp. 1134–1139. Lawrence Erlbaum, Mahwah, NJ (2003)

On the Role of AI in the Ongoing Paradigm Shift within the Cognitive Sciences

Tom Froese

Centre for Computational Neuroscience and Robotics, University of Sussex, UK
t.froese@sussex.ac.uk

Abstract. This paper supports the view that the ongoing shift from orthodox to embodied-embedded cognitive science has been significantly influenced by the experimental results generated by AI research. Recently, there has also been a noticeable shift toward enactivism, a paradigm which radicalizes the embodied-embedded approach by placing autonomous agency and lived subjectivity at the heart of cognitive science. Some first steps toward a clarification of the relationship of AI to this further shift are outlined. It is concluded that the success of enactivism in establishing itself as a mainstream cognitive science research program will depend less on progress made in AI research and more on the development of a phenomenological pragmatics.

Keywords: AI, cognitive science, paradigm shift, enactivism, phenomenology.

1 Introduction

Over the last two decades the field of artificial intelligence (AI) has undergone some significant developments (Anderson 2003). Good old-fashioned AI (GOFAI) has faced considerable problems whenever it attempts to extend its domain beyond simplified "toy worlds" in order to address context-sensitive real-world problems in a robust and flexible manner[1]. These difficulties motivated the Brooksian revolution toward an embodied and situated robotics in the early 1990s (Brooks 1991). Since then this approach has been further developed (e.g. Pfeifer & Scheier 1999; Pfeifer 1996; Brooks 1997), and has also significantly influenced the emergence of a variety of other successful methodologies, such as the dynamical approach (e.g. Beer 1995), evolutionary robotics (e.g. Harvey *et al.* 2005; Nolfi & Floreano 2000), and organismically-inspired robotics (e.g. Di Paolo 2003). These approaches are united by the claim that cognition is best understood as embodied and embedded in the sense that it emerges out of the dynamics of an extended brain-body-world systemic whole.

These developments make it evident that the traditional GOFAI mainstream, with its emphasis on perception as representation and cognition as computation, is being challenged by the establishment of an alternative paradigm in the form of

---

[1] For example: the commonsense knowledge problem (Dreyfus 1991, p. 119), the frame problem (McCarthy & Hayes 1969), and the symbol grounding problem (Harnard 1990).

M. Lungarella et al. (Eds.): 50 Years of AI, Festschrift, LNAI 4850, pp. 63–75, 2007.

embodied-embedded AI. How is this major shift in AI related to the ongoing paradigm shift within the cognitive sciences? [2] Section 2 analyzes the role of AI in the emergence of what has been called "embodied-embedded" cognitive science (e.g. Clark 1997; Wheeler 2005). Recently, there has also been a noticeable shift in interest toward "enactivism" (e.g. Thompson 2007; 2005; Di Paolo, Rohde & De Jaegher 2007; Torrance 2005), a paradigm which radicalizes the embodied-embedded approach by placing autonomous agency and lived subjectivity at the heart of cognitive science. How AI relates to this further shift is still in need of clarification, and section 3 provides some initial steps in this direction. Finally, section 4 argues that since many of the claims of enactivism are grounded in the phenomenological domain, its success as a major cognitive science research program depends less on progress in AI research and more on the development of a phenomenological pragmatics.

2   Toward Embodied-Embedded Cognitive Science

Much of contemporary cognitive science owes its existence to the founding of the field of AI in the late 1950s by the likes of Herbert Simon, Marvin Minsky, Allen Newell, and John McCarthy. These researchers, along with Noam Chomsky, put forth ideas that were to become the major guidelines for the computational approach which has dominated the cognitive sciences since its inception[3]. In order to determine the impact of AI on the ongoing shift from such orthodox computationalism toward embodied-embedded cognitive science, it is necessary to briefly consider some of the central claims associated with these competing theoretical frameworks.

2.1   Theories of Cognition

The paradigm that came into existence with the birth of AI, and which was essentially identified with cognitive science itself for the ensuing three decades and which still represents the mainstream today, is known as *cognitivism* (e.g. Fodor 1975). The cognitivist claim, that cognition is a form of computation (i.e. information processing through the manipulation of symbolic representations), is famously articulated in the Physical-Symbol System Hypothesis which holds that such a system "has the necessary and sufficient means for general intelligent action" (Newell & Simon 1976). From the cognitivist perspective cognition is essentially centrally controlled, disembodied, and decontextualized reasoning and planning as epitomized by abstract problem solving. Accordingly, the mind is conceptualized as a digital computer and cognition is viewed as fundamentally distinct from the embodied action of an autonomous agent that is situated within the continuous dynamics of its environment.

The cognitivist orthodoxy remained unchallenged until *connectionism* arose in the early 1980s (e.g. McClelland, Rumelhart *et al.* 1986). The connectionist alternative views cognition as the emergence of global states in a network of simple components,

---

[2] Whether any of the major changes in AI or cognitive science are paradigm shifts in the strict Kuhnian sense is an interesting question but beyond the scope of this paper. Here the notion is used in the more general sense of a major shift in experimental practice and focus.

[3] See Boden (2006) for an extensive overview of the history of cognitive science.

and promises to address two shortcomings of cognitivism, namely by 1) increasing efficiency through parallel processing, and 2) achieving greater robustness through distributed operations. Moreover, because it makes use of artificial neural networks as a metaphor for the mind, its theories of cognition are often more biologically plausible. Nevertheless, connectionism still retains many cognitivist commitments. In particular, it maintains the idea that cognition is essentially a form of information processing in the head which converts a set of inputs into an appropriate set of outputs in order to solve a given problem. In other words, "connectionism's disagreement with cognitivism was over the nature of computation and representation (symbolic for cognitivists, subsymbolic for connectionsists)" (Thompson 2007, p. 10), rather than over computationalism as such (see also Wheeler 2005, p. 75). Accordingly, most of connectionism can be regarded as constituting a part of orthodox cognitive science.

Since the early 1990s this computionalist orthodoxy has begun to be challenged by the emergence of *embodied-embedded* cognitive science (e.g. Varela, Thompson & Rosch 1991; Clark 1997; Wheeler 2005), a paradigm which claims that an agent's embodiment is constitutive of its perceiving, knowing and doing (e.g. Gallagher 2005; Noë 2004; Thompson & Varela 2001). Furthermore, the computational hypothesis has given way to the dynamical hypothesis that cognitive agents are best understood as dynamical systems (van Gelder 1998). Thus, while the embodied-embedded paradigm has retained the connectionist focus on self-organizing dynamic systems, it further holds that cognition is a situated activity which spans a systemic totality consisting of an agent's brain, body, and world (e.g. Beer 2000). In order to assess the importance of AI for this ongoing shift toward embodied-embedded cognitive science, it is helpful to first consider the potential impact of theoretical argument alone.

2.2 A Philosophical Stalemate

The theoretical premises of orthodox and embodied-embedded cognitive science can generally be seen as Cartesian and Heideggerian in character, respectively (e.g. Wheeler 2005; Dreyfus 2007; Anderson 2003). The traditional Cartesian philosophy accepts the assumption that any kind of being can be reduced to a combination of more basic atomic elements which are themselves irreducible. On this view cognition is seen as a general-purpose reasoning process by which a relevant representation of the world is assembled through the appropriate manipulation and transformation of basic mental states (Wheeler 2005, p. 38). Orthodox cognitive science adopts a similar kind of reductionism in that it assumes that symbolic/subsymbolic structures are the basic representational elements which ground all mental states[4], and that cognition is essentially treated as the appropriate computation of such representations. What are the arguments against such a position?

The Heideggerian critique starts from the phenomenological claim that the world is first and foremost experienced as a significant whole and that cognition is grounded in the skilful disposition to respond flexibly and appropriately as demanded by contextual circumstances. Dreyfus (1991, p. 117) has argued that such a position

---

[4] In contrast to the Cartesian claim that mental stuff is ontologically basic, orthodox cognitive science could hold that these constitutive elements are not basic in this absolute sense because they are further reducible to physical states. However, this change in position does not make any difference with regard to Heidegger's critique of this kind of reductionism.

questions the validity of the Cartesian approach in two fundamental ways. First, the claim of *holism* entails that the isolation of a specific part or element of our experience as an atomic entity appears as secondary because it already presupposes a background of significance as the context from which to make the isolation. From this point of view a reductionist attempt at reconstructing a meaningful whole by combining isolated parts appears nonsensical since the required atomic elements were created by stripping away exactly that contextual significance in the first place. As Dreyfus (1991, p. 118) puts it: "Facts and rules are, by themselves, meaningless. To capture what Heidegger calls significance or involvement, they must be *assigned relevance*. But the predicates that must be added to define relevance are just more meaningless facts". From the Heideggerian perspective it therefore appears that the Cartesian position is faced with a problem of infinite regress. Second, if we accept the claim of *skills*, namely that cognition is essentially grounded in a kind of skilful know-how or context-sensitive coping, then the orthodox aim of reducing such behaviour into a formal set of input/output mappings which specify the manipulation and transformation of basic mental states appears to be hopelessly misguided.

Judging from these philosophical considerations it seems that the Heideggerian critique of the Cartesian tradition could have a significant impact on the paradigm shift from orthodox toward embodied-embedded cognitive science. However, since the two approaches have distinct underlying constitutive assumptions (e.g. reductionism vs. holism), there exists no *a priori* theoretical argument which would force someone holding a Cartesian position to accept the Heideggerian critique from holism and skills. Similarly, it is not possible for the Cartesian theorist to prove that worldly significance can indeed be created through the appropriate manipulation and transformation of abstract and de-contextualized representational elements. The problem is that, like all rational arguments, both accounts of cognition are founded on a particular set of premises which one is at liberty to accept or reject. Thus, even if the development of a strong philosophical position is most likely a necessary factor in the success of the embodied-embedded paradigm, it is by itself not sufficient. In other words, there is a fundamental stalemate in the purely *philosophical* domain; a shift in constitutive assumptions cannot be engendered by argumentation alone.

2.3 An Empirical Resolution

It has often been proposed that this theoretical stalemate has to be resolved in the empirical domain of the cognitive sciences (e.g. Dreyfus & Dreyfus 1988; Clark 1997, p. 169; Wheeler 2005, p. 187). The authors of the Physical-Symbol System Hypothesis (Newell & Simon 1976) and the Dynamical Hypothesis (van Gelder 1998) are also in agreement that only sustained empirical research can determine whether their respective hypotheses are viable. Research in AI[5] is thereby awarded the rather privileged position of being able to help resolve theoretical disputes which have plagued the Western philosophical tradition for decades if not centuries. This

---

[5] It is worth noting that there are compelling arguments for claiming that the results generated by AI research are not "empirical" in the same way as those of the natural sciences, and that this is likely to weaken their impact outside the field. Nevertheless, it is still the case that AI can provide "valuable tools for re-organising and probing the internal consistency of a theoretical position" (Di Paolo, Noble & Bullock 2000).

reciprocal relationship between AI and theory has been captured with the slogan "understanding by building" (e.g. Pfeifer 1996; Pfeifer & Scheier 1999, p. 299).

In what way has AI research managed to fulfil this role? Dreyfus (1991, p. 119), for example, has argued that the Heideggerian philosophy of cognition has been vindicated because GOFAI faces significant difficulties whenever it attempts to apply its Cartesian principles to real-world situations which require robust, flexible, and context-sensitive behavior. In addition, he demonstrates that the Heideggerian arguments from holism and skills can provide powerful explanations of why this kind of AI has to wrestle with the frame and commonsense knowledge problems. In a similar vein, Wheeler (2005, p. 188) argues compellingly that the growing success of embodied-embedded AI provides important experimental support for the shift toward a Heideggerian position in cognitive science. He argues that Heidegger's claim that a cognitive agent is best understood from the perspective of "being-in-the-world" is put to the test by embodied-embedded AI experiments which investigate cognition as a dynamical process which emerges out of a brain-body-world systemic whole.

2.4 The Failure of Embodied-Embedded AI?

In light of these developments it seems fair to say that AI can have a significant impact on the ongoing shift from orthodox toward embodied-embedded cognitive science. However, while embodied-embedded AI has managed to overcome some of the significant challenges faced by traditional GOFAI, it has also started to encounter some of its own limitations. Considering the seemingly insurmountable challenge to make the artificial agents of current embodied-embedded AI behave in a more robust, flexible, and generally more life-like manner, particularly in the way that more complex living organisms do, Brooks (1997) was led to entertain the following sceptical reflections: "Perhaps we have all missed some organizing principle of biological systems, or some general truth about them. Perhaps there is a way of looking at biological systems which will illuminate an inherent necessity in some aspect of the interactions of their parts that is completely missing from our artificial systems. [...] I am suggesting that perhaps at this point we simply do not *get it*, and that there is some fundamental change necessary in our thinking" (Brooks 1997). Has the field of AI managed to find the missing "juice" of life in the past decade?

The existential philosopher Dreyfus, while mostly known in the field for his scathing criticisms of GOFAI, has recently referred to the current work in embodied-embedded AI as a "failure" (Dreyfus 2007). He points to the lack of "a model of our particular way of being embedded and embodied such that what we experience is significant for us in the particular way that it is. That is, we would have to include in our program a model of a body very much like ours". Similarly, Di Paolo (2003) argues that embodied-embedded robots, while in many respects an improvement over traditional GOFAI, "can never be truly autonomous. In other words the presence of a closed sensorimotor loop *does not* fully solve the problem of meaning in AI". These problems are even further amplified because, while embodied-embedded AI has focused on establishing itself as a vialbe alternative to the traditional computational paradigm, relatively little effort has been made to connect its experimental work with theories outside the field of AI, such as with theoretical biology, in order to address issues of autonomy and embodiment (Ziemke 2007). It thus seems that slowly there is an awareness growing in the field of embodied-embedded AI that something crucial is still missing in the current implementations of autonomous systems, and that this shortcoming is likely related to their particular manner of embodiment. But what could this elusive factor be? In order to answer this question we need to shift our focus back to recent developments in the cognitive sciences.

3 Further: The Shift Toward Enactivism

The enactive paradigm originally emerged as a part of embodied-embedded cognitive science in the early 1990s with the publication of the influential book *The Embodied Mind* (Varela, Thompson & Rosch 1991). It has recently distinguished itself by more explicitly placing autonomous agency in addition to lived subjectivity at the heart of cognitive science (e.g. Thompson 2007; Thompson 2005; Di Paolo, Rohde & De Jaegher 2007). How AI relates to this further shift in the cognitive sciences is still in need of clarification. This section provides some initial steps in this direction by considering how AI can contribute to the enactive account of how our bodily activity relates to the subjective mind at three interrelated "dimensions of embodiment": 1) bodily self-regulation, 2) sensorimotor coupling, and 3) intersubjective interaction (Thompson & Varela 2001). While the development of such fully enactive AI is a significant challenge to existing methodologies, it has the potential of providing a fresh perspective on some of the issues currently faced by embodied-embedded AI.

3.1 Bodily Self-regulation

This dimension of embodiment is central to the enactive account of autonomy. Since embodied-embedded AI has always been involved in extensive studies of autonomous systems (e.g. Pfeifer & Scheier 1999), it might seem that such research is particularly destined to relate to enactivism in a mutually informative manner. Unfortunately, things are not as straightforward; the enactive paradigm has a very different view of what constitutes autonomy when compared to most embodied-embedded AI (Froese, Virgo & Izquierdo 2007). Its approach can be traced to the notion of *autopoiesis*, a systems concept which originated in the theoretical biology of the 1970s (e.g. Maturana & Varela 1980). Enactivism defines an autonomous agent as a precarious self-producing network of processes which constitutes its own identity; the paradigmatic example being a living organism. Drawing from the bio-philosophy of Hans Jonas (1966), it is claimed that such an autonomous system, one whose being is its own doing, should be conceived of as an individual in its own right, and that this process of self-constitution brings forth, in the same stroke, what is other, namely its world (e.g. Thompson 2007, p. 153). In other words, it is proposed that the continuous reciprocal process, which constitutes the autonomous system as a distinguishable individual, also furnishes it with an intrinsically meaningful perspective on its environment, i.e. autonomy lies at the basis of *sense-making* (Weber & Varela 2002).

It follows from these considerations that today's AI systems are not autonomous in the enactive sense. They do not constitute their own identity, and the only "identity" which they can be said to possess is projected onto them by the observing researcher (Ziemke 2007). The popular methodology of evolutionary robotics, for example, presupposes that an "individual" is already defined by the experimenter as the basis for selection by the evolutionary algorithm, and in the dynamical approach to AI it is up to the investigator to distinguish which subpart of the systemic whole actually constitutes the "agent" (Beer 1995). The enactive notion of autonomous agency therefore poses a significant difficulty for current AI methodologies. Nevertheless, it is worth noting that AI researchers do not have to synthesize actual living beings in order for their work to provide some relevant insights into the dimension of bodily self-regulation. Following Di Paolo (2003), a first step would be to investigate artificial systems with some self-sustaining dynamic structures. In this manner embodied-embedded AI can move beyond its current focus on closed sensorimotor feedback loops by implementing systems which have a reciprocal link between internal organization and external behaviour. Indeed, there are signs that a shift toward more concern with bodily self-regulation is starting to develop. This is demonstrated by an increasing interest in homeostasis as a regulatory mechanism for investigating, for example, sensory inversion (e.g. Di Paolo 2003), the emergence of sensorimotor coupling (e.g. Ikegami & Suzuki forthcoming), behavioural preference (e.g. Iizuka & Di Paolo forthcoming), and active perception (e.g. Harvey 2004).

3.2 Sensorimotor Coupling and Intersubjective Interaction

Since sensorimotor embodiment is the research target of most current embodied-embedded AI, its results can have an impact on this aspect of enactivism. However, since the vast majority of such work is not concerned with how the constraints of constitutive autonomy are related to the emergence of sensorimotor behavior, it is not contributing to the enactive account of how an autonomous agent is able to bring forth its own cognitive domain. To become more relevant in this respect, the field needs to adapt its methodologies so as to deal with the enactive proposal that an agent's sense-making is grounded in the active regulation of ongoing sensorimotor coupling in relation to the viability of a precarious, dynamically self-sustaining identity (Weber & Varela 2002). This is an area which has been practically unexplored, although some promising work has begun (e.g. Ikegami & Suzuki forthcoming; Di Paolo 2003).

These considerations can be extended to the domain of intersubjective interaction, since this dimension of embodiment also involves distinctive forms of sensorimotor coupling (Thompson & Varela 2001). An enactive account of social understanding based on this continuity has recently been outlined by Di Paolo, Rohde and De Jaegher (2007). They make the important suggestion that the traditional focus on the embodiment of individual interactors needs to be complemented by an investigation of the interaction process that takes place between them. This shift in focus enables them to extend the enactive notion of sense-making into the realm of social cognition in the form of *participatory sense-making*. The development of such an account is important for embodied-embedded AI, because most of its current research remains limited to "lower-level" cognition. Exploring the domain of social interaction might provide it with the necessary means to tackle the problem of scalability (Clark 1997, p. 101), in particular because such inter-action can constitute new ways of sense-making that are not available to the individual alone. The challenge is to implement

AI systems that constitute the social domain by means of an interaction process that is essentially embodied and situated, as opposed to the traditional means of formalized transmissions of abstract information over pre-specified communication channels. Di Paolo, Rohde and De Jaegher (2007) review some initial work in this direction which demonstrates that "these models have the possibility to capture the rich dynamics of reciprocity that are left outside of traditional individualistic approaches".

3.3 A Fully Enactive AI?

It is debatable if AI research should be considered as enactive rather than embodied-embedded if it does not address some form of bodily self-regulation[6]. In this sense the authors of *The Embodied Mind* perhaps got slightly carried away when they referred to the emergence of Brooks's behaviour-based robotics as a "fully enactive approach to AI" (Varela, Thompson & Rosch 1991, p. 212). However, this is not to say that embodied-embedded AI does not have an impact on the shift toward enactivism, it does, but only to the extent that there is an overlap between the two paradigms. Its current influence is therefore by no means as significant as it has been on the shift toward embodied-embedded cognitive science. For example, Thompson's recent book *Mind in Life*, which can be considered as a successor to *The Embodied Mind*, does not even include AI as one of the cognitive science disciplines from which it draws its insights (Thompson 2007, p. 24). Indeed, at the moment it seems more likely that the influence will run more strongly from enactive cognitive science to AI instead. Its account of *autonomous agency*, for example, has the potential to provide embodied-embedded AI with exactly the kind of bodily organizational principle that has been identified as missing by Brooks (1997). In addition, the enactive notion of *sense-making*, as a biologically grounded account of how a system must be embodied in order for its encounters to be experienced as significant, can be used as a response to Dreyfus's (2007) vague requirement of "a detailed description of our body", which apparently has not even "a chance of being realized in the real world". Furthermore, there is a good possibility that the field's current restriction to "lower-level" cognition could be overcome in a principled manner by extending its existing research focus on sensorimotor embodiment to also include *participatory sense-making*.

Of course, it goes without saying that all of these aspects of enactivism are also open to further refinement through artificial modelling, and that some initial work in this direction has already begun. Nevertheless, for AI to have a more significant impact on the ongoing shift toward enactive cognitive science, it must address some considerable challenges that face its current methodologies. The field needs to extend its current preoccupation with sensorimotor interaction in the behavioural domain to include a concern of the constitutive processes that give rise to that domain in living systems. Maybe Brooks (1997) was right when he suggested that in order for AI to be more life-like perhaps "there is some fundamental change necessary in our thinking". At least such a change is indeed necessary for the development of a fully enactive AI.

---

[6] In a similar manner it could be argued that since recent work in enactive perception (e.g. Noë 2004) is more concerned with sensorimotor contingencies than with autonomous agency or lived subjectivity, such work might be more usefully classified as part of embodied-embedded cognitive science rather than enactivism proper. See also Thompson (2005).

4 From AI to Phenomenology

How can such fully enactive AI impact on the cognitive sciences? This section argues that, while clearly an important aspect, it is not sufficient to displace the orthodox mainstream. More than just having to make Heideggerian AI more Heideggerian, as Dreyfus (2007) proposes, Heideggerian cognitive science itself must become more Heideggerian by shifting its focus from AI to phenomenology, a shift which coincides with a movement from embodied-embedded cognitive science to enactivism.

4.1 An Empirical Stalemate

Two decades ago Dreyfus and Dreyfus (1988) characterized GOFAI as a project in which "the rationalist tradition had finally been put to an empirical test, and it had failed". Nevertheless, despite this supposed 'failure' no alternative has yet succeeded in fully displacing the orthodox mainstream in AI or cognitive science. While it could be argued that more progress in embodied-embedded or enactive AI will eventually remedy this situation, a more serious problem becomes apparent when we consider why this perceived 'failure' did not remove the orthodox framework from the mainstream. As Wheeler (2005, p. 185) points out, this did not happen for the simple reason that researchers are always at liberty to interpret practical problems as mere temporary difficulties which will eventually be eliminated through more scientific research and additional technological development. Accordingly, Wheeler goes on to conclude that a resolution of the standoff must await further empirical evidence.

However, while Wheeler's appeal to more experimental data is evidently useful when resolving theoretical issues within a particular approach, it is not clear whether it is also valid when deciding between different paradigms: you always already have to choose (whether explicitly or not) one paradigm over the others from which to interpret the data. Furthermore, this choice is significant because "the conceptual framework that we bring to the study of cognition can have profound empirical consequences on the practice of cognitive science. It influences the phenomena we choose to study, the questions we ask about these phenomena, the experiments we perform, and the ways in which we interpret the results of these experiments" (Beer 2000). Thus, since data is only meaningful in a manner which crucially depends on the underlying premises of the investigator, the current empirical stalemate in AI appears to be less due to a lack of empirical evidence and more due to the fact that the impact of an experiment fundamentally depends on an interpretive aspect[7]. In other words, in order for experimental *data* to be turned into scientific *knowledge* it first has to be *interpreted* according to (often implicitly) chosen constitutive assumptions. Moreover, our premises even ground the manner in which we distinguish between noise and data[8]. It follows from this that the major cause of the standoff in the philosophical domain also plays a significant role in the current empirical stalemate:

---

[7] Again, this is not to say that such experimental evidence has no effect. The point is simply that, while a necessary component, it is not *sufficient* for a successful paradigm shift.

[8] Consider, for example, the empirical fact that the fossil record shows long periods of stasis interspersed with layers of rapid phyletic change. Someone who believes that evolution proceeds gradually will treat this fact as irrelevant noise, while someone who claims that evolution proceeds as punctuated equilibria will view such a finding as supporting evidence.

both domains of enquiry require an interpretative action on the part of the observer. And, more importantly, while it is possible to influence this act of interpretation through research progress, its outcome cannot be fully determined by such external events since any kind of understanding always already presupposes interpretative activity[9]. In addition, the impact of this potential influence is also limited because the significance of such advances might not become apparent if one does not already hold the kind of constitutive assumptions required to understand them appropriately.

These considerations give a rather bleak outlook for the possibility of actively generating a successful paradigm shift in the cognitive sciences, and at this point it might seem relatively futile to worry about such abstract problems and just get on with the work. Indeed, considering the overall state of affairs this is in many respects a sensible and pragmatic course of action. Nevertheless, it is evidently the case that we choose a paradigm for our research. However, if rational argument combined with empirical data is still not sufficient to necessarily establish this choice, then what is it that determines which premises are assumed? And how can this elusive factor be influenced? The rest of this section provides a tentative answer to these questions by focusing on a crucial aspect of enactivism that has not been addressed so far.

4.2 A Phenomenological Resolution

The enactive account of autonomous agency as expressed in terms of systems biology is complemented by a concern with the first-person point of view, by which is meant the subjectively *lived experience* associated with cognitive and mental events (Varela & Shear 1999). Since the enactive framework incorporates both biological agency and phenomenological subjectivity, it allows the traditional mind-body problem to be recast in terms of what has recently been called the "body-body problem" (Hanna & Thompson 2003). On this view the traditional "explanatory gap" (Levine 1983) is no longer absolute since the concepts of subjectively lived body and objective living body both require the notion of living being. Though more work needs to be done to fully articulate the details, this reformulation of the hard problem of consciousness can be seen as one of the major contributions of enactivism (Torrance 2005).

Nevertheless, it is not yet clear how a concern with subjective experience could provide us with a way to move beyond the stalemate that we have identified in the previous sections. Surely enactivism is just more philosophical theory? However, to say this is to miss the point that it derives many of its crucial insights from a source that is quite distinct from standard theoretical or empirical enquiry, namely from careful *phenomenological* observations that have been gained through the principled investigation of the structure of our lived experience (see Ch. 2 in Thompson 2007 for an overview). But what about the insights from which Heidegger originally deduced his claims? If his analysis of the holistic structure of our "being-in-the-world" is one of the most influential accounts of the Husserlian phenomenological tradition, then why did it not succeed in convincing mainstream cognitive scientists? The regrettable answer is that, while his claims have sometimes been probed in the philosophical or

---

[9] From the enactive view this is hardly surprising (Varela, Thompson & Rosch 1991, p. 10-12), and it can ground this epistemological reflection in its biology of autonomy by claiming that a living system always constitutes its own perspective on the world. Indeed, at one point enactivism was actually called "the hermeneutic approach" (Thompson 2007, p. 24).

empirical domain, there have not been many sustained and principled efforts in orthodox cognitive science to verify their validity in the phenomenological domain.

If enactivism is to avoid this fate then it needs to focus less on the development of enactive AI, and more on the promotion of principled phenomenological studies. Indeed, according to Di Paolo, Rohde and De Jaegher (2007) the central importance of experience is perhaps the most revolutionary implication of enactivism since "phenomenologically informed science goes beyond black marks on paper or experimental procedures for measuring data, and dives straight into the realm of personal experience" such that, for example, "no amount of rational argument will convince a reader of Jonas's claim that, as an embodied organism, he is concerned with his own existence if the reader cannot see this for himself". Thus, enactivism implicates an element of personal practice. Similarly, Varela and Shear (1999) outline the beginnings of a project "where neither experience nor external mechanism have the final word", but rather stand to each other in a relationship of mutual constraints. They point out that the collection of phenomenological data requires a disciplined training in the skilful exploration of lived experience. Such an endeavour might already be worthwhile in itself, but in the context of the stalemate in the cognitive sciences it comes with an added benefit. In a nutshell this is because, while it is still the case that phenomenological data first has to be interpreted from a particular point of view before it can be integrated into a conceptual framework, generating such data also requires a change in our mode of experiencing. Moreover, this change in our experiential attitude is constituted by a change in our mode of *being*, and this in turn entails a change in our *understanding* (Varela 1976). Thus, it is this being, our everyday "Dasein", which determines how we interpret our world. Of course, since we are autonomous agents this does not mean that actively practicing phenomenology necessarily commits us to enactivism. But perhaps by changing our awareness in this manner we will be able to understand more fully the reasons, other than theory and empirical data, which are at the root of why we prefer one paradigm over another.

5   Conclusion

The field of AI has had a significant impact on the ongoing shift from orthodox toward embodied-embedded cognitive science mainly because it has made it possible for philosophical disputes to be addressed in an experimental manner. Conversely, enactivism can have a strong influence on AI because of its biologically grounded account of autonomous agency and sense-making. The development of such enactive AI, while challenging to current methodologies, has the potential to address some of the problems currently in the way of significant progress in embodied-embedded AI. However, if an alternative paradigm is to be successful in actually displacing the orthodox mainstream, then it is unlikely that theoretical arguments and empirical evidence alone are sufficient. For this to happen it will be necessary that a *phenomenological pragmatics* is established as part of the general methodological toolbox of contemporary cognitive science. This shift of focus from AI to phenomenology coincides with a shift from embodied-embedded cognitive science to enactivism. Unfortunately, however, most of our current academic institutions are not concerned with supporting phenomenology in any principled manner, and it will be one of the major challenges facing the realization of enactivism as the mainstream of cognitive science to devise appropriate ways of changing this. In this context, Terry Winograd's turn toward teaching Heidegger in computer science courses at Stanford when he became disillusioned with traditional GOFAI appears in a new light[10].

References

Anderson, M.L.: Embodied Cognition: A field guide. Artificial Intelligence 149(1), 91–130 (2003)

Beer, R.D.: A dynamical systems perspective on agent-environment interaction. Artificial Intelligence 72(1-2), 173–215 (1995)

Beer, R.D.: Dynamical approaches to cognitive science. Trends in Cognitive Sciences 4(3), 91–99 (2000)

Boden, M.A.: Mind as Machine: A History of Cognitive Science, vol. 2. Oxford University Press, Oxford, UK (2006)

Brooks, R.A.: Intelligence without representation. Artificial Intelligence 47(1-3), 139–160 (1991)

Brooks, R.A.: From earwigs to humans. Robotics and Autonomous Systems 20(2-4), 291–304 (1997)

Clark, A.: Being There. The MIT Press, Cambridge, MA (1997)

Di Paolo, E.A.: Organismically-inspired robotics: homeostatic adaptation and teleology beyond the closed sensorimotor loop. In: Murase, K., Asakura, T. (eds.) Dynamical Systems Approach to Embodiment and Sociality, Advanced Knowledge International, Adelaide, Australia, pp. 19–42 (2003)

Di Paolo, E.A., Noble, J., Bullock, S.: Simulation Models as Opaque Thought Experiments. In: Bedau, M.A., et al. (eds.) Proc. of the 7th Int. Conf. on the Synthesis and Simulation of Living Systems, pp. 497–506. The MIT Press, Cambridge, MA (2000)

Di Paolo, E.A., Rohde, M., De Jaegher, H.: Horizons for the Enactive Mind: Values, Social Interaction, and Play, Cognitive Science Research Paper, vol. 587, Sussex Uni., UK (2007)

Dreyfus, H.L.: Being-in-the-World. The MIT Press, Cambridge, MA (1991)

Dreyfus, H.L.: Why Heideggerian AI failed and how fixing it would require making it more Heideggerian. Philosophical Psychology 20(2), 247–268 (2007)

Dreyfus, H.L., Dreyfus, S.E.: Making a mind versus modelling the brain: artificial intelligence back at a branch-point. Daedalus 117(1), 15–44 (1988)

Fodor, J.A.: The Language of Thought. Harvard Uni. Press, Cambridge, MA (1975)

Froese, T., Virgo, N., Izquierdo, E.: Autonomy: a review and a reappraisal. In: Almeida e Costa, F. et al. (eds.), Proc. of the 9th Euro. Conf. on Artificial Life, Springer-Verlag, Berlin, Germany (in press, 2007)

Gallagher, S.: How the Body Shapes the Mind. Oxford University Press, New York (2005)

Hanna, R., Thompson, E.: The Mind-Body-Body Problem. Theoria et Historia Scientarum 7(1), 24–44 (2003)

Harnard, S.: The symbol grounding problem. Physica D 42, 335–346 (1990)

Harvey, I.: Homeostasis and Rein Control: From Daisyworld to Active Perception. In: Pollack, J., et al. (eds.) Proc. of the 9th Int. Conf. on the Simulation and Synthesis of Living Systems, pp. 309–314. The MIT Press, Cambridge, MA (2004)

---

[10] See Dreyfus (1991, p. 119).

Harvey, I., Di Paolo, E.A., Wood, R., Quinn, M., Tuci, E.A.: Evolutionary Robotics: A new scientific tool for studying cognition. Artificial Life 11(1-2), 79–98 (2005)

Iizuka, H., Di Paolo, E.A.: Toward Spinozist robotics: Exploring the minimal dynamics of behavioral preference. Adaptive Behavior (forthcoming)

Ikegami, T., Suzuki, K.: From Homeostatic to Homeodynamic Self. BioSystems (forthcoming)

Jonas, H.: The Phenomenon of Life: Toward a Philosophical Biology. Northwestern University Press, Evanston, Illinois (2001)

Levine, J.: Materialism and Qualia: The Explanatory Gap. Pacific Philosophical Quarterly 64, 354–361 (1983)

Maturana, H.R., Varela, F.J.: Autopoiesis and Cognition: The Realization of the Living. Kluwer Academic Publishers, Dordrecht, Holland (1980)

McCarthy, J., Hayes, P.J.: Some philosophical problems from the standpoint of artificial intelligence. In: Meltzer, B., Michie, D. (eds.) Machine Intelligence 4, pp. 463–502. Edinburg University Press, Edinburgh, UK (1969)

McClelland, J.L., Rumelhart, D.E., The PDP Research Group: Parallel Distributed Processing. In: Psychological and Biological Models, vol. 2, The MIT Press, Cambridge, MA (1986)

Newell, A., Simon, H.A.: Computer Science as Empirical Enquiry: Symbols and Search. Communications of the Association for Computing Machinery 19(3), 113–126 (1976)

Noë, A.: Action in Perception. The MIT Press, Cambridge, MA (2004)

Nolfi, S., Floreano, D.: Evolutionary Robotics. The MIT Press, Cambridge, MA (2000)

Pfeifer, R.: Building 'Fungus Eaters': Design Principles of Autonomous Agents. In: Maes, P., et al. (eds.) Proc. of the 4th Int. Conf. on the Simulation of Adaptive Behavior, pp. 3–12. The MIT Press, Cambridge, MA (1996)

Pfeifer, R., Scheier, C.: Understanding Intelligence. The MIT Press, Cambridge, MA (1999)

Thompson, E.: Sensorimotor subjectivity and the enactive approach to experience. Phenomenology and the Cognitive Sciences 4(4), 407–427 (2005)

Thompson, E.: Mind in Life. The MIT Press, Cambridge, MA (2007)

Thompson, E., Varela, F.J.: Radical embodiment: neural dynamics and consciousness. Trends in Cognitive Sciences 5(10), 418–425 (2001)

Torrance, S.: In search of the enactive: Introduction to special issue on enactive experience. Phenomenology and the Cognitive Sciences 4(4), 357–368 (2005)

van Gelder, T.: The dynamical hypothesis in cognitive science. Behavioral and Brain Sciences 21(5), 615–665 (1998)

Varela, F.J.: Not One, Not Two. The Co-Evolution Quarterly 12, 62–67 (1976)

Varela, F.J., Shear, J.: First-person Methodologies: What, Why, How? Journal of Consciousness Studies 6(2-3), 1–14 (1999)

Varela, F.J., Thompson, E., Rosch, E.: The Embodied Mind: Cognitive Science and Human Experience. The MIT Press, Cambridge, MA (1991)

Weber, A., Varela, F.J.: Life after Kant: Natural purposes and the autopoietic foundations of biological individuality. Phenomenology and the Cognitive Sciences 1, 97–125 (2002)

Wheeler, M.: Reconstructing the Cognitive World. The MIT Press, Cambridge, MA (2005)

Ziemke, T.: What's life got to do with it? In: Chella, A., Manzotti, R. (eds.) Artificial Consciousness, pp. 48–66. Imprint Academic, Exeter, UK (2007)

On the Information Theoretic Implications of Embodiment – Principles and Methods

Rolf Pfeifer[1], Max Lungarella[2,*], Olaf Sporns[3], and Yasuo Kuniyoshi[2]

[1] Dept. of Informatics, University of Zurich, Zurich, Switzerland
[2] Dept. of Mechano-Informatics, The University of Tokyo, Tokyo, Japan
[3] Dept. of Psychological and Brain Sciences, Indiana University, Bloomington (IN), USA
pfeifer@ifi.unizh.ch, maxl@isi.imi.i.u-tokyo.ac.jp,
osporns@indiana.edu, kuniyosh@isi.imi.i.u-tokyo.ac.jp

Abstract. Embodied intelligent systems are naturally subject to physical constraints, such as forces and torques (due to gravity and friction), energy requirements for propulsion, and eventual damage and degeneration. But embodiment implies far more than just a set of limiting physical constraints; it directly supports the selection and processing of information. Here, we focus on an emerging link between information and embodiment, that is, on how embodiment actively supports and promotes intelligent information processing by exploiting the dynamics of the interaction between an embodied system and its environment. In this light the claim that "intelligence requires a body" means that embodied systems actively induce information structure in sensory inputs, hence greatly simplifying the major challenge posed by the need to process huge amounts of information in real time. The structure thus induced crucially depends on the embodied system's morphology and materials. From this perspective, behavior informs and shapes cognition as it is the outcome of the dynamic interplay of physical and information theoretic processes, and not the end result of a control process that can be understood at any single level of analysis. This chapter reviews the recent literature on embodiment, elaborates some of the underlying principles, and shows how robotic systems can be employed to characterize and quantify the notion of information structure.

Keywords: Embodiment, Information Processing, Morphology, Materials.

1 Introduction

The stance taken here strongly differs from the still widely held traditional one of "cognition as computation" where intelligence is considered to be algorithmic and the result of abstract symbol manipulation. While this computational perspective has led to many important theoretical insights and applications, most of the emphasis has been on exclusively internal mechanisms of information processing. Contrasting the computational perspective, there has been a considerable amount of research demonstrating that cognition is embodied and best understood as a situated activity.

---

* Current affiliation: JST ERATO, The University of Tokyo, Tokyo, Japan.

M. Lungarella et al. (Eds.): 50 Years of AI, Festschrift, LNAI 4850, pp. 76–86, 2007.
© Springer-Verlag Berlin Heidelberg 2007

The extensive conceptual and empirical groundwork for embodied cognition laid within psychology, cognitive science, philosophy artificial intelligence, and robotics has been reviewed elsewhere [1–16]. Building on this body of empirical and theoretical work, here we address a specific set of issues surrounding the potential link between embodiment and information processing.

Our main thesis is that the interaction between physical and information processes is central for the emergence and development of intelligence. When talking about agents in the real world, it is important to realize that information is not just "out there", an infinite tape ready to be loaded and processed by the cognitive machinery of the brain. Instead, through physical (embodied) interactions with the environment, information structure (e.g., spatio-temporal correlations in a visual input stream, redundancies between different perceptual modalities, or regularities in sensory patterns that are invariant with respect to changes in illumination, size, or orientation) is actively induced in sensory inputs. In the context of this review, we will use the term information structure to refer to the structure in the sensory data typically induced by and meaningful with respect to some purposive or intended action such as grasping or walking. As suggested here, the presence of such structure might be essential for the acquisition of a broad range of cognitive and motor abilities such as multimodal sensory integration, cross-modal learning, perceptual categorization, reaching, object manipulation, language, and locomotion.

We first discuss a case study, categorization, illustrating the main concepts, and we formulate two pertinent principles. Subsequently, we expand on the notion of information structure and information self-structuring, and show how quantitative measures can be used to provide corroboration and theoretical groundwork. We will then briefly discuss the role of these ideas in learning and development and look at how dynamics can be exploited to structure sensory stimulation. Finally, we discuss the implications of the ideas developed in this chapter for theories of cognition and cognitive development.

2   Categorization in the Real World

For autonomous embodied agents acting in the real world (animals, humans, robots), perceptual categorization – the ability to make distinctions – is a hard problem. First, based on the stimulation impinging on its sensor arrays (sensation) the agent has to rapidly determine and attend to what needs to be categorized. Second, the appearance and properties of objects or events in the environment being classified vary continuously, e.g., owing to occlusions, and changes of distances and orientations with respect to the agent. And third, the environmental conditions (e.g., illumination, viewpoint, and background noise) fluctuate considerably. There is much relevant work in computer vision that has been devoted to extracting scale- and translation-invariant low-level visual features and high-level multidimensional representations for the purpose of robust perceptual categorization [17–19]. Following this approach, however, categorization often turns out to be a very difficult if not an impossible computational feat, especially when adequate information is lacking. A solution that can only be pursued by embodied agents, but is not available when using a purely computational (i.e., disembodied) approach, is that through their interaction with the environment, agents generate the sensory stimulation required to perform the proper categorization and thus drastically simplify the problem of mapping sensory stimulation onto perceptual categories. The most typical and effective way is through a process of sensory-motor coordination.

Because of its almost universal presence in behaving organisms, sensory-motor coordination has been widely studied in psychology, neuroscience, and robotics [20–31]. Studies indicate how sensory-motor coordination, for instance, simplifies category formation (for a review, see [30]), influences visual experience [25], and determines concept acquisition [32]. One demonstration of the idea of exploiting coordinated interaction with the environment is a study by Pfeifer and Scheier [23] in which it is shown that mobile robots can reliably categorize big and small wooden cylinders only if their behavior is sensory-motor coordinated. The artificial evolution experiments by Nolfi [26] and Beer [27] illustrate a similar point: the fittest agents, i.e., those that could most reliably find the category to which different kind of objects belonged, were those engaging in sensory-motor coordinated behavior. Intuitively, in these examples, the interaction with the environment (a physical process) creates additional (that is, previously absent) sensory stimulation which is also highly structured thus facilitating subsequent information processing. Computational economy and temporal efficiency are purchased at the cost of behavioral interaction, so to speak.

3 Information Self-structuring

The idea that the synergy between the world and the observer's actions plays a primary role for the emergence and development of cognition is much in tune with previous work on direct and active perception [33–35], animate, interactive, and enactive vision [36–38]. From an information theoretical point of view, embodied agents generate information structure in their sensory stimulation as they – actively – interact with the environment. It is important to note that in this process, the specific morphology (type and placement of the sensors and actuators) and the materials used unavoidably determine the resulting information structure. Because of the high density of touch sensors on the fingertips and because of the shape of the hand, for instance, grasping automatically leads to rich, structured tactile stimulation. The coordinated sensory-motor action of grasping induces stable patterns of stimulation characterized by correlations between the activities of receptor neurons within a sensor modality, as well as correlations between receptor neurons in different modalities (vision, touch, audition, and proprioception). Such correlations or statistical dependencies (which are instances of information structure) create redundancy across sensory channels, which may help reducing the effective dimensionality of the input, and which in turn – given the typically staggering number of possible configurations that the input system can assume – significantly simplify perception. We call this idea the "principle of information self-structuring" [28,29,39].

Theoretical studies and robot models provide quantitative evidence for the notion that self-generated motor activity can create information structure in sensory-motor data [23,24,26,28,29,39,40]. For instance, in [28] it is shown how a simple robot capable of saliency-based attentional behavior – an instance of an active vision system – self-structures the information present in its sensory and motor channels (Fig. 1). The results exposed in the article also demonstrate that sensory-motor coordination leads to a better embedding of the visual input into a low-dimensional space, as compared to un-coordinated behavior. Traditionally, such dimensionality reduction is seen as the result of internal processing of a neural architecture, for example through mechanisms in early visual processing that lead to efficient low-dimensional (sparse) encoding by exploiting input redundancies and regularities [41–43]. We suggest that the generation of structured information through embodied interaction provides an additional mechanism contributing to efficient neural coding. In this context we also point out a distinct advantage of using robotic devices rather than working with humans or animals. Robots allow for comprehensive recording and analysis of complete histories of sensory stimulation and motor activity, and enable us to conduct precisely controlled experiments while introducing systematic changes in body morphology, materials, and control architectures [44,45].

The theoretical concepts outlined in this section receive support from experiments with human subjects showing that most of our sensory experiences involve active (i.e., sensory-motor) exploration of the world (e.g., through manipulation or visual inspection) [25,37]. Such exploration promotes not only object recognition [46–48], but also, for instance, the learning of the three-dimensional structure of objects [49], and depth perception [50].

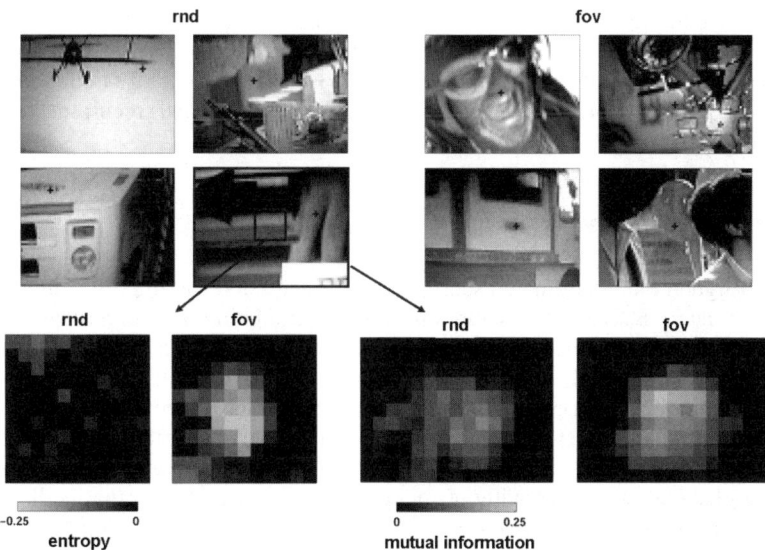

Fig. 1. Information structure in the visual field as a function of embodiment. Images show sample video frames obtained from a disrupted (a; "rnd", "low embodiment" – no sensory-motor coupling) and normally tracking (b; "fov", "high embodiment" – high sensory-motor coupling) and active vision system. Plots at the bottom show spatial maps of entropy and mutual information, expressed as differences relative to the background. There is a significant decrease in entropy (c) and an increase in mutual information (d) in the center of the visual field for the "fov" condition, compared to little change in the "rnd" condition. (Data replotted from [28]).

4 Learning and Development

There is an interesting implication of information self-structuring for learning. Information structure does not exist before the interaction occurs, but emerges only while the interaction is taking place. However, once it has been induced, learning can pick up on it such that next time around, the responsible sensory-motor information structure is more easily reactivated. It follows that embodied interaction lies at the root of a powerful learning mechanism as it enables the creation of time-locked correlations and the discovery of higher-order regularities that transcend the individual sensory modalities.

These ideas also extend to development. It is generally recognized that structured information and statistical regularities are crucial for perception, action, and cognition – and their development [4,32,41,51,52]. At a very young age, babies frequently use several sensory modalities for the purpose of categorization: they look at objects, grasp them, stick them into their mouths, throw them on the floor, and so on. The resulting intra- and intermodal sensory stimulation appears to be essential for concept formation [4,32,53,54]. As they grow older, infants can perform categorization based on the visual modality alone which implies that they must have learned something about how to predict sensory stimulation in one modality using the information available through another modality, for instance, the haptic from the visual one. By virtue of its continuous influence on the development of specialized neurons and their connections that incorporate consistent statistical patterns in their inputs, information structure plays a critical role in development. It is easier for neural circuits to exploit and learn sensory-motor patterns containing regularities and recurrent statistical features.

5 On Morphology, Dynamics, and Control

We have argued that coordinated sensory-motor interaction can impose consistent and invariant (that is, learnable) structure on sensory stimulation. It is important to realize that such information structure can also result from the dynamics of the interaction of a given morphology with the surrounding environment. Several studies with robots, for instance, indicate that computational processes involved in control can be partially subsumed (or taken over) by the morphological properties of the agent [55–59]. A paradigmatic example is provided by passive dynamic walkers which are robots – or rather mechanical structures without microprocessors or motors – that walk down a slope without control and actuation [56]. The walker's morphology (center of mass, length of the limbs, and the shape of the feet) and its materials are carefully designed so as to exploit the physical constraints present in its ecological niche (friction, gravity, inclination of the slope) for locomotion. To get the robot to learn to walk on level surfaces, one can use the mechanical design obtained during passive dynamic walking and endow it with actuators (e.g., located in the ankles or hips) [60]. The natural dynamics of the (body-environment) system provides the target for learning the control policy for the actuators by stabilizing the limit cycle trajectory that the robot follows – the dynamics structures the output of the angle sensors located in the joints, so to speak – and the robot learns to walk adaptively on flat ground within a relatively short period of time.

It is interesting to observe that as a consequence of the different data distributions resulting from different sensory morphologies a dependency exists between morphology, dynamics, and learning speed [60–62]. For example, by exploiting the non-homogenous arrangement of facets in the insect eye (denser in the front than on the side), the phenomenon of motion parallax can be "compensated away" and the adaptability of neural controller can be greatly improved [62]. We infer that the design of controller and morphology are, in a sense, inseparable, since the structure of both impacts information processing. However, while some progress has been made to optimize the design of robot controllers, robot morphology still largely remains a matter of heuristics. Future progress in the design of intelligent robots will require analytical tools and methodologies to exploit the interaction between morphology and computation [59].

The specific morphology of the body and the interaction of body and environment dynamics also shape the repertoire of preferred movements: a loosely hanging bouncing arm moves in a complex trajectory, but its control is extremely simple (the knowledge of how to move the limb seems to reside in the limb itself), whereas moving the hand in a straight path – a seemingly simple trajectory – requires a lot of control. It follows that part of the "processing" is done by the dynamics of the agent-environment interaction, and only sparse neural control needs to be exerted when the self-regulating and stabilizing properties of the natural dynamics can be exploited (see Fig. 2). The idea that brain, morphology, materials, and environment share responsibility in generating information structure has been called the "principle of ecological balance" [57] because there is a "balance" or task distribution between the different aspects of an embodied agent.

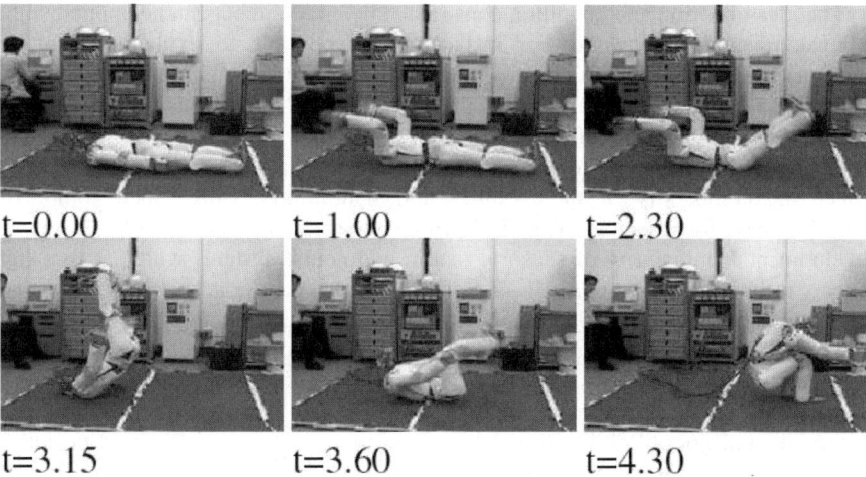

Fig. 2. Humanoid exploiting natural dynamics of body-environment interaction. Note that the robot is underactuated with respect to the ground which makes the equations of motion intractable analytically. By applying sparse but well-timed control actions the system transits from a lying (t=0) to a squatting position (t=4.30).

6 On the Interaction of Physical and Information Processes

The importance of the interaction between physical and information processes can hardly be over-estimated. The complexity of perceptual categorization in the real world, for instance, cannot be managed by computational means only. We have therefore stressed the significance of sensory-motor coordination. The principle of information self-structuring illustrates that physical interaction with the real world, in particular sensory-motor coordinated interaction, induces structured sensory stimulation, which, given the proper morphology, substantially facilitates neural processing, and hence sets the foundations for learning and development of perception and cognition in general.

We can take the idea of interaction of physical and information processes a step further by looking at the dynamics of embodied systems. We mentioned that because of the constraints provided by their embodiment, the movements of embodied systems follow certain preferred trajectories. It turns out that in biological agents such dynamics typically leads to rich and structured sensory stimulation. For example, as grasping is much easier than bending the fingers of the hand backwards, grasping is more likely to occur, and owing to the morphology (e.g., the high density of touch sensors on the fingertips), the intended sensory stimulation is induced. The natural movements of the arm and hand are – as a result of their intrinsic dynamics – directed towards the front center of the body. This in turn implies that normally a grasped object is moved towards the center of the visual field thereby inducing correlations in the visual and haptic channels which, as we pointed out earlier, simplify learning. So we see that an interesting relationship exists between morphology, intrinsic body dynamics, generation of information structure, and learning.

The idea of action and cognition constrained by embodiment can be applied within a developmental framework. For instance, it is possible to explain the infant's immaturity and initial limitations in morphology (e.g., wide spacing of photoreceptors in the retina), as unique adaptations to the environmental constraints of the ecological niche [63]. The specific effect of this arrangement is to filter out high spatial frequency information, and to make close objects most salient to the infant and hence reduce the complexity of the required information processing. Such complexity reduction may, for instance, facilitate learning about size constancy [64]. That is, the developmental immaturity of sensory, motor, and neural systems which at first sight appears to be an inadequacy, is in fact of advantage, because it effectively decreases the "information overload" that otherwise would most certainly overwhelm the infant [53,65]. A similar phenomenon occurs at the level of the motor system where musculo-skeletal constraints limit the range of executable movements and hence implicitly reduce the number of control variables. The neural system exploits such constraints and control is simplified by combining a rather small set of primitives (e.g., synergies [66] or force fields [67]), in different proportions rather than individually controlling each muscle.

Here, we have outlined a view of sensory-motor coordination and natural dynamics as crucial causal elements for neural information processing because they generate information structure. Our argument has revolved mainly around brain areas directly connected to sensory and motor systems. It is likely, however, that embodied systems operating in a highly coordinated manner generate information structure and statistical regularities at all hierarchical levels within their neural architectures, including effects on neural activity patterns far removed from the sensory periphery. This hypothesis leads to several predictions, testable in animal or robot experiments. For example, activations or statistical relationships between neurons in cortical areas engaged in sensorimotor processing should exhibit specific changes across different states of sensorimotor coordination or coupling. Increased structuring of information through embodiment would be associated with increased multimodal synchronization and binding, or more efficient neural coding.

7 Conclusion

The conceptual view of perception as an active process has gained much support in recent years (e.g., [25–29,38,57]). The work reviewed in this chapter provides additional evidence for this view and proposes a new link between embodiment and information. Perception cannot be treated as a purely computational problem that unfolds entirely *within* a given information processing architecture. Instead, perception is naturally embedded within a physically embodied system, interacting with the real world. Thus, the *interplay* between physical and information processes gives rise to perception. We identified specific contributions of embodiment to perceptual processing through the active generation of structure in sensory stimulation, which may pave the way towards a formal and quantitative analysis. The idea of inducing information structure through physical interaction with the real world has important consequences for understanding and building intelligent systems, by highlighting the fundamental importance of morphology, materials, and dynamics.

References

1. Anderson, M.L.: Embodied cognition: a field guide. Artif. Intelligence 149, 91–130 (2003)
2. Brooks, R.A.: New approaches to robotics. Science 253, 1227–1232 (1991)
3. Chiel, H., Beer, R.: The brain has a body: adaptive behavior emerges from interactions of nervous system, body, and environment. Trends in Neurosciences 20, 553–557 (1997)
4. Clark, A.: An embodied cognitive science? Trends in Cognitive Sciences 3(9), 345–351 (1999)
5. Glenberg, A.M., Kaschak, M.P.: Grounding language in action. Psychonom. Bull. and Rev. 9(3), 558–565 (2002)
6. Iida, F., Pfeifer, R., Steels, L., Kuniyoshi, Y. (eds.): Embodied Artificial Intelligence. Springer, Heidelberg (2004)
7. Lakoff, G., Johnson, M.: Philosophy in the Flesh: The Embodied Mind and its Challenge to Western Thought. Basic Books, New York (1999)
8. Lungarella, M., Metta, G., Pfeifer, R., Sandini, G.: Developmental robotics: a survey. Connection Science 15(4), 151–190 (2003)
9. Pfeifer, R., Scheier, C.: Understanding Intelligence. MIT Press, Cambridge, MA (1999)
10. Pfeifer, R., Bongard, J.C.: How the Body Shapes the Way we Think – A New View of Intelligence. MIT Press, Cambridge, MA (2007)
11. Smith, L., Thelen, E.: Development as a dynamic system. Trends in Cognitive Sciences 7(8), 343–348 (2003)

12. Smith, L., Gasser, M.: The development of embodied cognition: six lessons from babies. Artificial Life 11(1/2), 13–30 (2005)
13. Sporns, O.: Embodied cognition. In: Arbib, M. (ed.) Handbook of Brain Theory and Neural Networks, MIT Press, Cambridge, MA (2003)
14. Thompson, E., Varela, F.J.: Radical embodiment: neural dynamics and consciousness. Trends in Cog. Sci. 5(10), 418–425 (2001)
15. Wilson, M.: Six views of embodied cognition. Psychonom. Bull. Rev. 9(4), 625–636 (2002)
16. Ziemke, T.: Situated and embodied cognition. Cog. Sys. Res. 3(3), 271–554 (2002)
17. Edelman, S.: Representation and Recognition in Vision. MIT Press, Cambridge, MA (1999)
18. Palmeri, T.J., Gauthier, I.: Visual object understanding. Nat. Rev. Neurosci. 5, 291–304 (2004)
19. Riesenhuber, M., Poggio, T.: Neural mechanisms of object recognition. Current Opinion in Neurobiology 22, 162–168 (2002)
20. Dewey, J.: The reflex arc concept in psychology. Psychol. Rev. 3, 357–370 (1896)
21. Piaget, J.: The Origins of Intelligence. Routledge, New York (1953)
22. Edelman, G.M.: Neural Darwinism. Basic Books, New York (1987)
23. Pfeifer, R., Scheier, C.: Sensory-motor coordination: the metaphor and beyond. Robotics and Autonomous Systems 20, 157–178 (1997)
24. Lungarella, M., Pfeifer, R.: Robots as cognitive tools: an information-theoretic analysis of sensory-motor data. In: Proc. of 1st Int. Conf. on Humanoid Robotics, pp. 245–252 (2001)
25. O'Regan, J.K., Noe, A.: A sensorimotor account of vision and visual consciousness. Beh. Brain Sci. 24, 939–1031 (2001)
26. Nolfi, S.: Power and limit of reactive agents. Neurocomputing 49, 119–145 (2002)
27. Beer, R.D.: The dynamics of active categorical perception in an evolved model agent. Adaptive Behavior 11(4), 209–243 (2003)
28. Lungarella, M., Pegors, T., Bulwinkle, D., Sporns, O.: Methods for quantifying the informational structure of sensory and motor data. Neuroinformatics 3(3), 243–262 (2005)
29. Lungarella, M., Sporns, O.: Mapping information flow in sensorimotor networks. PLoS Computational Biology 2(10), 1301–1312 (2006)
30. Harnad, S.: Cognition is categorization. In: Cohen, H., Lefebvre, C. (eds.) Handbook of Categorization in Cognitive Science, Elsevier, Amsterdam (2005)
31. Poirier, P., Hardy-Vallee, B., DePasquale, J.F.: Embodied categorization. In: Cohen, H., Lefebvre, C. (eds.) Handbook of Categorization in Cognitive Science, Elsevier, Amsterdam (2005)
32. Gallese, V., Lakoff, G.: The brain's concepts: The role of the sensory-motor system in conceptual knowledge. Cog. Neuropsychol. 22(3/4), 455–479 (2005)
33. Gibson, J.J.: The Ecological Approach to Visual Perception. Houghton-Mifflin, Boston (1979)
34. Bajcsy, R.: Active perception. Proc. of the IEEE 76(8), 996–1005 (1988)
35. Beer, R.D.: The dynamics of active categorical perception in an evolved model agent. Adaptive Behavior 11(4), 209–243 (2003)
36. Ballard, D.: Animate vision. Artificial Intelligence 48, 57–86 (1991)
37. Churchland, P.S., Ramachandran, V., Sejnowski, T.: A critique of pure vision. In: Koch, C., Davis, J. (eds.) Large-Scale Neuronal Theories of the Brain, MIT Press, Cambridge, MA (1994)
38. Noe, A.: Action in Perception. MIT Press, Cambridge, MA (2004)

39. Lungarella, M., Sporns, O.: Information self-structuring: key principle for learning and development. In: Proc. of 4 thInt. Conf. on Development and Learning, pp. 25–30 (2005)
40. Metta, G., Fitzpatrick, P.: Early integration of vision and manipulation. Adaptive Behavior 11(2), 109–128 (2003)
41. Barlow, H.B.: The exploitation of regularities in the environment by the brain. Beh. Brain Sci. 24, 602–607 (2001)
42. Olshausen, B., Field, D.J.: Sparse coding of sensory inputs. Curr. Op. Neurobiol. 14, 481–487 (2004)
43. Simoncelli, E., Olshausen, B.: Natural image statistics and neural representation. Ann. Rev. Neurosci. 24, 1193–1216 (2001)
44. Webb, B.: Can robots make good models of biological behavior? Behavioral and Brain Sciences 24, 1033–1050 (2001)
45. Sporns, O.: What neuro-robotic models can tell us about neural and cognitive development. In: Mareschal, D. (ed.) Neuroconstructivism. Perspectives and Prospects, vol. II, Oxford University Press, Oxford, UK (2005)
46. Woods, A.T., Newell, F.: Visual, haptic, and cross-modal recognition of objects and scenes. J. of Physiology – Paris 98(1–3), 147–159 (2004)
47. Harman, K.L., Humphrey, G.K., Goodale, M.A.: Active manual control of object views facilitates visual recognition. Curr. Biol. 9(22), 1315–1318 (1999)
48. Vuong, Q.C., Tarr, M.J.: Rotation direction affects object recognition. Vis. Res. 44(14), 1717–1730 (2004)
49. James, K.H., Humphrey, G.H., Goodale, M.A.: Manipulating and recognizing virtual objects: where the action is. Canad. J. of Exp. Psych. 55(2), 113–122 (2001)
50. Wexler, M., van Boxtel, J.J.A.: Depth perception by the active observer. Trends in Cognitive Sciences 9(9), 431–438 (2005)
51. Gibson, E.J., Pick, A.D.: An Ecological Approach to Perceptual Learning and Development. Oxford University Press, Oxford (2000)
52. Lewkowicz, D.J.: Perception of serial order in infants. Dev. Sci. 7(2), 175–184 (2004)
53. Bahrick, L.E., Lickliter, R., Flom, R.: Intersensory redundancy guides the development of selective attention, perception, and cognition in infancy. Curr. Dir. Psychol. Sci. 13(3), 99–102 (2004)
54. Lewkowicz, D.J.: The development of intersensory temporal perception: an epigenetic systems/limitations view. Psychol. Bull. 126(2), 281–308 (2000)
55. Raibert, M.H.: Legged Robots that Balance. MIT Press, Cambridge, MA (1986)
56. Collins, S., Ruina, A., Tedrake, R., Wisse, M.: Efficient bipedal robots based on passive-dynamic walkers. Science 307, 1082–1085 (2005)
57. Pfeifer, R., Iida, F., Bongard, J.C.: New robotics: design principles for intelligent systems. Artificial Life 11(1/2), 99–120 (2005)
58. Geng, T., Porr, B., Worgotter, F.: Fast biped walking with a sensor-driven neuronal controller and real-time online learning. Int. J. of Robotics Research 25(3), 243–259 (2006)
59. Paul, C., Lungarella, M., Iida, F.(eds.): Morphology, control, and passive dynamics. Robotics and Autonomous Systems 54(8), 617–718 (2006)
60. Lungarella, M., Berthouze, L.: On the interplay between morphological, neural, and environmental dynamics: a robotic case-study. Adaptive Behavior 10(3/4), 223–241 (2002)
61. Tedrake, R., Zhang, T.W., Seung, H.S.: Stochastic policy gradient reinforcement learning on a simple 3D biped. In: Proc. of 10 th Int. Conf. on Intelligent Robots and Systems, pp. 3333–3338 (2004)

62. Lichtensteiger, L., Pfeifer, R.: An optimal sensory morphology improves adaptability of neural network controllers. In: Hochet, B., Acosta, A.J., Bellido, M.J. (eds.) PATMOS 2002. LNCS, vol. 2451, pp. 850–855. Springer, Heidelberg (2002)
63. Bjorklund, E., Green, B.: The adaptive nature of cognitive immaturity. American Psychologist 47(1), 46–54 (1992)
64. Bjorklund, E.: The role of immaturity in human development. Psychological Bulletin 122(2), 153–169 (1997)
65. Turkewitz, G., Kenny, P.: Limitations on input as a basis for neural organization and perceptual development: a preliminary theoretical statement. Dev. Psychobio. 15(4), 357–368 (1982)
66. Bernstein, N.: The Co-ordination and Regulation of Movements Oxford: Pergamon. Pergamon, Oxford (1969)
67. Mussa-Ivaldi, F.A., Bizzi, E.: Motor learning through the combination of motor primitives. Phil. Trans. Roy. Soc. Lond. B 355, 1755–1769 (2000)

Development Via Information Self-structuring of Sensorimotor Experience and Interaction

Chrystopher L. Nehaniv[1], Naeem Assif Mirza[1], and Lars Olsson[2]

[1] Adaptive Systems Research Group, University of Hertfordshire, College Lane, Hatfield, Hertfordshire AL10 9AB, U.K.
{C.L.Nehaniv, N.A.Mirza}@herts.ac.uk
[2] Netemic Ltd., Prospect Studios, 67 Barnes High Street, London SW13 9LE, U.K.
lo@abstractvoid.se

Abstract. We describe how current work in Artificial Intelligence is using rigorous tools from information theory, namely *information distance* and *experience distance* to organize the self-structuring of sensorimotor perception, motor control, and experiential episodes with extended temporal horizon. Experience is operationalized from an embodied agent's own perspective as the flow of values taken by its sensors and effectors (and possibly other internal variables) over a temporal window. Such methods allow an embodied agent to acquire the sensorimotor fields and control structure of its own body, and are being applied to pursue autonomous scaffolded proximal development in the zone between the familiar experience and the unknown.

1 Introduction: Information Self-structuring in Ontogeny

Modern Artificial Intelligence (AI) research has increasingly focused on adaptive, embodied agents with rich sensing capabilities situated in complex environments, that develop in their capabilities over the course of their "lifetimes" (ontogeny) [1, 2]. In our and related research particular attention is paid to the process of autonomous self-structuring in response to a history of self-motivated interaction with a rich environment. The aim is to investigate in artificial agents mechanisms of motivation, learning, development, and temporal awareness with inspiration drawn from biology, psychology, philosophy, engineering, and mathematics.

In this article we review a class of methods for discovering relationships between any and all sensors and actuators that an agent has access to. The methods use the measure of information distance based on Shannon information theory [3] and capture the degree to which the time-varying nature of a variable may be predictable from another. These measures have been used in robots to autonomously discover sensorimotor maps from unknown sensors grounded in interaction with the environment and to discover fundamental control laws for unknown actuators [4, 5] thus gaining mastery over one's own embodiment (which may well be changing). Related classical geometric and statistical methods have also been

M. Lungarella et al. (Eds.): 50 Years of AI, Festschrift, LNAI 4850, pp. 87–98, 2007.

used in simulation to discover sensorimotor relationships [6] and the related structure of space via sensing and acting [7]. Further, the information-theoretic and related methods have been used to characterize behaviour of robots from the robot's perspective [8, 9] and also to measure, geometrically, how one sensorimotor experience differs from another [10]. The self-structuring of the sensory and motor competencies is enabled by the tight coupling of the agent with the environment [11, 2, 5] and the agent can directly base action on its own history of interaction with the environment (including the social environment) to make this possible [12].

2   Information Distance Measures

2.1   Sensors as Information Sources

An agent situated and acting in an environment will have access to many external and internal variables any of which can be modeled as random variables changing over time. These can be thought of as generalized "sensory" inputs, from sources having any character at all (whether sensory, motor, or internal), such as, e.g., registration on sensory surfaces (activations of retinal neurons in vision or cochlear hairs in hearing, readings coming from spatially distributed tactile sensors such as skin and whiskers, etc.), proprioception, motor values sent to effectors, physiological variables, other internal states, etc. Consider one such random variable $\mathcal{X}$ changing with time, taking value $x(t) \in X$, where X is the set of its possible values. Time is taken to be discrete (i.e. t will denote a natural number) and $\mathcal{X}$ takes values in a finite set or "alphabet" $X = \{x_1, \ldots, x_m\}$ of possible values[1].

2.2   Information Distance

For any pair of jointly distributed random variables ("sensors") $\mathcal{X}$ and $\mathcal{Y}$ the *conditional entropy* $H(\mathcal{X}|\mathcal{Y})$ of $\mathcal{X}$ given $\mathcal{Y}$ is the amount of uncertainty (in bits) that remains about the value $\mathcal{X}$ given that the value of $\mathcal{Y}$ is known, and is given by

$$H(\mathcal{X}|\mathcal{Y}) = -\sum_{x \in X} \sum_{y \in Y} p(x, y) \log_2 \frac{p(x, y)}{p(y)},$$

where $p(x, y)$ is the joint distribution of $\mathcal{X}$ and $\mathcal{Y}$.[2]

The *information distance* between $\mathcal{X}$ and $\mathcal{Y}$ is then defined as

$$d(\mathcal{X}, \mathcal{Y}) = H(\mathcal{X}|\mathcal{Y}) + H(\mathcal{Y}|\mathcal{X}).$$

---

[1] The approach generalizes to continuous time and value sets with appropriate changes.

[2] The methods require the assumption of approximate local stationarity of the joint distribution of random variables representing the sensorimotor variables over a temporal window and that this distribution can be estimated closely enough by sampling the sensorimotor variables.

Crutchfield [13] shows that this satisfies the mathematical axioms of equivalence, symmetry and the triangle inequality and so is a *metric*. (See the Appendix for a visual proof of this fact.) Thus d defines a geometric structure on any space of jointly distributed information sources, such as the sensorimotor variables of an embodied agent.

The metric space geometric structure is advantageous as it potentially allows one to exploit the highly developed advanced techniques of mathematical analysis and geometry in the control of behaviour.

2.3   Experience Distance

Given the above definitions we can operationalize an agent's *experience* from time t over a temporal horizon of h time units as $E(t, h) = (\mathcal{X}^1_{t,h}, \ldots, \mathcal{X}^N_{t,h})$ where $\mathcal{X}^1, \ldots, \mathcal{X}^N$ is the set of all sensorimotor (or other) variables available to the agent and each $\mathcal{X}^i_{t,h}$ is the random variable estimated from the values of $\mathcal{X}^i$ over a window of length h beginning at time t ($1 \leq i \leq N$).

We can then define a metric, the *experience metric D*, on experiences of temporal horizon h as

$$D(E, E') = \sum_{k=1}^{N} d(\mathcal{X}^k_{t,h}, \mathcal{X}^k_{t',h}),$$

or, alternatively, the *cross-modal experience metric D'*, as

$$D'(E, E') = \sum_{i=1}^{N} \sum_{j=1}^{N} d(\mathcal{X}^i_{t,h}, \mathcal{X}^j_{t',h}),$$

where $E = E(t, h)$ and $E' = E(t', h)$ are experiences of an agent at time t and t' over horizon h and d is the information distance. That D (and similarly D') is a metric follows from the fact that the metric axioms hold component-wise, since d is a metric.

As experiences are collected, they can be placed in a *metric space of experience* using either of these experience metrics. Experiences generated from similar behaviour as judged from the human observer's perspective generally turn out to be nearby from the robot's perspective in terms of the experience metric in such a space [14].

This operational notion of experience facilitates the application of information-theoretic methods to sensorimotor variables in a way that is directly related to *embodiment*. Such an agent-centric approach already brings these rigorous methods closer to a type of Shannon information that is meaningful for perception and action, however it is possible to go much further and develop a rigorous notion of *relevant information* specific to particular organisms and agents, by relating action, information and utility – see [15].

3 Development of Artificial Cortex: Using Information Theory as a Means for Self-organizing Sensorimotor Structures Grounded in Experience

How can raw, uninterpreted information from unknown sensors come to be used by a developing embodied agent with no prior knowledge of its motor capabilities?

In nature, cognitive structures appear to be organized in the course of evolution and also in the course of development so as to reflect information-theoretic relations arising in the interaction of sensors, actuators, and the environment (including the social environment). That wiring economy for intracortical connectivity of functionally related neural processing structures yields evolutionary fitness has been proposed as a general principle giving rise to topographic structure of cortical maps (see review in [16]) and permitting "extraordinary speed in representing ecologically significant information" [17].

We have applied the information distance metric to develop and reconstruct "retinotopic" and cortex-like sensorimotor maps for robots as they engage in interaction with real-world environments (see Figure 1, and [4, 5] for details). Information distance (rather than mutual information or other measures such as Hamming distance) appears to lead to the best structured cortex-like maps of sensorimotor variables, especially for cross-modal integration [18]. This power might be due to information distance's metric nature, which allows natural geometrization of information sources (which could possibly also give raise to wiring economy), coupled with the capacity of information distance to detect relations between information sources that are informationally, but not necessarily linearly (nor even continuously), correlated.

Even in brain areas much older than the cortex, such as the superior colliculus in mammals (the area homologous to the optic tectum in other vertebrates), cross-modal alignment of visual, auditory, and tactile and somatosensory maps is evident [19]. For instance, in the ferret or barn owl such visual and auditory maps are aligned in this region in proximity to neural pre-motor control circuitry allowing the animal to combine sensory modalities in guiding action, e.g. combining or using either of visual and auditory information in neural maps to guide head movements and eye saccades in orienting toward prey, or, e.g. in reaching in primates; moreover maps are maintained and aligned over the course of development, which may be activity and sensory stimulation dependent – see [20, 21].

In artificial embodied agents, such sensory fields that are constructed on the basis of information distance (see preceding section) methods [4] can then be used to autonomously discover sensorimotor laws (Figure 2), e.g. optical or tactile flow and visually guided movement [5]. The particular embodiment and environment experienced and the changes in it can shape the sensorimotor maps generated in this way, as well as drive their dynamic unfolding and adaptation in ontogeny [4, 5].

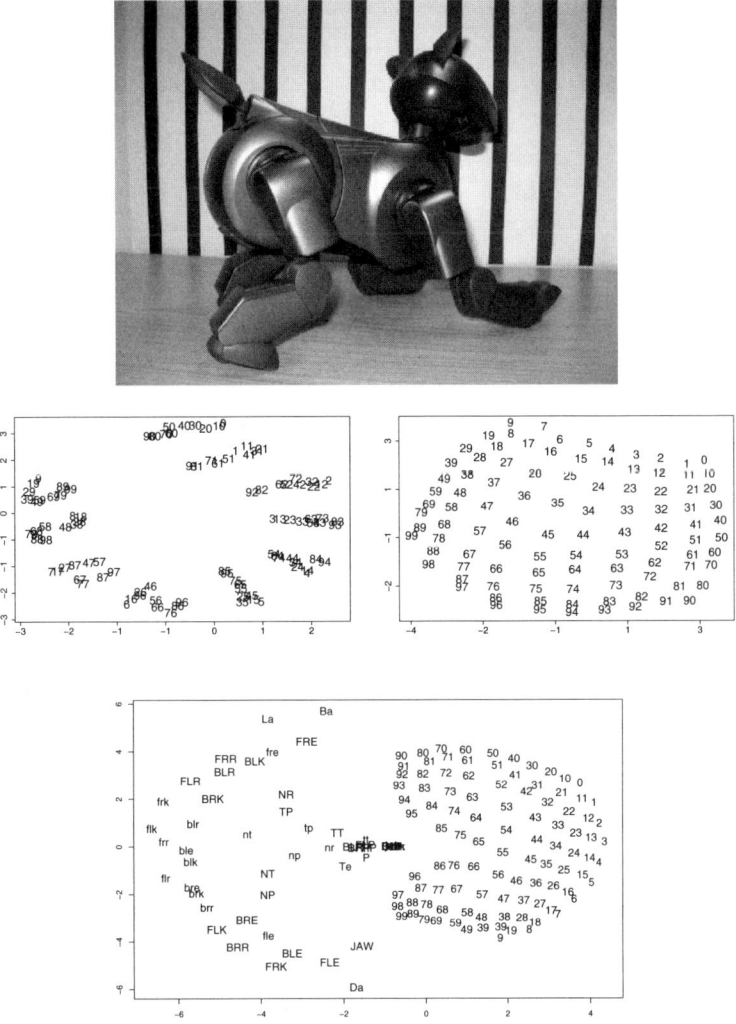

Fig. 1. Sony Aibo in striped environment develops impoverished distinctions between sensors, but further development may allow distinctions to unfold. Top: Robot moving in the striped environment. In the remaining subfigures, points represent individual information sources (sensors or actuators of the robot) plotted using the information distance (and collapsed into two dimensions). Middle left: Sensory organization of the vision sensors (pixels in the visual field) developed in impoverished environment reflects only similar sensory experience of visual receptors from the same columns. Middle right: Sensory organization of vision sensors after moving to richer visual environment reflects their topographical layout. Bottom: Cortex-like "Aibunculus" sensorimotor organization – analogous to the somatosensory "homuncular" cortical maps – recovered based on self-structuring during agent-environment interaction using information distance, discovering visual field (numbered sensors, clustered "retinotopically" and arrayed to the right) and left-right bilateral body symmetry along an anterior-posterior axis.

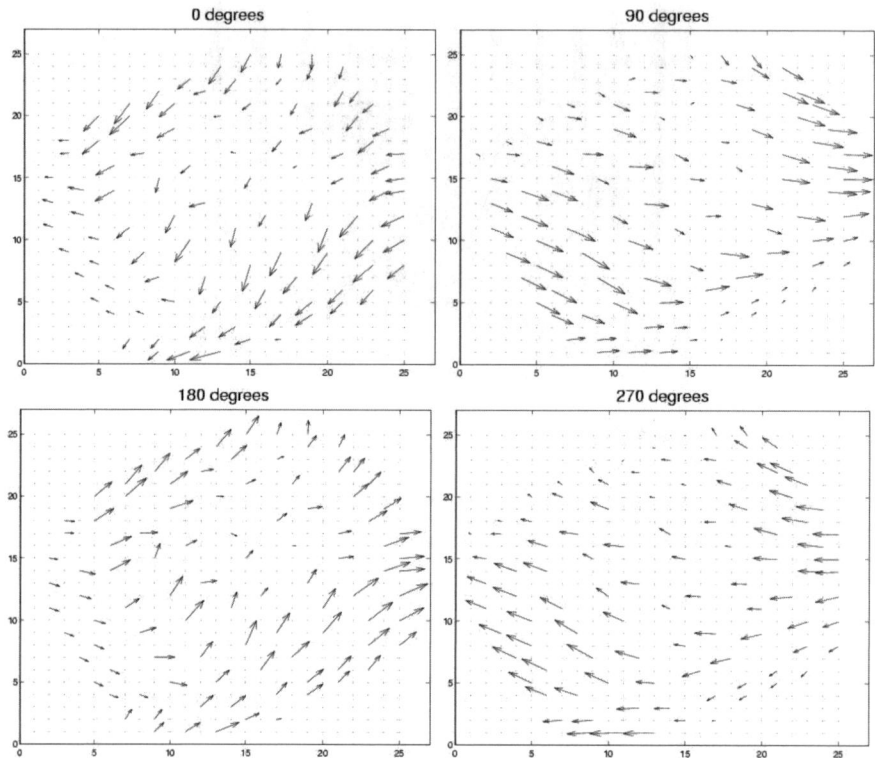

Fig. 2. Sensory fields (in this case a two-dimensional visual field), sensory flows on the fields (the regularities in value shifts due to actions - in this case head movements), and sensorimotor laws (describing the flows resulting from actions) are autonomously derived from experience based on bottom-up information self-structuring and used to control visually guided movement [5]. Figure shows discovered sensorimotor regularities in sensory flows induced by motions of the Aibo's head in various directions, where 0 degrees is up, 180 degrees is down, 270 degrees is right, and 90 degrees is left from the robot perspective.

Alignment of multimodal information sources is demonstrated using the Aibo robot for red, green, blue color channels in vision via *entropy maximization* and *information distance* self-structuring, and this combination of information distance and entropy maximization is shown by far to out-perform other metrics and measures for sensory map construction (see [18] for details). Combining multimodal sensory integration with pre-motor control based on alignment of sensory and body maps is a next natural target for such methods.

4   Temporally Extended Experience and Interaction Histories

How can embodied agents develop in response to extended experiences at various scales in time?

Generally, in AI so far the *temporal horizon* [22] has either been limited to the immediate time-scale (traditional behaviour-based reactivity), short-term modulation (affective computing and robotics), or, if longer term, then has generally remained ungrounded (case-based reasoning or symbolic planning) and not susceptible to developmental change in ontogeny. *Autobiographic agents* dynamically construct and reconstruct their experiences in a process of remembering while actively engaged in interaction with the rest of the world [23].

Using extensions of the information metric to experiential episodes of various temporal horizons (see section 2.3), it is possible to impose geometric order on a robot's temporally extended sensorimotor experiences, at various temporal scales [24]. The structure of these dynamic spaces of experiences provides an agent-centric enactive representation of interaction histories with the environment (including the social environment), grounded in the temporally extended sensorimotor experience and used in generating action [8, 10, 25].

Potentially an agent can act using this dynamically growing, developing space of experiences to return to familiar experiences, predict the effect of continuing on a current behavioural trajectory, and explore at the boundary of what is already mastered (cf. Vygotsky's notion of "zone of proximal development"). By using temporally extended experiences to guide action and interaction, we will have the beginnings of post-reactive robotics and grounded episodic intelligence in artificially constructed enactive agents that grow, develop, and adapt their cognitive structures with a broader temporal horizon.

This possibility is explored in our experiments where a robot uses actions determined by a history of interaction to develop the capability to play the early learning game "peekaboo" of iteratively seeing/revealing and not-seeing/hiding the face with an interaction partner [10, 12]. The architecture uses experience distance (based on information distance) to compare experiences and to place them in a metric space. Actions are chosen based on proximity in this space and motivational value of experience. (See Figure 3, and [12] for details.) Peekaboo not only has inherent simple narrative temporal and rule structure [26], but is also believed to be important in providing scaffolding to young infants in developing social expectations and primary intersubjectivity [27]. By forming expectations and selecting action based on previous temporally extended experiences, the agent is able to develop the capacity to engage in practice of temporally complex skills such as social play, and to re-engage in them when similar experience arises again. It should also be possible to explore at the geometric boundary of already mastered skills and familiar behaviour in the experience metric space, which grows and changes dynamically with the lifelong history of interaction.

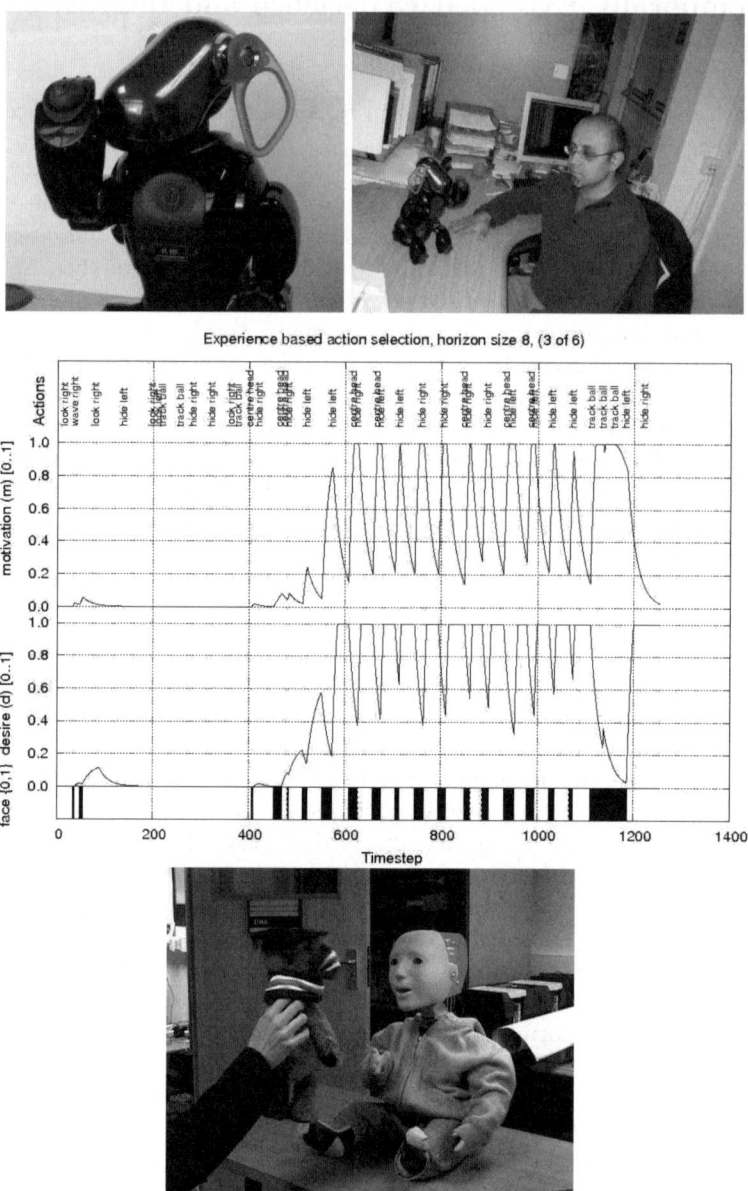

Fig. 3. Top left: Aibo hides face in the autonomous development of the ability to engage in a 'peekaboo' turn-taking game. Top right: Aibo engaging in interaction games with human partner based on interaction history and informationally structured space of experiences. Center: Dynamics of internal variables and actions selected as face is seen or not, with iterations by black/white pattern at bottom of panel indicative of 'peekaboo'-style interaction. Bottom: Interaction games with another platform, the KASPAR child-sized humanoid robot built at University of Hertfordshire.

It may be that using the information and experience distance metrics to organize sensorimotor and episodic experience might capture relations that, in natural organisms, are reflected in the topologies arising from such self-organizing principles as spike time dependent plasticity (cf. [28, 29]) that structure neural connections in development and spatiotemporal sensorimotor pattern learning.

5   Summary and Outlook

Information methods can guide the autonomous organization and structuring of sensorimotor data, lead to the autonomous detection of sensorimotor laws, and underpin the acquisition of sensorimotor control starting with raw uninterpreted sensory data and unknown actuators. Similarly, by extending the methods to encompass sensorimotor flow during particular temporally extended intervals, episodic experience can be operationalized for an embodied system. The geometry of experiences is organized by their information-theoretic structure, and is proposed as a basis for achieving development in robots that grow up, re-engaging in familiar activity, exploring at the boundary of what is already developed, controllable, and mastered. This includes not only sensorimotor experience of static environments, but also interaction histories in dynamic environments involving social interaction or changing embodiment.

Acknowledgments

We are grateful to Daniel Polani, Kerstin Dautenhahn, and René te Boekhorst for collaboration on various aspects of the work surveyed here. This work was partially conducted within the EU Integrated Project RobotCub (Robotic Open-architecture Technology for Cognition, Understanding, and Behaviours), funded by the EC through the E5 Unit (Cognition) of FP6-IST under Contract FP6-004370.

References

[1] Lungarella, M., Metta, G., Pfeifer, R., Sandini, G.: Developmental robotics: A survey. Connection Science 15(4), 151–190 (2003)
[2] Lungarella, M., Sporns, O.: Information self-structuring: Key principles for learning and development. In: ICDL 2005. Proc. of 4th International Conference on Development and Learning, pp. 25–30 (2005)
[3] Shannon, C.E.: A mathematical theory of communication. Bell Systems Technical Journal 27, 379–423, 623–656 (1948)
[4] Olsson, L., Nehaniv, C.L., Polani, D.: Sensory channel grouping and structure from uninterpreted sensor data. In: 2004 NASA/DoD Conference on Evolvable Hardware, Seattle, Washington, USA, June 24-26, 2004, pp. 153–160. IEEE Computer Society Press, Los Alamitos (2004)
[5] Olsson, L., Nehaniv, C.L., Polani, D.: From unknown sensors and actuators to actions grounded in sensorimotor perceptions. Connection Science 18(2), 121–144 (2006)

[6] Pierce, D., Kuipers, B.: Map learning with uninterpreted sensors and effectors. Artificial Intelligence 92, 169–229 (1997)

[7] Philipona, D., O'Regan, J.K., Nadal, J.P.: Is there something out there? Inferring space from sensorimotor dependencies. Neural Computation 15(9) (2003)

[8] Mirza, N.A., Nehaniv, C.L., Dautenhahn, K., te Boekhorst, R.: Using sensory-motor phase-plots to characterise robot-environment interactions. In: CIRA 2005. Proc. of 6th IEEE International Symposium on Computational Intelligence in Robotics and Automation, pp. 581–586. IEEE Computer Society Press, Los Alamitos (2005)

[9] Kaplan, F., Hafner, V.: Mapping the space of skills: An approach for comparing embodied sensorimotor organizations. In: ICDL 2005. Proceedings of the 4th IEEE International Conference on Development and Learning, pp. 129–134. IEEE Computer Society Press, Los Alamitos (2005)

[10] Mirza, N.A., Nehaniv, C.L., Dautenhahn, K., te Boekhorst, R.: Interaction histories: From experience to action and back again. In: ICDL 2006. Proceedings of the 5th IEEE International Conference on Development and Learning, Bloomington, IN, USA, IEEE Computer Society Press, Los Alamitos (2006)

[11] te Boekhorst, R., Lungarella, M., Pfeifer, R.: Dimensionality reduction through sensory-motor coordination. In: Kaynak, O., Alpaydin, E., Öja, E., Xu, L. (eds.) ICANN 2003 and ICONIP 2003. LNCS, vol. 2714, pp. 496–503. Springer, Heidelberg (2003)

[12] Mirza, N.A., Nehaniv, C.L., Dautenhahn, K., te Boekhorst, R.: Grounded sensorimotor interaction histories in an information theoretic metric space for robot ontogeny. Adaptive Behavior 15(2), 167–187 (2007)

[13] Crutchfield, J.: Information and its metric. In: Lam, L., Morris, H.C. (eds.) Nonlinear Structures in Physical Systems - Pattern Formation, Chaos and Waves, pp. 119–130. Springer-Verlag, New York (1990)

[14] Mirza, N.A., Nehaniv, C.L., Dautenhahn, K., te Boekhorst, R.: Using temporal information distance to locate sensorimotor experience in a metric space. In: CEC 2005. Proc. of 2005 IEEE Congress on Evolutionary Computation, Edinburgh, Scotland, UK, 2-5 September 2005, vol. 1, pp. 150–157. IEEE Computer Society Press, Los Alamitos (2005)

[15] Nehaniv, C.L., Polani, D., Olsson, L., Klyubin, A.S.: Information-theoretic modeling of sensory ecology. In: Laubichler, M., Müller, G.B. (eds.) Modeling Biology: Structures, Behavior, Evolution. The Vienna Series in Theoretical Biology, MIT Press, Cambridge (in press)

[16] Chklovskii, D.B., Koulakov, A.A.: Maps in the brain: What can we learn from them? Annu. Rev. Neurosci. 27, 369–392 (2004)

[17] Diamond, M.E., Petersen, R.S., Harris, J.A.: Learning through maps: Functional significance of topographic organization in the primary sensory cortex. J. Neurobiol. 41, 64–68 (1999)

[18] Olsson, L., Nehaniv, C.L., Polani, D.: Measuring informational distances between sensors and sensor integration. In: Artificial Life X, MIT Press, Cambridge (2006)

[19] King, A.J.: Sensory experience and the formation of a computational map of auditory space in the brain. BioEssays 21(11), 900–911 (1999)

[20] Knudsen, E.I., Brainard, M.S.: Creating a unified representation of visual and auditory space in the brain. Annual Review of Neuroscience 18, 19–43 (1995)

[21] King, A.J.: The superior colliculus. Current Biology 14(4), R335–R338 (2004)

[22] Nehaniv, C.L., Dautenhahn, K., Loomes, M.J.: Constructive biology and approaches to temporal grounding in post-reactive robotics. In: McKee, G.T.,

Schenker, P., (eds.) Sensor Fusion and Decentralized Control in Robotics Systems II (September 19-20, Boston, Massachusetts), Proceedings of SPIE. vol. 3839 pp. 156–167 (1999)

[23] Dautenhahn, K., Christaller, T.: Remembering, rehearsal and empathy - towards a social and embodied cognitive psychology for artifacts. In: Ó Nualláin, S., McKevitt, P., Mac Aogáin, E., (eds.) Two Sciences of the Mind: Readings in Cognitive Science and Consciousness. John Benjamins North America Inc. pp. 257–282 (1996)

[24] Nehaniv, C.L.: Sensorimotor experience and its metrics. In: Proc. of 2005 IEEE Congress on Evolutionary Computation, Edinburgh, Scotland, UK, 2-5 September 2005, vol. 1, pp. 142–149. IEEE Computer Society Press, Los Alamitos (2005)

[25] Nehaniv, C.L., Mirza, N.A., Dautenhahn, K., te Boekhorst, R.: Extending the temporal horizon of autonomous robots. In: Murase, K., Sekiyama, K., Kubota, N., Naniwa, T., Sitte, J. (eds.) AMiRE 2005. Proc. of the 3rd International Symposium on Autonomous Minirobots for Research and Edutainment, pp. 389–395. Springer, Heidelberg (2006)

[26] Bruner, J.S., Sherwood, V.: Peekaboo and the learning of rule structures. In: Bruner, J., Jolly, A., Sylva, K. (eds.) Play: Its Role in Development and Evolution, pp. 277–285. Penguin, New York (1975)

[27] Rochat, P., Querido, J.G., Striano, T.: Emerging sensitivity to the timing and structure of protoconversation in early infancy. Developmental Psychology 35(4), 950–957 (1999)

[28] Song, S., Miller, K.D., Abbott, L.F.: Competitive Hebbian learning through spike-timing-dependent synaptic plasticity. Nature Neuroscience 3, 919–926 (2000)

[29] Izhikevich, E.M.: Polychronization: Computation with spikes. Neural Computation 18, 245–282 (2006)

Appendix. Short Proof that Information Distance Satisfies the Axioms for a Metric on Information Sources

We can give a short proof that d is really a metric: Specifically, d is a metric since it satisfies the following axioms for every three (jointly distributed) information sources $\mathcal{X}$, $\mathcal{Y}$ and $\mathcal{Z}$:

1. $d(\mathcal{X}, \mathcal{Y}) = 0$ if and only if $\mathcal{X}$ and $\mathcal{Y}$ are equivalent (equivalence).
2. $d(\mathcal{X}, \mathcal{Y}) = d(\mathcal{Y}, \mathcal{X})$ (symmetry).
3. $d(\mathcal{X}, \mathcal{Y}) + d(\mathcal{Y}, \mathcal{Z}) \geq d(\mathcal{X}, \mathcal{Z})$ (triangle inequality).

Proof: In the first condition, "equivalent" means "informationally equivalent", i.e. that knowing the value of $\mathcal{X}$ completely determines the value of $\mathcal{Y}$, and vice versa. This can only be the case exactly when both of the conditional entropies are zero. The second condition is trivial from the symmetry of the expression $H(\mathcal{X}|\mathcal{Y}) + H(\mathcal{Y}|\mathcal{X})$. To see that the triangle inequality holds, draw a "Venn diagram" visualization for the entropies of the three random variables $\mathcal{X}$, $\mathcal{Y}$, $\mathcal{Z}$ (see Fig. 4). Now the quantity $d(\mathcal{X}, \mathcal{Y})$ corresponds to the "double crescent" region (i.e. excluding the overlap) for $\mathcal{X}$ and $\mathcal{Y}$ representing the sum of their (non-negative) conditional entropies in bits. Now it is obvious that the double crescent for $\mathcal{X}$ and $\mathcal{Y}$ together with the double crescent for $\mathcal{Y}$ and $\mathcal{Z}$ cover the one for the pair $\mathcal{X}$ and $\mathcal{Z}$, and, since for all the variously shaded regions the corresponding entropies are non-negative, it follows from the covering that the inequality holds.   □

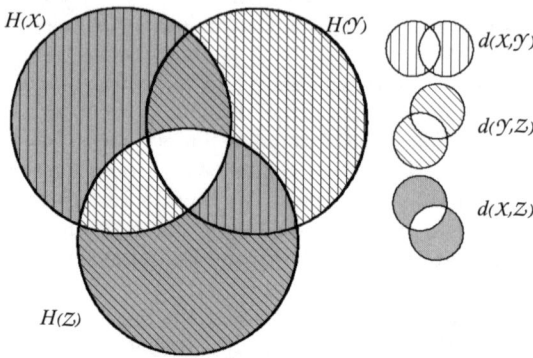

Fig. 4. Visual Proof of the Triangle Inequality for Information Distance. Visualization of the entropies H of three information sources modeled as random variables $\mathcal{X}$, $\mathcal{Y}$ and $\mathcal{Z}$, with the variously shaded double-crescent regions showing, for each pair of variables, the sum of these conditional entropies, which gives their information distance. Right: Three information distances are visualized as double-crescent regions in the key. Here the left crescent for the information distance $d(\mathcal{X}, \mathcal{Y})$ from $\mathcal{X}$ to $\mathcal{Y}$ represents the conditional entropy $H(\mathcal{X}|\mathcal{Y})$ and the right crescent represents the conditional entropy $H(\mathcal{Y}|\mathcal{X})$; similarly, the other double-crescent regions corresponding to $d(\mathcal{Y}, \mathcal{Z})$ and $d(\mathcal{X}, \mathcal{Z})$ are shown. Left: Venn diagram visualization for the entropies of the three information sources. The triangle inequality holds since the double-crescent region for $d(\mathcal{X}, \mathcal{Z})$ is completely covered by those for $d(\mathcal{X}, \mathcal{Y})$ and $d(\mathcal{Y}, \mathcal{Z})$.

How Information and Embodiment Shape Intelligent Information Processing

Daniel Polani[1], Olaf Sporns[2], and Max Lungarella[3]

[1] Dept. of Computer Science, University of Hertfordshire, Hatfield, UK
[2] Dept. of Psychological and Brain Sciences, Indiana University, Bloomington (IN), USA
[3] Artificial Intelligence Laboratory, University of Zurich, Zurich, Switzerland
d.polani@herts.ac.uk, osporns@indiana.edu, lunga@ifi,unizh.ch

Abstract. Embodied artificial intelligence is based on the notion that cognition and action emerge from interactions between brain, body and environment. This chapter sketches a set of foundational principles that might be useful for understanding the emergence ("discovery") of intelligence in biological and artificial embodied systems. Special emphasis is placed on information as a crucial resource for organisms and on information theory as a promising descriptive and predictive framework linking morphology, perception, action and neural control.

1 Introduction

Artificial Intelligence (AI) strives to understand what "thinking" is by building intelligent entities capable of perceiving, understanding, and manipulating the world surrounding them. The choice of the physical and computational substrates necessary to realize such entities remains a matter of debate. In the early years of electronic computation, one had several different competing approaches to implement processes of thought electronically: cybernetics, systems theory, neural networks, analog machines, and the von Neumann architecture. The classical framework of AI eventually was built on top of the model proposed by von Neumann which emerged as winner. With the success of the von Neumann concept, the algorithmic view of AI prevailed. Intelligence became synonymous with rule-based processing of symbolic information, within a computational architecture that existed independently of its physical implementation. Such a functionalist stance explicitly divorced intelligence from its material or biological substrate. Intelligent systems were targeted at implementing mechanisms derived from the reconstruction of models of human self-inspection or engineered based on technological principles oriented at achieving well-defined and specific goals. In other words, AI became an essentially "platonic" endeavour directed at the top-down design of symbol processing intelligent systems.

While some of the major challenges of AI became reachable (e.g., human-competitive chess-playing software), success was too fragmented. Moreover, there was quite some uncertainty as to what degree one could actually project a phenomenon (e.g., intelligent control) that nature had managed to "engineer" (in fact, evolve) on its own onto a human-designed architecture. Natural solutions have to be *always* viable, i.e., provide stable even if non-optimal solutions in the face of

M. Lungarella et al. (Eds.): 50 Years of AI, Festschrift, LNAI 4850, pp. 99–111, 2007.

uncertainty, noise or incomplete input, or unpredictable changes of context. While viability might seem an incidental property that distinguishes artificial from natural systems, it also fundamentally counteracts the top-down construction philosophy of artificial systems. If this property is taken seriously, emergence of viable solutions for intelligent agents cannot be separated from a permanent entrenchment of the agent in the real world. In particular, the agent's controller is not developed in a platonic world and planted into the agent, but needs to provide the agent with adequate behaviours during its entire lifetime.

Interestingly, a direct danger to the enterprise of platonic "universal" intelligence is posed by the concerns expressed by theorems of the "no free lunch" type [1]. Essentially, such theorems state that finding optima efficiently (and thus efficient learning) is impossible in arbitrary worlds. We do have, however, an existence proof for consistent emergence of intelligence – namely in the biological realm. Biological intelligence appears in as distant species as, say, humans and octopuses; eye evolution reappears in 40-60 different independent lines of descent and often repeats central morphological motifs in remotely related species [2]. In other words, while "no free lunch" type considerations are important for an understanding of abstract "platonic" models, they probably are of lesser relevance for the emergence of intelligence in real-world scenarios. In fact, the world is not arbitrary, but intricately structured and constrained by a subtly intertwined set of properties, e.g., symmetries, continuities, and smoothness. Intelligence is thus *fundamentally* a result of embodied interaction which exploits structure in the world [3]. Two questions remain: how embodiment actually manages to drive the emergence of intelligence under the constraints of uninterrupted viability, as there is no intelligent designer? Is it possible to formulate an architecture-invariant concept that captures the essence of (neural or morphological) computation? If such a concept could be found, one could then apply it to the informational dynamics of an agent acting in its environment. This would yield a computational analogue of what the Carnot-machine is for thermodynamics: by realizing a cycle of information exchange between system and environment, it would provide a consistent framework from which the laws of information processing could be derived given the constraints governing the flow into and out of the system.

Such a perspective would change the way we look at intelligent information processing. Instead of primarily constructing algorithms that solve a particular given task (as in the conventional approach), the phenomenon of intelligent information processing would emerge from an informationally balanced dynamics, without intervention or guidance from an external intelligent designer. Intelligence would be "discovered" rather than engineered (i.e., evoked rather than constructed). For this purpose, it is necessary to identify and formulate suitable quantitative principles. Here, we suggest that Shannon's measure of information [4] (and any quantities derived from it) is a main candidate to help us define such a framework. In the following, we discuss the state-of-the-art of this view on intelligence and how it points towards future perspectives for AI.

2 Information as a Guiding Principle

First attempts to relate information theory to the control of cybernetic systems were done by Ashby [5] who proposed the principle of "requisite variety" (that is, the idea that for control to be possible, the available variety of control actions must be equal or greater than the variety of disturbances). Around the same time, it was suggested that the organization of sensory and perceptual processing might be explained by principles of informational economy (e.g., redundancy reduction [6] and sparse coding [7]). Order and structure (i.e., redundancy) in sensor stimulation were hypothesized to greatly simplify the task of the brain to build up "cognitive" maps or working models of the surrounding environment. In AI, with the increasing dominance of the algorithmic as opposed to the cybernetic approach, the use of information theory was neglected for a long time (applications were typically limited to quantifications of classification strengths, such as in algorithms to generate decision trees). One problem lay in the fact that it was not clear how to make systematic use of information theory to design or understand intelligent systems. In view of the lack of progress, Gibson [8] suspected that the failure of information theory was intrinsic, because in its original form it considers a sender whose intention is to communicate with a receiver, while – so Gibson's argument – the environment of an agent has no intent of informing the agent.

Within the resurgence of neural networks, a major step ahead was taken by Linsker who proposed the principle of maximum information preservation (Infomax; [9]). His objective was to identify principles that would help narrow down architectures plausible for biological settings, e.g., the early processing stages of the visual system. The underlying tenet was the following: due to the intricate processes of the higher-level stages, earlier stages cannot predict what information is going to be used later. The most unbiased hypothesis is thus to assume that earlier stages maximize the total information throughput. In other words, everything that is processed in the later stages of the vision system has to pass through these early stages. This hypothesis was applied to a feed-forward network making some general architectural assumptions, namely a layered structure and localized receptive fields of the neurons arranged in two-dimensional sheets. Maximization of the amount of information preserved at each processing stage caused the neurons' receptive fields to exhibit response profiles similar to the filters found in the vision systems of many organisms. The Infomax principle provides a powerful and general mathematical framework for self-organization of an information processing system that is independent of the rule used for its implementation.

Another dramatic illustration of the central importance of information for living systems comes from work on bioenergetics. Surprisingly, information transmission and processing are metabolically expensive. For example, the blowfly retina consumes 10% of the energy used by the resting fly and, similarly, the human brain accounts for up to 20% of the oxygen consumption at rest [10]. If metabolic cost of a particular informational resource (i.e., neural information processing capacity) is limiting, there is not only a good chance that neural circuits have evolved to reduce its metabolic demands, but also that it will be exploited to a significant degree (and sometimes close to its limit) by a biological system [11]. These results indicate that "information" is almost as important as energy [12]. Motivated by this dominant role of information in living systems, we will therefore suggest to entirely focus on information "metabolism" as *the* single principle driving the emergence and formation of intelligence.

The obvious caveat is that the complexity of living systems may make it hard to pinpoint one single universal principle guiding the emergence of a class of phenomena such as intelligence. Many types of selection pressures, driven by a variety of resource requirements or by other factors (such as sexual selection) act on an organism. Why can we expect that it is sufficient to concentrate on the information "metabolism" to understand the emergence of intelligence? Although present space does not permit a discussion of this question at full length, we wish to reemphasize a few important arguments supporting this view. Not only do sensors and neural structures, as mentioned above, consume a considerable amount of energy in living agents, but also it is known that in living beings sensors and the neural substrate operate close to the theoretically optimal level of information processing. Information is thus a resource of primary importance for a living being and one can expect available capacities to be fully exploited (if not fully exploited, these capacities will degenerate away during evolution). In addition, any further constraints arising from other selection pressures can be factored into the trade-off between available information capacity and the particular interaction dynamics and embodiment.

But perhaps the strongest indicator that universal principles may play a role in the emergence of intelligence is the fact that natural intelligence arises in so many different forms and guises. Species as remotely related and with drastically different sensorimotor and neural substrates as mammals, birds, and octopuses all exhibit a high degree of intelligence. It is hard to believe that evolution would "reinvent the wheel" of intelligent information processing for each of these branches – much more plausible is the assumption of a small number of universal principles giving rise to appropriate evolutionary pressures. Intelligence is, after all, about the right strategy of processing and utilizing information. Therefore, in the following, we will concentrate on the role of *information* in the emergence of intelligence in embodied systems, to the exclusion of any other possible candidate concepts. As we will see, even this restricted set of assumptions provides a rich set of possible paths towards both an understanding of natural as well as the construction of artificially intelligent embodied agents.

3   Information and Embodiment

Once we accept the idea that information is a resource of major importance for living organisms, how does it help direct our attempts to understand the emergence of intelligence in biology and to create intelligence in artificial (embodied) systems?

3.1 Structure of Information

First of all, it turns out that for a living being information is not – as its use as a measure for communication effort might insinuate – a "bulk" quantity. In fact, information is, in general, intricately structured. The information bottleneck formalism [13] illustrates this point most lucidly. Of all the information impinging on a living agent, only a fraction is of true significance to the choice of the agent's actions. This is demonstrated in a striking way by experiments identifying what environmental cues humans are actually conscious of. In controlled settings, for instance, one can show that such cues are surprisingly impoverished (*cf.* phenomenon of change blindness; e.g., [14]). In other words, for a living agent information can be separated into a part that is used (and perhaps recorded), and a part which remains unused.

Such a "split" makes particular sense in the light of the above hypothesis of information being a central resource for living agents. In this case, one would expect evolution to drive up brain power, thus the capacity for information processing until the brain's metabolic costs would outweigh the gains. Thus, a natural limit exists on the amount of information that an agent can process. In other words, information processing must be governed by a *parsimony principle*: only selected components of information should be extracted and processed from the environment, namely those which make the best use of resources with respect to the acquisition and processing cost they entail. Such selected components constitute the *relevant information* available to the agent. It turns out that in typical scenarios, relevant information can be massively reduced while incurring only in moderate losses in overall performance [15]. The performance costs of an agent acting in its environment thus induce structure on information itself by separating relevant from irrelevant information. Information is hence imbued with a "semantic" flavor.[1] But a performance cost profile is not the only factor that provides information with structure. On an even more primordial level, already the *embodiment* of the agent, before the inclusion of any external (evolutionary or task-dependent) performance measures, imposes structure on the *information flows* [3,12,16–19].

To formalize these intuitions, we can express the interaction of the agent with its environment as a causal Bayesian network model of the perception-action loop and consider the information flows arising in the given system [12,20]. The causal Bayesian model allows quantifying the dynamics of the agent as a family of probability distributions, capturing different types of mutual information and information flow quantities in the system. It is found that a given embodiment of an agent favours particular information flows. As a thought experiment, consider, for example, a legged robot where each leg contains a movement sensor. Evidently, one can expect to find the movement sensor mainly reflecting information about the movement of the particular leg it is mounted on, and only to a minor degree that of the other legs. This qualitative intuition can be made precise in a quantitative way and implies the existence of individual, largely separate information flows for the different legs of the robot.

The power of the method extends far beyond this simplistic model and furthermore allows for natural decompositions of information flows. It hence provides a quantitative, theoretically well-founded formalism for characterizing how exactly embodiment induces a bias on *what* information an embodied agent will process and in *which* way. We note that this bias is prior to any concrete goals and tasks the agent may have.

In addition, the "information view" also abstracts away the information processing architecture – which may explain why different species can solve similar tasks using

---

[1] Semantics was intentionally omitted in the original formulation of information by Shannon, but its absence in the purist interpretation of information theory was long felt to be limiting the potential of information theory to understanding intelligence.

entirely differing brain "hardware." It further relates to the principle of "degeneracy" [21], i.e., the capacity of a system to achieve the same function through many different structural arrangements – a principle found also across individual brains of the same species that are known to differ significantly in terms of their structural components and interconnections, while generating similar perceptual and cognitive performance. Degeneracy fits naturally within an evolutionary framework: as long as brains manage to evolve means to accomplish concrete information processing tasks, it is of minor relevance which part of the brain achieves the task and what its detailed architecture is. The abstractive power of information theory promotes the isolation of necessary from fortuitous aspects of the information processing architecture. It indicates, *ab initio*, what forms of information processing are favoured, prior to any other "implementational" constraints determined by mechanical, biological, and other factors.

Note that on a long time scale, embodiment itself is subject to evolution. Once a concrete embodiment is established, some information flows are reinforced while others are suppressed. It follows that evolution can be seen to operate indirectly on the structure of information flows and even envisage models under which information-theoretic criteria may direct evolution in a Baldwin-like fashion in an environment providing otherwise little short-term fitness structure [22–24]. To simplify the discussion, in what follows, we will restrict ourselves to the case in which the embodiment is given *a priori*.

If the main hypothesis is right that embodiment generates – beyond any concrete implementational substrate – a preferred mode of information processing, then for any given embodied system there should be *natural* controllers. Such controllers would include particular pattern detectors, elementary behaviours and filters, as well as utility (performance) profiles appropriate for the given embodiment [25,26]. The properties of such natural controllers emerge from the *complete* perception-action loop. Note that, at the same time, the time scales characterizing changes of environment, morphology ("hardware"), or controller ("software") are vastly different. Thus, in this picture, the apparent Cartesian duality between body and mind put forward by the classical view evaporates into a mere matter of time scales. In particular, this view suggests that the canonical development of a controller for an embodied system (both in biology and engineering) would first involve starting from the natural information flows emerging from the agent's embodiment, before any concrete tasks and goals are addressed. This is a major deviation from the conventional philosophy which states that the overall control of an embodied agent is attained by a "skillful" combination of partial strategies that achieve individual subgoals. Rather, it makes clear in a mathematically transparent and computationally accessible way how embodiment imposes *a priori* constraints on suitable controllers.

This view provides a plausible argument why nature is able to discover viable solutions for the control of such a variety of "hardware" realizations. It also leads to a novel perspective of how the robot designers could go about designing a controller that is particularly suitable to a given hardware and that could be adapted on-the-fly to any changes of the underlying hardware. Thus, the skeptic's distrust of viewing an agent's life as merely a sequence of goal-driven behaviours maximizing some utility function is vindicated. While a weak notion of "goal" may still exist, the natural way of looking at one's world is prior to all this: it is a basic fact that an agent has a body.

Only then, goals may pop into place. In particular, goals can be shifted adaptively, while the agent is still equipped with a more or less established sensorimotor repertoire.

3.2 Virtue Out of Necessity: Parsimony as Organizational Principle

We start this section by noting that the capacity of perception-action loops to structure information only emerges if the information processing capability is *limited*. Indeed, a system with unlimited information processing capability would have no need to structure information and could process each event as an individual, indivisible, incomparable symbol out of its infinite alphabet of symbols. Thus, it has been proposed that intelligence arises from the need to compress (e.g., sensor data streams can be compressed by identifying regularities and patterns in them; see http://prize.hutter1.net/).

How little information is sometimes necessary to achieve a good chance of survival is exemplified by species of bacteria that can switch randomly from a "safe" hibernating state to a "risky" wake state in which they have the opportunity to reproduce, but are also exposed to possible attacks, e.g., by antibiotics. Recent research indicates that such bacteria do not employ any sensory information to evaluate whether it is safe to switch state, but switch randomly between wake and hibernating states [27]. In information-theoretical treatments of a related scenario (a bacterial colony), it is possible to show that the information processing capacity necessary for survival of an agent can, under certain circumstances, be reduced to zero if fitness requirements are only moderately reduced from the maximum [16]. This observation suggests an *information parsimony* principle: for a given level of required performance, an intelligent agent could aim to minimize its use of information processing resources and the associated expenditures.

A closer look reveals that information parsimony is essentially a "dual" formulation of the Infomax principle: instead of Infomax's view that the agent will maximize the information throughput through a given system, information parsimony emphasizes that, for a given level of performance, the agent will strive to *minimize* the necessary information processing capacity. While both views are mathematically equivalent in the limit of a stationary system, the difference in the formulations essentially emphasizes the time scales of development and evolution. Infomax assumes a given "hardware" for which the throughput is then maximized (i.e., development), while information parsimony assumes a certain performance level (fitness) which needs to be achieved with the least possible informational effort (i.e., evolution).

In both cases, an agent should strive to make use of its informational capacity to its fullest, and use as much as possible of the available information to its advantage; when such information cannot be made use of, evolution will reduce the unused information processing capacity in the long run. One of the most striking examples from biology is the loss of eye function in blind cave animals [28]. In our hypothesis, the parsimony principle will not be limited to this prominent case, but extends to all levels of information processing by the agent. The rigorous formulation of the information parsimony principle lends additional plausibility to approaches to generate intelligent agent controls based on minimal dynamical systems (e.g., [22,29]), and promises additional insights into what lies behind the emergence of apparent intelligence (i.e., organisms doing the right thing) with seemingly very limited abilities.

Additional quantitative principles can be formulated using information-theoretical notions. For instance, the decomposition of information flows (such as the multivariate bottleneck principle [30]) can lead to the emergence of concept detectors (e.g., in a world imbued with a chemical gradient, such decomposition would produce through self-organization detectors for concepts such as direction, parity, long-term or short-term timers [12]). While such low-level concepts seem to emerge in a bottom-up fashion, the question arises in how far the information decomposition view can tackle concept formation in the context of AI. A long-standing challenge, concept formation is being looked at by a large body of work from different viewpoints. Among these, the informational view promises to provide a coherent, far-reaching framework (on a practical level, methods such as independent component analysis or multivariate bottleneck methods are known methods to decompose data into "natural" components [30,31]). In particular, it offers a coherent currency by which the cost of abstraction (relevant to concept formation) can be dealt with universally throughout the system. So, in this view, we propose that concept formation becomes quasi a by-product of the necessity of managing the complexity of processing information. According to this picture, limited informational resources require a decomposition of incoming information flows into largely independent subcomponents of lower complexity which are then handled individually.

The above principles are not arbitrary but arise naturally from the toolbox of information theory. We thus do not just have pockets of isolated quantities, but a whole network of interrelated principles: in fact, information can be seen as forming a language and as such, it allows formulating, quantifying and relating different aspects of information processing in a quantitative, non-metaphorical manner. In summary, the introduction of independent component analysis as a systematic machine learning tool, as well as models such as the regular and multivariate information bottleneck moved us away from the "bulk" picture of information towards a picture of "structured" information in a precise information-theoretical sense. Another important insight is that embodiment is increasingly recognized as a driving factor after decades of an almost Cartesian split in AI research, separating computational processes from a body which was merely regarded as a translation device between the environment and an internal model of the world [22]. One of the most visible expressions of this "paradigm shift" is the increasing interest in morphological computation [3,32,33]. Today, such novel perspectives are joined by a variety of powerful new theoretical and experimental tools which have led to the development of a candidate framework recruiting Shannon information for the study of intelligence (and routes for its emergence) in embodied systems.

4 Outstanding Research Issues

The view that information theory can provide a comprehensive approach for understanding the emergence of intelligence in biological systems and for producing intelligent behavior in artificial systems – while anticipated for decades (e.g., Ashby [5])

– has only recently begun to crystallize. The core issue is to which extent information processing found in biological organisms can be understood in terms of information-theoretical principles. We wish to emphasize once again that information theory does not just apply to high-level overall performance measures of an agent but reaches down to lowest levels of biological information exchange (e.g., the genetic and the neural code [34,35]). We hypothesize that whatever frame is adopted to understand and model biological and artificial information processing, a suitable formalism needs to provide descriptive and predictive power. While a comprehensive (quantitative) framework may remain out of reach for still some time, the adoption of an information-theoretical perspective holds significant promise, as it allows the investigation of a considerable number of relevant issues under a unifying theme. In this section, we would like to discuss several future issues of interest from a higher-level perspective.

One major issue relates to what drives the *evolution* of perception-action loops. We suggest that a major force in their evolution is the selection pressure due to information processing requirements. Going beyond perception-action loops, information theory may even allow the formulation of more general statements about the informational characteristics of evolving systems. For instance, the "principle of ecological balance" [3] (called "complexity monotonicity thesis" in a different context [36]) states that there is a correlation between sensory, motor and information processing capacity. Although a good chance exists that fundamental information-theoretic principles can be identified supporting this thesis, its universality is not entirely obvious. From the observation of living organisms one expects the (potential) sensorial capacity to exceed the motor capacity by far, and memory and the total information processing capacity to be much higher than the sensorial capacity. The question is whether these relations are incidental or universal, and, if the latter is indeed the case, whether this universality can be expressed quantitatively.

Intimately related to this issue is the question of whether concepts such as *relevant information* [15] (i.e., information requirements deriving from external tasks or fitness criteria) could yield a powerful enough drive to instigate an "arms race" between the information necessary to achieve a goal, the information captured by the sensors for this purpose, the required processing capacity of the brain, and the actuator processing capacity necessary to carry out the tasks. This poses central questions concerning the relationship between the co-evolution of brain, morphology, and control and the emergence of complex systems responsive to relevant (structured) information. Complex systems typically contain a high amount of non-repetitive and non-random structure. In particular, the amount of structure of nervous systems can be characterized by a measure of neural complexity which assesses in an information-theoretical context the interplay of highly segregated and highly integrated neural processes [37]. The presumed increase of neural complexity over evolution may simply reflect intense selection pressure on neural structures to efficiently deal with informational challenges posed by co-evolving body morphology, sensorimotor information flows, and eco-niches. As more and more information structure is generated, there is an increased need for neural structures to extract and integrate information, which in turn drives complexity to higher levels.

Two closely related issues are the ones of open-ended evolution and development, that is, how to create systems that do not stop evolving or do not stop learning "interesting" things. There have been various attempts at designing intrinsic motivation systems that capture the drive towards novel and curious situations. They are either based on the notion of *empowerment*, a measure of the power of the agent to modify its environment in a way that is detectable by the agent itself [25], *homeokinesis*, i.e., selection of action sequences which maintain sensorial predictability [24], predictability of action sequences [38], or the maintenance of an abstract cognitive variable, the learning progress, which has to be kept maximal [39]. It is interesting to note that on the physical side, all these approaches seem in one way or another to relate to the notion of maximum entropy production – a principle believed to be relevant in guiding the dynamics in many complex systems (e.g., [40]). In fact, some of the above principles are aptly formulated in the language of information. It thus is natural to explore possible avenues to unify these different but "similar-minded" approaches.

5 Final Remarks

If we, at the end of this chapter, take a step back and, as a final reflection, consider the issues from a bird's eye view, where does this place us? Compared to other sciences, AI is a strangely hybrid creature. For instance, engineering sciences (or more engineering-oriented branches of computer science, such as software engineering) are typically constructive: starting from a more or less uncontested basis, a "code of practice" for the creation of state-of-the-art artifacts is developed. The basis may occasionally be revised to encompass novel developments (e.g., in software engineering: object-orientation or extreme programming), but the task is mostly about improving the paths from a firm basis to an envisaged artifact.

At the other end of the spectrum, we have sciences such as physics which attempt to model observed phenomena. In such sciences, the *foundations*, and not the constructive aspects of a system are at the core of the issues. The physicists aim to simplify and minimize the assumptions behind their models while attempting to capture as many essential features as possible of the examined phenomena. The universal descriptive and predictive power of models such as Maxwell's theory, relativity and quantum theory or thermodynamics are striking and, in fact, one of the mysteries of science.[2]

Biology, on the one hand, resembles physics, as it studies real natural phenomena. On the other hand, it incorporates elements of reverse engineering, as it attempts to disentangle the intricate mechanisms underlying living beings, all of which have historical origins as products of evolution. Unlike physics, however, in biology the reduction of phenomena to few simple principles has until recently only been

---

[2] Construction was historically not part of the physicist's agenda. This "pure" agenda has begun to change in recent years with the advent of massive computational power, introducing a new branch, *computational physics*, into the picture.

achieved in a few areas (such as the universality of the genetic code); the complexity of biological phenomena, and their place outside of the range of validity of powerful theoretical models such as equilibrium thermodynamics makes it difficult to impose more or less universal organizing principles on the vast collection of empirical data. While evolution provides a universal theory for all living organisms, the nature of this theory is quite different from the sort of theories that serve as the foundations of physics.

AI lies in between physics and biology (in modern AI, biomechanics, material science, neuroscience come also into play and are increasingly superseding the role of psychology and linguistics which dominated classical AI; see [3]). AI belongs to the realm of engineering, and rightly so, because it strives to *construct* intelligent systems. In many aspects, engineering approaches to AI have proven efficient and powerful. However, there is also a universalistic aspiration in AI. Not unlike physics, AI aims to find fundamental principles underlying the emergence of intelligence. This goal is fueled by the observed power of biological systems which achieve intelligence at many different levels, quite unlike the engineered intelligent systems which are usually optimized for one particular task. Biology under the Darwinian stance is "engineered without an engineer", successfully reinventing wheels (eyes, and other "goodies", actually) again and again in an extremely large and complex space – a strong indication that some universal set of principles are at work. A satisfying picture of AI should aim (and hopefully will be able) to isolate and exploit such principles. Consider thus, the grand goal of AI, the one of understanding how intelligence can be "engineered without engineer": it lies between the constructive view of the engineering sciences, the "first principles" view of physics, and the biological view. The latter one is particularly opaque, since any fundamental principles may be buried in volumes of fortuitous historic accidents or restrictions of the biological substrate.

As discussed in this chapter, a primary candidate for building a suitable framework is provided by a suitable adaptation of information theory to the information processing task posed to *embodied* agents: they thus may turn out to serve as the "Carnot-machine" for intelligent information processing. It is striking that information theory which was developed by Shannon essentially as a response to an engineering challenge, not only provides a different way of looking at probability theory (which later was used extensively in AI), but also found to be intimately related to the physical field of thermodynamics. The well-understood formalism of the latter, however, reached its limits in the "exotic" non-equilibrium states of biology. We conjecture that information theory will play an important role in linking the convoluted world of biological information processing, a physics-like set of fundamental principles for intelligent information processing, and the goal of engineering an intelligent system, all in the service of getting closer to the grand vision of artificial intelligence.

Acknowledgements

The authors would like to thank Rolf Pfeifer and the anonymous reviewer for very helpful and clarifying comments.

References

1. Wolpert, D.H., MacReady, W.G.: No free lunch theorems for optimization. IEEE Trans. Evol. Comp. 1, 67–82 (1996)
2. van Salvini-Plawen, L., Mayr, E.: On the evolution of photoreceptors and eyes. In: Hecht, M.K., Steere, W.C., Wallace, B. (eds.) Evol. Biol. 10, 207–26 (1977)
3. Pfeifer, R., Bongard, J.C.: How the Body Shapes the Way We Think. MIT Press, Cambridge, MA (2007)
4. Shannon, C.E.: A mathematical theory of communication. Bell System Tech. J. 27, 379–423 (1948)
5. Ashby, R.W.: Requisite variety and its implications for the control of complex systems. Cybernetica 1, 83–99 (1958)
6. Attneave, F.: Some informational aspects of visual perception. Psychol. Rev. 61, 183–193 (1954)
7. Barlow, H.B.: Possible principles underlying the transformation of sensory messages. In: Rosenblith, W.A. (ed.) Sensory Communication, pp. 217–234. MIT Press, Cambridge, MA (1961)
8. Gibson, J.J.: The ecological approach to visual perception. Houghton Mifflin, Boston (1979)
9. Linsker, R.: Self-organization in a perceptual network. Computer 21(3), 105–117 (1988)
10. Laughlin, S.B., de Ruyter van Steveninck, R.R., Anderson, J.C.: The metabolic cost of neural information. Nat. Neurosci. 1(1), 36–41 (1998)
11. de Ruyter van Steveninck, R.R., Laughlin, S.B.: The rate of information transfer at graded-potential synapses. Nature 279, 642–645 (1996)
12. Klyubin, A., Polani, D., Nehaniv, C.: Representations of space and time in the maximization of information flow in the perception-action loop. Neural Comp. 19(9), 2387–2432 (in press)
13. Tishby, N., Pereira, F.C., Bialek, W.: The information bottleneck method. In: Proc. of 37th Ann. Allerton Conf. on Communication, Control and Computing, pp. 368–377 (1999)
14. O'Regan, K., Alva, N.: A sensorimotor account of vision and visual consciousness. Behav. Brain Sci. 24, 939–1031 (2001)
15. Polani, D., Nehaniv, C., Martinetz, T., Kim, J.T.: Relevant information in optimized persistence vs. progeny strategies. In: Rocha, et al. (eds.) Proc. of 10 th Int. Conf. on Artificial Life, pp. 337–343 (2006)
16. Klyubin, A.S., Polani, D., Nehaniv, C.L.: Organization of the information flow in the perception-action loop of evolved agents. In: Proc. of 2004 NASA/DoD Conf. on Evolvable Hardware, pp. 177–180 (2004)
17. Lungarella, M., Pegors, T., Bulwinkle, D., Sporns, O.: Methods for quantifying the information structure of sensory and motor data. Neuroinformatics 3(3), 243–262 (2005)
18. Lungarella, M., Sporns, O.: Mapping information flow in sensorimotor networks. PLoS Computational Biology 2, 1301–1312 (2006)
19. Ay, N., Polani, D.: Information flows in causal networks. Santa Fe Institute Working Paper 06-05-014 (2006)
20. Klyubin, A.S., Polani, D., Nehaniv, C.L.: All else being equal be empowered. In: Capcarrère, M.S., Freitas, A.A., Bentley, P.J., Johnson, C.G., Timmis, J. (eds.) ECAL 2005. LNCS (LNAI), vol. 3630, pp. 744–753. Springer, Heidelberg (2005)
21. Tononi, G., Sporns, O., Edelman, G.M.: Measures of degeneracy and redundancy in biological networks. Proc. Natl. Acad. Sci. USA, 3257–3262 (1999)

22. Prokopenko, M., Gerasimov, V., Tanev, I.: Evolving spatiotemporal coordination in a modular robotic system. In: Proc. of 9th Int. Conf. on the Simulation of Adaptive Behavior (2006)
23. Sporns, O., Lungarella, M.: Evolving coordinated behavior by maximizing information structure. In: Rocha, et al. (eds.) Proc. of 10th Int. Conf. on Artificial Life, pp. 323–329 (2006)
24. Der, R.: Self-organized acquisition of situated behavior. Theory in Bioscience 120, 1–9 (2001)
25. Klyubin, A.S., Polani, D., Nehaniv, C.L.: Empowerment: A universal agent-critic measure of control. In: Proc. of IEEE Congress of Evolutionary Computation, pp. 128–135 (2005)
26. Tedrake, R., Zhang, T.W., Seung, H.S.: Stochastic policy gradient reinforcement learning on a simple 3D biped. In: Proc. of 10th Int. Conf. on Intelligent Robots and Systems, pp. 3333–3338 (2004)
27. Kussell, E., Leibler, S.: Phenotypic diversity, population growth, and information in fluctuating environments. Science 309, 2075–2078 (2005)
28. Yamamoto, Y., Stock, D.W., Jeffery, W.R.: Hedgehog signaling controls eye degeneration in blind cavefish 431, 844–847 (2004)
29. Beer, R.: The dynamics of active categorical perception in an evolved model agent. Adaptive Behavior 11(4), 209–243 (2003)
30. Friedman, N., Mosenzon, O., Slonim, N., Tishby, N.: Multivariate information bottleneck. In: Proc. of 17th Conf. on Uncertainty in Artificial Intelligence, pp. 152–161. Morgan Kaufmann Publishers, San Francisco (2001)
31. Comon, P.: Independent component analysis. In: Proc. Int. Signal Processing Workshop on Higher-order Statistics, Chamrousse, France, pp. 111–120 (1991)
32. Paul, C., Lungarella, M., Iida, F. (eds.): Morphology, dynamics and control. Special issue of Robotics and Autonomous Systems 54(8), 617–718 (2006)
33. Pfeifer, R., Gomez, G., Iida, F.: Morphological computation for adaptive behavior and cognition. Int. Cong. Series 1291, 22–29 (2006)
34. Adami, C.: Introduction to Artificial Life. Springer, Heidelberg (1998)
35. Zhaoping, L.: Theoretical understanding of the early visual processes by data compression and data selection. Network: Computation in Neural Systems 17(4), 301–334 (2006)
36. Bosse, T., Sharpanskykh, A., Treur, J.: On the complexity monotonicity thesis for environment, behaviour and cognition. In: Int. Conf. on Complex Systems, 1728 (2006)
37. Tononi, G., Sporns, O., And Edelman, G.M.: A measure for brain complexity: relating functional segregation and integration in the nervous system. Proc. Natl. Acad. Sci. USA 91, 5033–5037 (1994)
38. Schmidhuber, J.: Curious model-building control systems. In: Proc. Int. Joint Conf. on Neural Networks pp. 1458–1463 (1991)
39. Oudeyer, P.-Y., Kaplan, F., Hafner, V.V.: Intrinsic motivation systems for autonomous mental development. IEEE Trans. Evol. Comp. 11(1), 265–286 (2006)
40. Dewar, R.: Maximum entropy production and the fluctuation theorem. J. Phys. A: Math. Gen. 38, 371–381 (2005)

Preliminary Considerations for a Quantitative Theory of Networked Embodied Intelligence

Fabio P. Bonsignorio

Department of Mechanics – University of Genova
Heron Robots srl, Via R.C.Ceccardi 1/18
16121 Genova, Italy
fabio.bonsignorio@heronrobots.com, bonsignorio@dimec.unige.it

Abstract. This paper exposes and discusses the concept of 'networked embodied cognition', based on natural embodied neural networks, with some considerations on the nature of natural collective intelligence and cognition, and with reference to natural biological examples, evolution theory, neural network science and technology results, network robotics. It shows that this could be the method of cognitive adaptation to the environment most widely used by living systems and most fit to the deployment of artificial robotic networks. Some preliminary ideas about the development of a quantitative framework are shortly discussed. On the basis of the work of many people a few approximate simple quantitative relations are derived between information metrics of the phase space behavior of the agent dynamical system and those of the cognition system perceived by an external observer.

Keywords: embodiment, intelligent agents, intelligence, information, entropy, complexity, dynamical systems, network, emergence.

1 Introduction

In nature there are many kinds of loosely coupled networks of intelligent embodied situated agents, largely varying in terms of quantity of agents and cognitive and adaptive capacity (i.e. of computational needs) of each agent. These networks of embodied situated intelligent agents have evolved - are emerged - on top of the very complex network of networks of the ecological relations between all the living beings of our planet after billion of years of natural evolution.

At an extreme we can observe that these networks of relations rely on the deeper network of quantomechanical interactions originating all the matter, suggesting that the same simple network paradigm could explain a wide range of phenomena from quantum to macroscopic level. The world could be seen as composed of many networks at many scales.

It seems that this could be the method of cognitive adaptation to the environment most widely used by living systems and most fit to the deployment of artificial robotic networks. These ideas have some analogy with the visions of Piaget [33,34] ('knowing is know how', his view of the learning process in the children sees it as an embodied learning process where 'ability' precedes 'knowledge'), Merleau-Ponty [35]

M. Lungarella et al. (Eds.): 50 Years of AI, Festschrift, LNAI 4850, pp. 112–123, 2007.

(empathy is based on mimicking, I know what you think because I can do the same things you can do), Bateson [36] (the importance of systemic 'relations' between living beings from which meaning emerges) and Marx [37] (collective learning of the masses through the collective 'praxis'), as well with recent neurobiology research results ('mirror neurons').

In particular Piaget's developmental vision of the learning process in the children, is very inspiring and shows many analogies with much work in cognition and robotics.

Piaget models the children's learning in four stages:

1. sensorimotor stage: children experience the world through interaction and senses and learn object permanence
2. preoperational stage: acquisition of motor skills
3. concrete operational stage: children begin to think logically about concrete events)
4. formal operational stage: development of abstract reasoning.

This vision is somehow completed by the vision of an 'ecology of mind' where information is seen as embodied in the structure of the relations between the entities in the environment proposed by Bateson.

In the last twenty years, starting from the pioneering work of R. Brooks [4] many prototypes exhibiting 'emerging' intelligence have been demonstrated showing how the road to the realization of the intelligent artifacts which constitutes the long term goal of AI research could be different from previously popular symbolic paradigm usually referred to as Good Old Fashioned Artificial Intelligence.

In the natural domain the most widely used method of 'intelligence', computation and 'cognition' seems to be 'embodied' biological neural networks.

A very simplified model of a biological neural network – not considering, usually, the plasticity of natural examples - is constituted by artificial neural networks, a schematic model of natural neural systems. In the original model of an artificial neural network (given by Rosenblatt's 'perceptron' [20], proposed in 1958) the computation is based on the triggering of an output signal when a threshold of a sum of weighted connection values is reached. Although today most current neural network algorithms are more sophisticated as they do not use thresholds, but rather continuous valued squashing functions, they are still an approximation of their natural counterparts.

There are several important results concerning artificial neural networks which suggest some general remarks. While a single layered perceptron have some environment mapping limits, the 'multilayered perceptron' mapping capabilities are remarkable. Hornik et alias[21], and Funahashi [22](1989) have demonstrated that an artificial neural network with an input layer, an output layer and a number of 'hidden' layers (a 'multilayered perceptron') can approximate within a given error boundary any real valued vector function, by means of an appropriate learning procedure applied to the weights on the connections.

This has led to many successful applications in automatic learning in AI and robotics.

If we consider the process of cognitive adaptation to the environment as a learning process in the sense of AI learning theory, whose final result is the 'fit' behaviour of the (living) agent, we can draw some interesting conclusions.

We can suppose that it is always possible to build a neural network approximating (in the sense of probabilistic approximate learning) any environment of given complexity, for instance measured by the Vapnik-Chervonenkis dimension of the simpler learning machine which can learn it, or more generally Kolgomorov (or Chaitin) complexity [2,3].

This can be interpreted by saying that a learning system for a physical (embodied) agent based on a neural network can be taught to interact effectively in (almost) any environment, or 'to know' (almost) any environment.

Several learning procedures for artificial neural networks have been demonstrated.

Particle swarm optimization (Kennedy and Eberhart,[23]) allow to tune the weights of connections by means of a swarm of agents in the environment: if we assume that any agent has an (almost) identical neural network and (almost) perfect communication between the agents this allow a collective learning (tuning of the weights) of the multi agent system.

Other approaches are genetic algorithms (imitating genetic natural evolution), evolutionary programming, reinforcement learning.

Ant algorithms mimic the ant colonies learning process based on external storage of information through pheromones paths (mathematically modeled by Millonas, [24]).

The importance of 'embodiment' is well shown by the MIT biped passive walker and by theoretical investigation by many people, for example by Pfeifer and Iida [5,6], which make clear that part of the 'computation' needed by control, intelligence and cognition are in fact performed by the physical morphology of the agent and by its physical relations within the environment.

A general schema seems to emerge.

The unit of (intelligent/cognitive/computational) adaptation to the environment is constituted by loosely coupled groups of neural networks embedded, or more properly 'embodied', into physical agents sensing and acting cooperatively in the physical environment.

The weights of the connection are determined in part by biological inheritance (modeled by genetic algorithms optimization), in part through social cooperative exploration (modeled by particle swarm optimization) and individual tuning (modeled by reinforcement learning).

The information can be maintained in part inside the neural networks of the individual of the group (communicated from agent to agent for instance by means of bees' waggle-dance or human language) in part externally (ant pheromone paths and human libraries). In part in the morphology of the agent body itself.

It could be interesting to notice that neural network themselves can be regarded as a massively parallel networked morphological computing system.

2 Some Theoretical Questions

The observation of natural intelligent systems and the practice of robotics research and engineering lead us to think that 'intelligence' (and 'meaning' if not 'consciousness') are 'emerging' characteristics springing from the evolution of loosely coupled networks of intelligent 'embodied' and 'situated' agents.

As we noticed above natural neural network themselves could be regarded as embodied and situated computing systems.

On the other end we miss a quantitative comprehensive theory which allows us to model the interplay between the agents' 'morphology', in other words their mechanical structure and the emerging of 'intelligence' and 'meaning'.

The main concept are that information processing – 'computing' – is a general characteristics of any material system, while 'intelligence' and 'meaning' should be syncronisation processes within networks of autonomous agents.

In particular we should be able to explain:

1. How the dynamics of an (embodied) agent is related to its information/computing capabilities (morphological computation)
2. How information/computing capabilities behave in a multi body agent system
3. How 'intelligence' and 'meaning' emerge from networks of embodied agent

The next section aims to give an answer to the first two questions that, following a modeling tradition coming from Boltzmann, are actually seen as two aspects of the general problem of information processing in multibody systems.

These considerations, as shown in [7,13,16], have some intriguing connection to the old questions raised by the 'Maxwell's demon', [19].

The relations shown below require, possibly, a more rigorous demonstration, on the other end they spread some light on the relations involved by the computing capabilities of dynamical systems.

The following section 4 shows a possible way to explain the emerging of intelligence and meaning as a collective process in highly networked systems [9], as it could be suggested for instance by the work on semiotic dynamics and the talking heads experiment by Steels [26,28].

3 On the Information Metrics of Intelligent Moving Bodies

Although it may seem strange only in recent times the classical results from Shannon theory, [1], have been applied to the modeling of control systems.

An interesting approach to the measure of the complexity of a dynamical system is given by Brudno in [8], here we prefer to start from Shannon entropy as it seems easier to link it to the system dynamics.

In this discussion we see, for simplicity, an embodied agent as a controlled dynamical system and we will show how the algorithmic complexity of the control program is related to the phase space 'footprint' of the dynamical system.

In [7] Shannon theory is applied to the modeling of controlled systems and statistical information metrics based definitions of controllability and observability are derived.

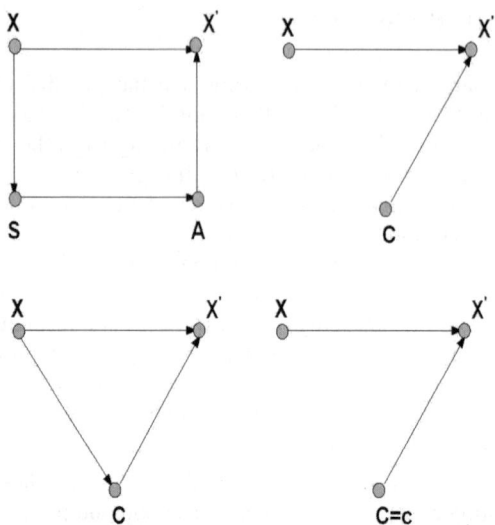

Fig. 1. Directed acyclic graphs representing a control process. (Upper left) Full control system with a sensor and an actuator. (Lower left) Shrinked Closed Loop diagram merging sensor and actuator, (Upper right) Reduced open loop diagram. (Lower right) Single actuation channel enacted by the controller's state C=c. The random variable X represents the initial state, X' the final state. Sensor is represented by state variable S and actuator is represented by state variable A.

With reference to figure 1, from [7], the dependencies of the random variable of a vertex are given by equation (1)

$$p(x_1, x_2, ..., x_n) = \prod_{i=1}^{N} p(x_i \mid \pi[x_i]) \qquad \pi[x_1] = 0 \qquad (1)$$

with $\pi[x_i]$ parents of X_i, i=1,2,3..N.

In the control graph of figure 1, the random variable X represents the initial state of the system to be controlled, and whose values x $\in$ X follow a probability distribution $p_X(x)$.

The initial state X is controlled to a final state X' by means of a sensor represented by state variable S and an actuator represented by state variable A.

Relation (1), like (2) and (3), follows from the assumption that the controller's variables can be connected by a bayesian network (a directed acyclic graph).

For simplicity sensor and actuator, represented by state variable S and A, are merged in the controller represented by state variable C.

$$p(x, x', c)_{closed} = p_X(x) p(c \mid x) p(x' \mid x, c) \qquad (2)$$

$$p(x, x', c)_{open} = p_X(x) p_C(c) p(x' \mid x, c) \qquad (3)$$

In equation (3) $p(x'|x,c)$ is an actuation matrix giving the probability that the controlled system in state $X=x$ is moved by means of the actuation to the state $X'=x'$, given the controller state $C=c$.

We define the mutual information $I(X,C)$ giving the amount of information that the controller is able to extract from the system state X.

$$I(X;C) = \sum_{x \in X, c \in C} p_{XC}(x,c) \log \frac{p_{XC}(x,c)}{p_X(x) p_C(c)} \qquad (4)$$

From the definition follow:

$$I(X;C) \geq 0$$
$$I(X;C) > 0, \; closed-loop \qquad (5)$$
$$I(X;C) = 0, \; open-loop$$

All logarithms are in base two (the logarithm base is not essential for our purposes)

The (Shannon) entropy of the initial state is then defined as in (6):

$$H(x) = -\sum_{x \in X} p_x(x) \log p_x(x) \qquad (6)$$

It can be demonstrated that the action of a controller can be decomposed into a series of individual conditional actuations depending on the internal state values of the controller, represented by C, that we call 'subdynamics', see [7].

It is shown that for a specific subdynamics, identified by 'c', holds

$$\Delta H^{c}_{closed} = H(X \mid c) - H(X' \mid c) \qquad (7)$$

The reason of decomposing a closed loop control action into a set of conditional actuations can be justified as it follows somehow from intuition that a closed loop controller, after the state estimation, the arrow from X to C in figure 1, can be seen as a set of open loop controllers acting on the set of estimated states, the arrow from C to X' in figure 1, see [7].

The left side of equation (7) gives the variation of entropy for a closed loop controller applying the control law for a given state 'c'.

From our perspective the most important result in [7] is given by (8)

$$\Delta H_{closed} \leq \Delta H^{max}_{open} + I(X;C) \qquad (8)$$

where ΔH^{max}_{open} is defined as the maximum optimal entropy for open loop over any input distribution:

$$\Delta H^{max}_{open} \triangleq \max_{p_X(x) \in P, c \in C} \Delta H^{c}_{open} \qquad (9)$$

This relation says that the variation of entropy of the (closed) controlled loop is bounded by the mutual information between the controller C and the state X, defined in equation (4).

The left side of equation (8) is the variation of entropy for all the given 'subdynamics', values of 'c', of the controller, it can be roughly seen as an integral of the variation of entropy of equation (7).

We assume in equation (10) that the 'information contribution' of the controller is roughly given by $I(X,C)$, this seems in accordance with the results shown in [10] and [12].

$$\Delta H_{controller} \cong \Delta H_{closed} - \Delta H_{open}^{max} \le I(X;C) \tag{10}$$

In [14] it is demonstrated that, where $K(X)$ is the Kolmogorov complexity [2], of X:

$$0 \le \sum_X f(x)K(X) - H(X) \le K(f) + o(1) \tag{11}$$

In this context Kolgomorov complexity can be interpreted as the length of the control program. $K(f)$ is the Kolgomogorov complexity of f, a function introduced instrumentally in order to show that $K(X) \approx H(X)$.

For not too Kolgomorov complex f functions, with the assumptions of equation (10) and considering that the variation of entropy can be roughly equated to $K(X)$ from equation (11), we can derive:

$$K(X) \approx \Delta H_{closed} - \Delta H_{open}^{max} \tag{12}$$

Since it is:

$$H_{shannon} = \frac{1}{k_b} H_{boltzmann} = \frac{1}{k_b} \log W \tag{13}$$

Relation (11) can be rewritten;

$$K(X) \approx \log W_{closed} - \log W_{open}^{max} \tag{14}$$

And finally:

$$K(X) \overset{+}{\le} \log \frac{W_{closed}}{W_{open}^{max}} \tag{15}$$

The equation (15) bounds the algorithmic complexity of the control program (the intelligence of the agent, in a simplified view) to the phase space volume of the controlled agent versus the phase space volume of the non controlled system.

From a qualitative standpoint (at least) this relations explains why a simpler walker like the MIT biped or the one described in [15] can be controlled with a 'short' program, while other walkers (like the Honda Asimo or the Sony Qrio) which don't have a limit cycle and show a larger phase space 'footprint' require more complex control systems.

4 A Self Organizing Network Model for Embodied Agent Communication Pragmatics Coevolution

Assuming that, whatever method we choose to quantify the relations between the dynamical characteristics of a system and the information about itself and the surrounding environment it can carry, every piece of matter does process information, we need to devise a mechanism to explain the emerging of intelligence and meaning.

The communication between intelligent networked agents could be seen as a pragmatic activity made possible by the self-organization of coevolving 'situated' and 'embodied' cognitive processes, physically distributed among the inter-communicating agents, motivated and initiated by physical finalized interactions with the environment.

The self organizing coevolutive model of communication and cognition, proposed here, considers the building of the model of the environment by a number of networked physical agents, a 'swarm', situated in an environment and interacting with it, as a collective learning process represented by the growth of a network of nodes mapping, for simplicity and at least, the already identified sensory motor correlations. The idea that a learning system based on some evolutionary process could show intelligence is not new and it was actually already raised by Turing in a famous 1950's paper, [31], here the evolution of the model is seen as a collective process performed by the whole network of networks of agents, this concept is strongly affected by Bateson's concept of an 'ecology of mind'.

The random activity of the network of moving and interacting physical situated agents allows the system to identify the regular patterns in the variables (sensors and actuator propioceptors data flow) and to connect them in a model of the environment.

After that the first node has been created, new nodes (highly correlated groups of sensors and actuators) are attached preferentially to the previous one, according to their 'fitness'.

We assume (for simplicity) that the 'cognitive network ' can be accessed by all the agents which were coevolving it and in fact share (constitute) it.

As a consequence the resulting representation is inherently shared (again as a simplification) between all the agents.

In this conceptual framework 'models of the environment' are shared very naturally, while it is needed a concept of 'self'.

By means of random action generation the nodes which have no consequences on the individual agent are labeled by that agent as 'non self'.

This mimic the basic behavior of the human and mammals immune system (which can be regarded by itself as a 'cognitive system'), see [30].

In order to develop a formal model of such a system, we should possibly try to leverage on the recent fast progress in statistical physics of evolving networks, some possibly relevant results are recalled below. Interest has focused mainly on the structural properties of random complex networks in communications, biology, social sciences and economics. A number of giant artificial networks of such a kind came into existence recently. This opens a wide field for the study of their topology, evolution, and complex processes occurring in them.

Such networks possess a rich set of scaling properties. A number of them are scale-free and show striking resilience against random breakdowns. In spite of large sizes of these networks, the distances between most their vertices are short , a feature known as the 'small world' effect.

It is known that growing networks self-organize into scale-free structures through the mechanism of preferential linking. Numerous networks, e.g., collaboration networks, public relations nets, citations of scientific papers, some industrial networks, transportation networks, nets of relations between enterprises and agents in financial markets, telephone call graphs, many biological networks, food and ecological webs, metabolic networks in cell etc., can be modeled in this way.

Here we suggest that evolving self organizing networks can (could) model the collective knowledge of a network of intelligent (artificial) autonomous agents. An 'object/process' in the environment is modeled by a node of the network with many links generated by a fitness process within a coevoluted learning process.

The shared model of the environment (including the agents) is seen as a growing network of nodes, where each node represents a set of statistically connected sensor and actuators.

This model is initially very coarse and is progressively refined and adapted as new nodes are added.

The approach to cognition shortly described here is somehow a generalization of the semiotic dynamics approach [26], although there are some differences.

In the, standard, semiotic dynamics model every agent maintains its own look up table associating the 'names' to the 'objects' it knows, updating it by means of the 'naming game'. Like in the model we propose here the multi agent systems collectively evolve towards an equilibrium condition where all the agents share the same dictionary (in a simplify view), but we assume here that all the sensory motor coordinations and higher level structures are shared and evolved collectively and that the 'self' perception springs from the continuing physical interaction of the single agent with the environment and the other agents.

We basicly assume that the model of the environment is distributed among all the agents and depends on the (co) evolution of their interaction between them and the environment in time.

In this perspective it is interesting to notice that in [27] the mathematical model of the collective behaviors of systems like that described in [28] are based on the theory of random acyclic graphs which is the basis of many network system physics formalizations.

In [27] the network of agents, where each word is initially represented by a subset of three or more nodes with all (possible) links present, evolves towards an equilibrium state represented by a fully connected graph, with only single links.

The statistical distribution of node connection, necessary to determine the information managing capability of the network through equation (6) and to link to equation (15) can be obtained from equations derived in the statistical physics of networks domain.

The probability Πi that a new node will connect to a node i already present in the network is a function of the connectivity k_i and on the fitness η_i of that node, such that:

$$\Pi_i = \frac{\eta_j k_i}{\sum_j \eta_j k_j} \tag{16}$$

A node i will increase its connectivity k_i at a rate that is proportional to the probability that a new node will attach to it, giving:

$$\frac{\partial k_i}{\partial t} = m \frac{\eta_i k_i}{\sum_j k_j \eta_j} \tag{17}$$

The factor m accounts for the fact that each new node adds m links to the system.

In [29] it is shown that the connectivity distribution, i.e. the probability that a node has k links, is given by the integral:

$$P(k) = \int_0^{\eta_{max}} d\eta \, \frac{\partial P\left(k_\eta(t) > k\right)}{\partial t} \propto \int d\eta \rho(\eta) \left(\frac{m}{k}\right)^{\frac{C}{\eta}+1} \tag{18}$$

where $\rho(\eta)$ is the fitness distribution and C is given by:

$$C = d\eta \rho(\eta) \frac{\eta}{1 - \beta(\eta)} \tag{19}$$

We define a proper η_i function which can be essentially a performance index of the effectiveness of sensory motor coordination and which control the growth of the network.

From equation (18) we can derive the expression for the Shannon entropy of the network:

$$HN = -\sum_{k=1}^{\infty} P(k) \log P(k) \tag{20}$$

It must be noticed that the concept model described here actually identify a large class of possible models, the short formal discussion above only aims to show that a networked system like that envisioned here, whose idea is strongly influenced by Bateson's concept of an 'ecology of mind', can actually manage information into the structure of its internal relations, as it can be shown starting from equation (20).

Moreover, this discussion should, possibly, be linked to the one in section 3, as the networks of agents we are considering here are actually embodied and situated dynamical systems, which do have a phase space representation.

It is thought that it could be interesting to elaborate this concepts with some further quantitative considerations on the related information metrics.

In particular it could be of some interest looking at the information flow, possibly with reference to coarse and fine grained entropy, see [13,14].

This could help some initial steps towards a comprehensive quantitative theory which should be eventually able to predict the results of experiments like those in [32] and to give guidance to the development of new improved intelligent artifacts.

5 A Research Program for Networked Embodied Intelligence?

If we share the vision, supported from empirical and theoretical work,[10,11,12, 26, 27], that 'intelligence' and 'meaning' are 'emerging' processes springing from loosely coupled networks of 'embodied' and 'situated' agents, we have in front of us the hard task of creating a theory able to explain these phenomena from a quantitative standpoint.

This paper support the idea that to chase such a theory (probably the work of a generation) we should look to the classical work of Shannon and Kolmogorov in order to better understand the relations between morphology and computation – embodiment - and to the statistical physics of network systems in order to explain the situatedness and the emerging of intelligence and meaning.

References

1. Shannon, C.E.: The Mathematical Theory of Communication. Bell Sys. Tech. J. 27, 379, 623 (1948)
2. Kolmogorov, A.N.: Three approaches to the quantitative definition of information. Problems Inform. Transmission 1(1), 1–7 (1965)
3. Chaitin, G.J.: On the length of programs for computing finite binary sequences: statistical considerations. J. Assoc. Comput. Mach. 16, 145–159 (1969)
4. Brooks, R.: A Robust Layered Control System for A Mobile Robot. IEEE Journal of Robotics and Automation (1986)
5. Pfeifer, R.: Cheap designs: exploiting the dynamics of the system-environment interaction. Three case studies on navigation. In: Conference on Prerational Intelligence — Phenomonology of Complexity Emerging in Systems of Agents Interacting Using Simple Rules. Center for Interdisciplinary Research, University of Bielefeld, pp. 81–91 (1993)
6. Pfeifer, R., Iida, F.: Embodied artificial intelligence: Trends and challenges. In: Iida, F., Pfeifer, R., Steels, L., Kuniyoshi, Y. (eds.) Embodied Artificial Intelligence. LNCS (LNAI), vol. 3139, pp. 1–26. Springer, Heidelberg (2004)
7. Touchette, H., Lloyd, S.: Information-theoretic approach to the study of control systems. Physica A 331, 140–172 (2004)
8. Brudno, A.A.: Entropy and the Complexity of the Trajectories of a Dynamical System. Transactions of the Moscow Mathematical Society, Moscow (1983)
9. Bak, P., Tang, C., Wiesenfeld, K.: Self-Organized Criticality: An Explanation of 1/f Noise. Phys. Rev. Letter 59(4), 381–384 (1987)
10. Gomez, G., Lungarella, M., Tarapore, D.: Information-theoretic approach to embodied category learning. In: Proc. of 10th Int. Conf. on Artificial Life and Robotics, pp. 332–337 (2005)
11. Philipona, D., O' Regan, J.K., Nadal, J.-P, Coenen, O.J.-M.D.: Perception of the structure of the physical world using unknown multimodal sensors and effectors. In: Advances in Neural Information Processing Systems (2004)
12. Olsson, L., Nehaiv, C.L., Polani, D.: Information Trade-Offs and the Evolution of Sensory Layouts. In: Proc. Artificial Life IX (2004)
13. Gacs, P.: The Boltzmann Entropy and Randomness Tests. In: Proc. 2nd IEEE Workshop on Physics and Computation (PhysComp 1994), pp. 209–216 (1994)

14. Gruenwald, P., Vitanyi, P.: Shannon Information and Kolmogorov Complexity. IEEE Transactions on Information Theory (2004)
15. Garcia, M., Chatterjee, A., Ruina, A., Coleman, M.: The Simplest Walking Model: Stability, Complexity, and Scaling, Transactions of the ASME. Journal of Biomechanical Engineering 120, 281–288 (1998)
16. Lloyd, S.: Use of mutual information to decrease entropy: Implication for the second law of thermodynamics. Phys. Rev. A 39(10), 5378–5386 (1989)
17. Lloyd, S.: Measures of Complexity: A Non exhaustive List. IEEE Control Systems Magazine (2001)
18. Wiener, N.: Cybernetics: or Control and Communication in the Animal and the Machine. MIT Press, Cambridge, MA (1948)
19. Maxwell, J.C.: Theory of Heat. Appleton, London (1871)
20. Rosenblatt, F.: The Perceptron: A Probabilistic Model for Information Storage and Organization in the Brain, Cornell Aeronautical Laboratory. Psychological Review 65(6), 386–408 (1958)
21. Hornik, K., Stinchcombe, M., White, H.: Multilayer feedforward networks are universal approximators. Neural Networks 2, 359–366 (1989)
22. Funahashi, K.: On the approximate realization of continuous mappings by neural networks. Neural Networks 2, 183–192 (1989)
23. Kennedy, J., Eberhart, R.: Particle swarm optimization. In: Proc.of the IEEE Intl. Conf. On Neural Network, vol. 4, pp. 1942–1948. IEEE, Washington DC, USA (1995)
24. Millonas, M.M.: Swarms, Phase transitions, and Collective Intelligence. In: Langton, C.G. (ed.) Artificial Life III. Santa Fe Institute Studies in the Sciences of the Complexity, vol. XVII, pp. 417–445. Addison-Wesley, Reading (1994)
25. Albert, R., Barabasi, A.L.: Statistical physics of complex networks. Rev. Mod. Phys. 74, 47–97 (2002)
26. Steels, L.: Semiotic dynamics for embodied agents. IEEE Intelligent Systems 32–38 (2006)
27. Baronchelli, A., Felici, M., Caglioti, E., Loreto, V., Steels, L.: Sharp Transitions towards Shared Vocabularies in Multi-Agent Systems, arxiv.org/pdf/physics/0509075 (2005)
28. Steels, L.: The Talking Heads Experiment. In: Words and Meanings, vol. 1. Laboratorium, Antwerpen (1999)
29. Bianconi, G., Barabasi, A.L.: Competition and multiscaling in evolving networks, arXiv:cond-mat/0011029 (2000)
30. Morpurgo, D., Serenità, R., Seiden, P., Celada, F.: Modelling thymic functions in a cellular automaton. International Immunology 7/4, 505–516 (1995)
31. Turing, A.M.: Computing machinery and intelligence. Mind 59, 433–460 (1950)
32. Lungarella, M., Sporns, O.: Mapping Information Flow in Sensorimotor Networks. PLOS Computational Biology 2(10), 1301–1312 (2006)
33. Piaget, J.: Introduction à l'Épistémologie Génétique (in French), Paris (1950)
34. Piaget, J., Inhelder, B.: The Growth of Logical Thinking from Childhood to Adolescence. Basic Books, New York (1958)
35. Merleau-Ponty, M.: Phenomenology of Perception (in French). Gallimard, Paris (1945)
36. Bateson, G.: Steps to an Ecology of Mind. University Of Chicago Press, Chicago (1972)
37. Marx, K.: Capital, vol. I (in German), Hamburg (1867)

A Quantitative Investigation into Distribution of Memory and Learning in Multi Agent Systems with Implicit Communications*

Roozbeh Daneshvar, Abdolhossein Sadeghi Marascht, Hossein Aminaiee, and Caro Lucas

Center of Excellence for Control and Intelligent Processing, School of Electrical and Computer Engineering, University of Tehran, Tehran, Iran
Faculty of Psychology, Tabriz University, Tabriz, Iran
roozbeh@daneshvar.ir, abdolhossein.sadeghi.marascht@gmail.com,
aminaiee@gmail.com, lucas@ipm.ir
http://cipce.ut.ac.ir

Abstract. In this paper we have investigated a group of multi agent systems (MAS) in which the agents change their environment and this change has the potential to trigger behaviors in other agents of the group in another time or another position in the environment. The structure makes it possible to conceptualize the group as a super organism incorporating the agents and the environment such that new behaviors are observed from the whole group as a result of the specific distribution of agents in that environment. This distribution exists in many aspects like a super memory (or even a super brain) that exists in the environment and is not limited to memories of the individuals. There is a distributed decision making that is done by the group of agents which, in a higher level consists of both individual and group decision makings, and can be viewed as emergent rather than consciously planned. As the agents change the environment, they decrease the error for the group and hence a distributed learning is also forming between the agents of the group. This implicit learning is related to the implicit memory existing in the environment. These two interpretations of memory and learning are assessed with experimental results where two robots perform a task while they are not aware of their global behavior.

Keywords: Multi Agent Systems, Implicit Communication, Distributed Memory, Distributed Learning, Complex Systems.

1 Introduction

When the agents in a MAS communicate with each other, the communication media is not limited to facilities for direct communication as there are also approaches for indirect communication between agents. One of these media for

---

* Any kind of military uses from the content and approaches of this article is against the intent of the authors.

M. Lungarella et al. (Eds.): 50 Years of AI, Festschrift, LNAI 4850, pp. 124–133, 2007.

indirect communication is the environment the agents are located in. The agents change their common environment and this change has the potential to trigger behaviors in other agents. When an agent fires behaviors of another agent (even if this is a tacit process), a kind of communication between two agents has taken place while this communication might happen in another position and different times. This method known as stigmergy was first introduced in [1] by Pierre-Paul Grass by observing the termites nest building behavior. The surrounding space which the agents are located in is a group of common (physical or non physical) elements that perform the role of a media for agent communications. When the state of an agent triggers behaviors of another agent, a kind of communication has formed between the two agents.

We have investigated the elements that make a MAS intelligent and we have considered how we can enhance a system without necessarily enhancing the individuals. We have considered the use of agents with bounded rationality that make simple decisions according to their perceived state of the environment and these simple decision makings lead to higher level decision makings in the system level. The agents are unconscious about their non-intentional behaviors while it is the essence of stigmergy that the consequences of behaviour affect subsequent behaviour [2]. These systems are inspired by the natural organic systems where there is a greater brain (or a super brain, in other words) making decisions that do not exist in individuals. The meta structure existing in groups of natural creatures or cells has the ability of containing a super memory that does not exist in individuals (like ants and their pheromone trails in an ant colony) and it has the abilities of data transformations like reactions of ants to pheromone trails in the environment and meta decision makings like the performance of an ant colony comparing the abilities of ants. Environmental changes can affect behaviors in two ways: qualitative effect in which the agents choice of action may be affected and quantitative effect in which the selected action may be unchanged, but the exact position, strength, frequency, latency, duration, or any other parameter of the action may be affected [3]. As an instance of this idea we can name scaffolding which means exploitation of external structure that is introduced in [4].

Many applications of this technique have been developed before. In [2] a group of mobile robots gather 81 randomly distributed objects and cluster them into one pile in which coordination of the agents movements is achieved through stigmergy (in an insect society individuals work as if they were alone while their collective activities appear to be coordinated [5]) and as another example, in [6] we have a group of RoboCup soccer players with emergent behaviors and some virtual springs that connect them together, each spring representing decision concern. The environment (potential field) acts as a super brain guiding the action of each player without the awareness of how the different concerns have been fused into that action. A self-organizing model of group formation in three-dimensional space is presented in [7]. It has also investigated the collective memory in such animal groups in which the previous history of group structure influences the collective behaviour exhibited as individual interactions change during the transition of a group from one type of collective behaviour to another. An external memory is used in [8] for reinforcement learning agents. The agents have the observation from the environment augmented by the memory bits and its output consists of the original actions in the environment augmented by actions that change the state of the memory.

2   Memory and Learning

Memory is the process by which the knowledge is encoded, stored and later retrieved. This is based on an ability to recall or recognize previous experience and depends on the persistence of learning in a state that can be revealed at a later time. It is related to any persistent effect of experience over time and can be used in the present. There are various definitions for memory as memory can be very broadly defined as any persistent effect of experience [9] memory is the process of maintaining information over time [10]. Also [11] adds two elements of usage and the time to the definition of memory which memory refers to the process by which past experience and learning can be used in a present time. [12] also refers to these elements and in his view, memory is the way in which we record the past and later utilize it so that it can affect the present situation. Other scientists use more features in description of memory: [13] used the concept of permanent change in behavior. In this approach, memory is a process that results in a relatively permanent change in behavior.

In a definition learning is described as a relatively permanent change in behavior that results from practice [14]. In [11], memory and learning have features in common. In his view, learning is defined as any relatively permanent change in the behavior, thoughts and feelings of an organism (human or other animal) that results from prior experiences. In other words, it is concerned with registering and storing information. Although there are some common features between memory and learning, some scientists try to distinguish them. According to [15] learning is the process of acquiring new information, while memory refers to the persistence of learning in a state that can be revealed at a later time. In [16] learning is described as a process by which we acquire knowledge about the world, while memory is the process by which that knowledge is encoded, stored and later retrieved ([16], [15] and [17]). The distinction between memory and learning is also accepted in [17]. In that approach, Learning is a relatively permanent change in an organism's behavior as a result of experiences. The study of learning and memory therefore requires the creation of behavioral measures to evaluate such behavioral changes.

2.1   Distributed Nature of Learning and Memory in Brain

As an ultra complex system, human brain uses a highly distributed strategy for storing information. There is no unique structure or region for learning and memory in the brain system and functionality of learning and memory in human brain depends on a complex co-operation of different regions. Learning and memory are complex phenomena that are formed by different subsets while these subsets are distributed in various structures and sub structures in the brain. As an instance, we know that the structural sites of explicit and implicit learning and memory are not the same as there are strong evidences about brain system; that learning and memory are not stored in a unitary structure or area (studies in the field of memory and learning have indicated that these phenomena are supported by a complex network of structures). Various kinds of memories are stored in various brain networks. In addition studies on patients with damaged brains have shown that lesions in different regions of the brain cause specific memory defects. It has been found that different representations of a unitary concept or phenomenon are stored in separate locations in the brain. When we think about that phenomenon, the mental concept of that phenomenon is an integrative product of the set of representation in multiple and distinct sites of the brain. In addition, we know that implicit memory is stored in a separate circuitry system including cerebellum and amigdala. All these facts indicate that learning and memory related information is stored in a highly distributed manner in brain ([16], [15], [13], [18], [19], [20] and [21]).

Evolutionarily, one advantage of this highly distributed strategy is clear in case of brain damage. Since different parts and bits of the information about concepts are stored in many different regions of the brain, in case of any lesion or damage, only one or some parts of information would be lost and the person can yet have other information intact and accessible [18].

3   Distributed Memory

We can specify a state for the environment which is specified with different configurations of parameters. These parameters are factors that are changed in the environment (like position of objects and agents in the space) and hence the state is defined of positions in the environment. The objects in the environment are distributed elements which specify the state of the whole environment. This state is configured with elements that are dispersed in the space and the whole forms a state which plays the role of an implicit and tacit memory[1]. We define the state of the environment as:

$$State = f(S_1, S_2, ..., S_i, ..., S_n) \tag{1}$$

in which S_i is the state of $Element_i$ in an environment containing n distinct elements. The states of elements are specified by a group of their attributes which are atomic and are not compound of other factors. When each of the S_i

---

[1] When the elements/objects of the environment show the state and have the potential ability to trigger the future behaviors of agents, they perform as memory elements that can be read by agents (when the agents perceive the position and state of elements and show respective behaviors) and they can also be changed (when the actions of the agents change the positions and states of elements) which seems that a new value is stored for that element.

determiners of state is changed, the whole state of the environment is changed and hence a new value is assigned to the distributed environment. The main point is that the changes in the environment are preserved such that the value is the same as the last value in future accesses made by the agents (even if the access is done in another time). If one of the elements is changed, then we have S_i' instead of S_i and the state of environment in Eq. 1 changes to $State' = f(S_1, S_2, ..., S_i', ..., S_n)$. In this case the value of the distributed memory is changed to a new one and this new state of the environment remains the same while the set of atomic states exists. The main point for a super memory is that it is a mixture of elements which their values are capable of affecting behaviors of agents in the group. This means that changing the elements which their values do not influence the behaviors of agents are not considered as a part of the state of the environment.

4   Distributed Learning

When an action is applied to the environment, the state changes to a new one:

$$State_B = f(State_A, Action_i) \qquad (2)$$

We consider a total error value for each state of the environment which is defined as

$$TotalError(State_A) = Error(State_A) + \sum_{i=1}^{n} \delta_i \times Err(Agent_i, State_A) \qquad (3)$$

in which $Error(State_A)$ is the error value associated to that state of the environment regardless of the agents, n is the number of agents, δ_i is a weight for the effect of error for $Agent_i$ on the total error and $Err(Agent_i, State_A)$ is the individual error of $Agent_i$ in $State_A$. The error value for each of the agents depends on the task that is under progress by the agents.

If we define Δ as an array of actions, $State_B = f(State_A, \Delta)$ means that if the chain of actions in Δ are applied to environment (with the same order as in Δ), the final state of environment is $State_B$. If we have a decrease for individual error as shown in Eq. 4

$$Err(Agent_i, State_B) \leq Err(Agent_i, State_A) \qquad (4)$$

or a decrease in error value of the state as in Eq. 5

$$Error(State_B) \leq Error(State_A) \qquad (5)$$

then we have a decrease in the value of total error as in Eq. 6.

$$TotalError(State_B) \leq TotalError(State_A) \qquad (6)$$

A change is made in the environment which can be either temporary or permanent and this change causes less error values both for the agents and for the task. This decrease in the total error value can be considered as a form of learning that happens implicitly in the environment in a distributed manner.

5   Experimental Results

An experiment is also set to assess the proposed interpretation of the environment. The robots used for testing are e-puck robots[2] as shown in Fig. 1. In this task, the robots are positioned behind an object (the initial positions and orientations are acceptable in a range and need not to be absolutely accurate) and they push an object toward a constrained path and the goal is that they move the object to a position at the end of the path without hitting the walls[3].

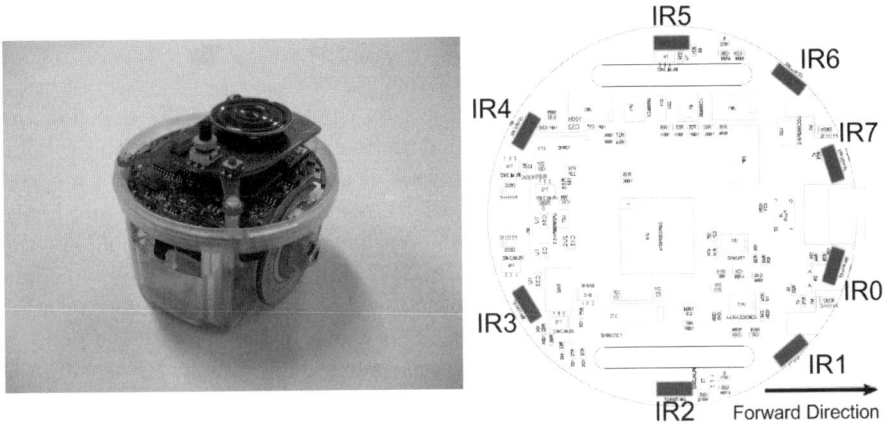

Fig. 1. A view of the e-puck robots which are used for experimental results (left) and plan of sensors on the robot (right)

5.1   Test Conditions and Algorithms

For this test we have two robots that are in charge of moving an object on a trajectory. The trajectory is surrounded by two walls and the robots and the object are between these two walls. The robots have only a sense of the surrounding objects and walls and they have no direct communication together. They perform actions according to their perceptions and the built-in rules while a higher level of behavior arises from the group of two robots (they are making decisions locally and are not aware of the global emerging behavior).

The robots use eight Infra Red sensors that are distributed around the body (the distribution is not uniform) and each robot is able to acquire a limited perception of surrounding objects by these sensors. The locations of these sensors are demonstrated in a map in Fig. 1.

The robots have two IR sensors in the front part of the body. These two sensors are used to ensure that the robots keep contact with the object at all times. When

---

[2] http://www.e-puck.org
[3] For this test we only concentrated on the simple behaviors of the robots and we did not add a constraint for detecting the end of the experiment while it can be added with additional external tools or algorithms.

the received signal from the left/IR7 sensor (ξ_7) is below a threshold, it means that the direction of the robot is not perpendicular to the surface of the object and hence the robot should turn left. The amount of this turning is such that both front sensors will sense the object at the desired orientation. The same procedure also goes for the right/IR0 sensor. The threshold (Δ) used for sensing the object in front of the robot is equal to %85 of the maximum value of signal that can be perceived by IR sensors. The algorithm used for pushing the object is as follows:

```
While (ξ7 < Δ) rotate right;
While (ξ0 < Δ) rotate left;
Go forward;
```

We have used two robots to push the object on the trajectory. The speed of the robots in this experiment is a factor that defines the behaviors of the robots. Whenever each of the robots senses the wall, it pushes the object with a speed faster than normal. This causes the object to rotate to the other side and hence the distance from the wall is increased. We have defined a threshold on the IR5/IR2 signals for robots. When the IR5/IR2 signals ξ_5/ξ_2 become more than a threshold (T), the velocity set point of left/right wheels of the robot (positive values of V) are set to maximum speed. This is while the velocity set point is equal to %50 of motors maximum speed on a normal movement. By sensing the wall, the robots change their behaviors to a temporary state. This ensures that they increase their distance with the wall. Concepts of sensing the wall and the respective behaviors are as follows:

```
If (ξ5 > T) // left wall is sensed
    VLeft=%100;
else
    VLeft=%50;

If (ξ2 > T) // right wall is sensed
    VRight=%100;
else
    VRight=%50;
```

5.2   Environment as a Common Memory

We define the state of the system for this problem as Eq. 7

$$State = S_{Object} \times S_{Robot1} \times S_{Robot2} \tag{7}$$

When the object moves, the state of the object and hence the state of the environment are changed. This is the memory element that lasts in the environment and agents can perceive it (when a robot finds the position of the object by sensors, this is a memory access process and the perceived data is gained from a stored piece of information in the environment). By a quantitative demonstration, the value of stored information in this distributed memory is specified by a function (for this case we have excluded the state of the robots from the overall state):

$$Value = g(X_{Object}, Y_{Object}, \theta_{Object}) \tag{8}$$

This value remains the same while none of the values related to mentioned elements are changed (the only factors that are capable of changing the value of memory for this problem are the actions made by robots). The agents make decisions according to the perceived data and their behaviors change according to the piece of information they have sensed from the common environment.

5.3 An Implicit Learning Process

For this task we can consider various factors for defining the total error value. As the agents have only one option (pushing the object), we omit the individual errors of agents and only consider the total error existing in the environment (the robots are not aware of these errors).

$$Total Error(State_A) = Error(State_A) \tag{9}$$

We can consider the distance from the target as one of the error factors

$$Error(State_A) = h(\Delta X_{State_A}, \Delta Y_{State_A}, \Delta \theta_{State_A}) \tag{10}$$

and as an instance we can specify the error function as

$$h(\Delta X, \Delta Y, \Delta \theta) = \alpha_1 \cdot \sqrt{(\Delta X)^2 + (\Delta Y)^2} + \frac{\alpha_2 \cdot \Delta \theta}{\sqrt{(\Delta X)^2 + (\Delta Y)^2}} \tag{11}$$

which means that the total error is commensurate to the distance of the object from the goal and its difference in angle to the final orientation[4] (a sample of the experiment is shown in Fig. 2).

When the object moves toward the trajectory (from (X1,Y1) to (X2,Y2)), it is getting closer to the final position and assuming a small value for

$$\sqrt{(\Delta X_2)^2 + (\Delta Y_2)^2} < \sqrt{(\Delta X_1)^2 + (\Delta Y_1)^2} \tag{12}$$

and hence accodrding to the error function, if the object is far enough from the target, we can ignore the orientation and the value of error in the new state is less than the value in previous state as shown in Eq. 13 (if the object gets closer to the target, the orientation plays an important role for the error value).

$$Error(State_2) < Error(State_1) \tag{13}$$

As a permanent error in the environment is decreased by previous changes, we can interpret this change in the state as an implicit learning process during this object pushing procedure.

---

[4] We can also define other error values like the distance of the object from the walls that is decreased by proper actions of agents. For this case we have only considered the simple error value by the distance from the final target.

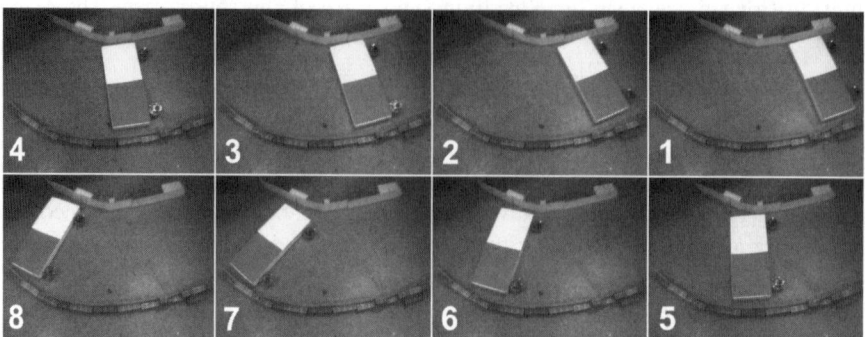

Fig. 2. Two robots are pushing the box in a curved path. When gets too close to one of the wall, the robot beside that wall compensates this deviaton by pushing the box with a faster speed (pictures are sorted from right to left, top to bottom).

6   Conclusions and Future Works

In this article we considered multi agent systems in which the environment played a significant role. The effects of making changes in the environment by the agents were under investigation and an interpretation of a distributed memory and a distributed learning in these systems was proposed. The proposed interpretations were assessed through experimental results of an object pushing task performed by two mobile robots. It was shown that how the memory and learning exist in the proposed example.

For future works, the concepts of distributed memory and distributed learning can be investigated with other experiments such as for the same experiment the number of robots may increase to more than two robots (while minor changes in the algorithms are necessary). On the other hand, the concept can be scaled up and be considered for other tasks in which a group of agents cooperate and their communication is implicit and is done via the common environment. The agents can use the potential facilities of the environment as implicit distributed memory elements which will be useful for tasks in which the communication between agents is not possible/reliable or when the environment is too complex.

Acknowledgments. We would like to appreciate AI and Robotics Lab. in University of Tehran ECE for providing facilities for experiments of this article.

References

1. Grasse, P.P.: La reconstruction du nid et les coordinations inter-individuelles chez bellicositermes natalensis et cubitermes sp. la theorie de la stigmergie: Essai d'interpretation des termites constructeurs. Insectes Sociaux, 41–83 (1959)
2. Beckers, R., Holland, O., Deneubourg, J.: From local actions to global tasks: Stigmergy and collective robotics. In: Brooks, R., Maes, P. (eds.) Artificial Life IV, pp. 181–189. MIT Press, Cambridge (1994)

3. Holland, O., Melhuish, C.: Stigmergy, self-organization, and sorting in collective robotics. Artif. Life 5(2), 173–202 (1999)
4. Clark, A.: Being There: Putting Brain, Body, and World Together Again. MIT Press, Cambridge (1998)
5. Theraulaz, G., Bonbeau, E.: A brief history of stigmergy. Artif. Life 5(2), 97–116 (1999)
6. Sharbafi, M., Daneshvar, R., Lucas, C.: New implicit approach in coordination between soccer simulation footballers. In: Fifth Conference on Computer Science and Information Technologies (CSIT 2005), Yerevan, Armenia (2005)
7. Couzin, I.D., Krause, J., James, R., Ruxton, G.D., Franks, N.R.: Collective memory and spatial sorting in animal groups. J. Theor. Biol. 218(1), 1–11 (2002)
8. Peshkin, L., Meuleau, N., Kaelbling, L.P.: Learning policies with external memory. In: ICML 99: Proceedings of the Sixteenth International Conference on Machine Learning, pp. 307–314. Morgan Kaufmann Publishers Inc. San Francisco, CA, USA (1999)
9. O'Reilly, R.C., Munakata, Y., Mcclelland, J.L.: Computational Explorations in Cognitive Neuroscience: Understanding the Mind by Simulating the Brain. The MIT Press, Cambridge (2000)
10. Matlin, M.W.: Cognition, 6th edn. John Wiley and Sons, Inc. Chichester (2005)
11. Sternberg, R.: Psychology. In search of the Human Mind, 3rd edn. Harcourt (2001)
12. Gleitman, H.: Psychology, 3rd edn. W.W. Norton and Company (1991)
13. Kolb, B., Whishaw, I.: Fundamentals of Human Neuropsychology, 3rd edn. Freeman (1990)
14. Atkinson, R., Atkinson, R., Smith, E., Bem, D., Hoeksema, N.: Hilgard's Introduction to Psychology. 12 th edn. Harcourt Brace (1996)
15. Gazzaniga, M., Ivry, R., Mangun, G.: Cognitive Neuroscience. The Biology of the mind. W.W. Norton and Company (1998)
16. Kandel, E., Schwartz, J., Jessell, T.: Principles of Neural science, 4th edn. Mc Graw Hill, New York (2000)
17. Kolb, B., Whishaw, I.: An Introduction to Brain and Behavior. Worth Publishers (1990)
18. Baddeley, A.: Human Memory: Theory and Practice. Lawrence Erlbaum, Mahwah (1990)
19. Steward, O.: Functional Neuroscience, vol. 23. Springer, Heidelberg (2000)
20. Purves, D., Augustine, G.J., Fitzpatrick, D., Hall, W.C., Lamantia, A.S., McNamara, J.O., Williams, S.M.: Neuroscience. Sinauer Associates, Inc. (2004)
21. Andrewes, D.G.: Neuropsychology: From Theory to Practice. Psychology Press (2001)

AI in Locomotion: Challenges and Perspectives of Underactuated Robots

Fumiya Iida[1,2], Rolf Pfeifer[2], and André Seyfarth[3]

[1] Computer Science and Artificial Intelligence Laboratory,
Massachusetts Institute of Technology, 32 Vassar Street, Cambridge, MA 02139, USA
[2] Artificial Intelligence Laboratory, Department of Informatics,
University of Zurich, Andreasstrasse 15, CH-8050 Zurich, Switzerland
[3] Locomotion Laboratory, University of Jena,
Dornburger Strasse 23, D-07743 Jena, Germany

Abstract. This article discusses the issues of adaptive autonomous navigation as a challenge of artificial intelligence. We argue that, in order to enhance the dexterity and adaptivity in robot navigation, we need to take into account the decentralized mechanisms which exploit physical system-environment interactions. In this paper, by introducing a few underactuated locomotion systems, we explain (1) how mechanical body structures are related to motor control in locomotion behavior, (2) how a simple computational control process can generate complex locomotion behavior, and (3) how a motor control architecture can exploit the body dynamics through a learning process. Based on the case studies, we discuss the challenges and perspectives toward a new framework of adaptive robot control.

1 Introduction

Navigation is one of the most fundamental functions of adaptive autonomous systems and it has been a central issue of artificial intelligence. As in most of the other topics of AI research, navigation has been traditionally treated as a "sense-think-act" process, where the problem is generally decomposed into three sub-processes of (1) identifying the situation, i.e. mapping the sensory stimulation on to an internal representation, the world model, (2) planning an action based on this world model, and (3) executing the physical action. In this framework, the navigation problem was nicely formulated by engineering terms as exemplified by the Simultaneous Localization and Map Building [1]. Although for many tasks these systems perform well, a considerable number of issues remain to be solved if compared to biological systems that routinely exhibit adaptive locomotion and navigation tasks in complex environments with great ease and robustness.

The studies of physiology and biomechanics revealed that animals' navigation capabilities generally rely on highly distributed mechanisms: object recognition and large-scale planning in the brain, reflexes and basic motor signals in peripheral nervous circuitry, and adaptive musculoskeletal dynamics in the mechanical level, for example. Although the decentralized nature of navigation mechanisms

M. Lungarella et al. (Eds.): 50 Years of AI, Festschrift, LNAI 4850, pp. 134–143, 2007.

was previously formulated by the so-called Behavior Based Approach [2,3] without explicit internal representation of the world, this approach generally does not explicitly discuss the physics of system-environment interactions, which makes the navigation capabilities still highly limited to relatively simple tasks such as obstacle avoidance and basic target following [4,5].

The computational framework of adaptive control architectures often ignores the fact that every behavior is the result of system-environment interaction, and it is implicitly assumed that computational processes and physical ones are independent problems. There are, however, a number of aspects where the computational processes have to take system-environment interactions into account as discussed in the field of embodied artificial intelligence [6,7,8]. For the navigation problem in particular, there are the following three main reasons. Firstly and most importantly, capabilities and limits of navigation are largely influenced by how robotic systems interact physically with environment. The well-designed mechanical structures are prerequisite for maximum forward speed, maneuverability, and energy efficiency for the locomotion in complex dynamic environment. Secondly, motor control architectures are highly related to the way how the system interacts physically with the environment. With a good mechanical design, computational process of motor control can be significantly simplified as demonstrated by Passive Dynamic Walkers [9,10,11], for example. And thirdly, because the dynamics of physical system-environment interaction are often highly nonlinear, the computational processes such as route planning cannot make decisions arbitrarily, but they need to take the physical constraints into account. For example, as demonstrated later in this article, the physical constraint of underactuated locomotion systems need to exploit the dynamics of hopping behavior in order to traverse rough terrains.

This article introduces three projects of locomotion machines with a special focus on underactuated systems, i.e. the systems that exploit passive dynamics for their behavioral functions. These case studies demonstrate how behavioral performances such as rapid movement, behavioral diversity, and complex behavior patterns can be improved in underactuated robotic systems by taking advantage of the interplay between material properties, body structures and dynamics, and adaptive control processes. Based on these case studies we will speculate further challenges and perspectives of robot navigation.

2   "Cheap Design" for Locomotion

Complex mechanical structures are a fertile basis of animals' adaptive behavior. Likewise, well-designed structures and mechanical properties of robot body are an important prerequisite, which make the robotic systems capable of achieving many behavioral variations for the purpose of adaptive behavior in complex dynamic environment. Exploiting physical constraints of the systems' own body and ecological niche is essential not only for energy efficient, rapid behavior with high maneuverability, but also simplified control, as nicely formulated by the principle of "cheap design" [6][8]. This section explains how physical

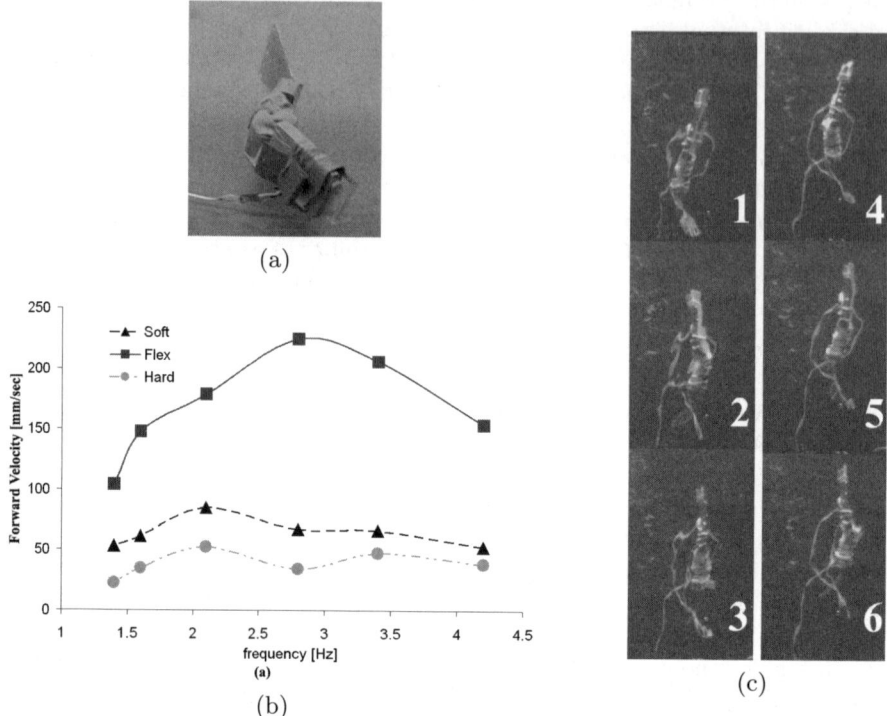

Fig. 1. Behavioral dynamics of the fish robot. (a) Photograph of the fish robot. (b) Forward velocity of three types of tail fins made of different materials (Square plot: flexible fin, triangle plot: soft fin, and circle plot: hard fin). (c) Time-series photographs of a typical forward swimming of the fish robot.

system-environment interactions can be exploited to achieve locomotion functions through a case study of an underwater locomotion robot. This "cheap" underwarter locomotion nicely demonstrates how locomotion capabilities are dependent on the physical system-environment interaction induced by mechanical design.

The fish robot has one single degree-of-freedom of actuation: it can basically wiggle its tail fin back and forth. The motor connects a rigid frontal plate and the elastic tail fin (Figure 1(a)). With this body structure, simple motor oscillation drives this fish robot not only in the forward direction, but also right, left, up, and down by exploiting fluidic friction and buoyancy [12]. Turning left and right is achieved by setting the zero-point of the wiggle movement either left or right at a certain angle. The buoyancy is such that if it moves forward slowly, it will sink (move down). The forward swimming speed is controlled by wiggling frequency and amplitude. If it moves fast and turns, its body will tilt slightly to one side which produces upthrust so that it will move upwards. For these behavioral variations, therefore, control of forward speed plays an important role. It is generated by the fluid dynamics as the elastic fin interacts with the environment. If the robot has inappropriate material properties in the tail fin, the locomotion performance is significantly degraded. Figure 1(b) shows how the forward velocity is related to the oscillation frequency of the motor and the material properties of tail fin.

This case study provides a nice illustration of how a computational process of motor control is related to the mechanical structure of the robot. The locomotion function is a consequence of physical system-environment interaction, i.e. the interaction between the frontal plate, the tail fin and the fluid, and the actuation simply induces the particular dynamic interaction. As a result, the control architecture can be very simple (oscillation of one motor). Another notable implication is the fact that the material properties of the robot body become important control parameters when motor control exploits the system-environment interaction. By changing the material property of the tail fin only, the same kinematic movement of the motor can result in fast or slow forward velocity.

3   Body Dynamics for Behavioral Diversity

Physical interaction is important not only in underwater locomotion but also for locomotion on the hard terrain. In this section we introduce a biped robot, which demonstrates two gait patterns - walking and running - by exploiting the dynamics induced by elastic legs interacting with the ground. This case study shows how behavioral diversity can be generated through a particular body structure and its dynamics.

Inspired from biomechanical models of human legs [13,14,15], each leg of this biped robot has one servomotor at the hip and two passive joints at the knee and the ankle (Figure 2(a)). Four springs, which are used to mimic the biological muscle-tendon systems, constrain the passive joints. Three of the springs are connected over two joints: they correspond to the biarticular muscles in the biological systems (i.e. two springs attached between the hip and the shank, another one between the thigh and the heel). Essentially, biarticular muscles induce more complex dynamics because the force exerted on each spring is not only dependent on the angle of a single joint but also the angle of the other joint. Interestingly, however, this unique design of the elastic legs enables the system to induce two different gait patterns, walking and running, by using a basic oscillation of the hip motors.

Despite the simplicity of the motor control, the leg behavior of walking is surprisingly similar to that of human [16]: As shown in Figure 2(c,e), during a stance phase, the body trajectory exhibits multiple peaks in vertical movement, the knee joint exhibits multiple flexion and extension movements, and the ankle joint rotates rapidly at the end. We found that these characteristics of joint trajectories are common also in human walking behavior. With the same configuration of the body design, this robot is also capable of running by varying the spring constants and a few motor oscillation parameters. As shown in

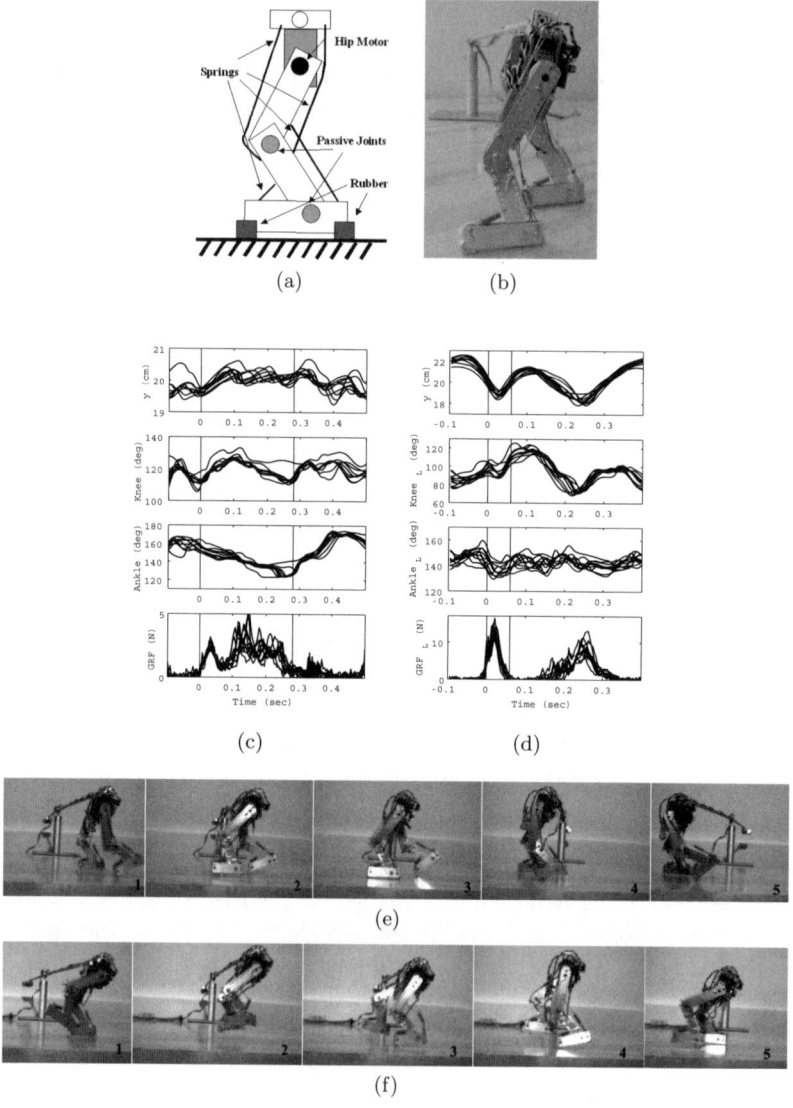

Fig. 2. Dynamic biped walking and running. (a) Schematic illustration and (b) photograph of the biped robot. This robot consists of a joint controlled by a servomotor (represented by a black circle) and three leg segments which are connected through two passive joints (gray circles). Four tension springs are attached to the segments and rubber materials are implemented at the two ground contact points of the foot body segment. The robot is attached to a pole to restrict rotational movement of the body. (c)Walking and (d) running dynamics are illustrated in terms of the vertical movement of body, knee joint angle, ankle joint angle, and vertical ground reaction force GRF (from top to bottom) which are aligned by the stance phase of 10 steps (the stance phase is indicated by two vertical lines in the figures). Time-series trajectories of the robot (c) walking and (d) running.

Figure 2(d,f), the robot shows a clear flight phase of about 0.1 second, resulting from the complex dynamics of the body and joint trajectories significantly different from those of walking [17].

This case study demonstrated how different kinds of behavioral patterns can be essentially generated by the body dynamics which are necessary in the adaptive locomotion scheme. By carefully designing elastic body structures, behavioral diversity can be not only achieved by the computational processes of motor control, but also significantly influenced by the dynamics induced by the interactions with simple motor action and the ground reaction force.

4 Control and Learning Through Body Dynamics

As shown in the previous sections, the use of body dynamics has a great potential to significantly improve the physical locomotion performances by using very simple control. However, a fundamental problem in control of underactuated systems lies in the fact that the desired behavior is always dependent on the environmental conditions. When the conditions are changed, the same motor commands result in a different behavior, and it is difficult to find the new set of motor commands in the new environment. In other words, the behavior is coupled with environmental properties, which the system could actually take advantage of. For example, the velocity curves of the fish robot are dramatically changed in rapid water flow or turbulence, and the biped walking and running is no longer possible in insufficient ground friction or a soft surface. In this sense, a dynamic adaptive controller is an essential prerequisite for underactuated systems.

This section introduces a case study of a hopping one-legged robot that learns motor control in order to traverse rough terrains [18]. This robot consists of one servomotor at the hip joint and two limb segments that are connected through a passive joint and a linear tension spring (Figure 3(a,b)). Although, on a level ground, this underactuated system exhibits periodic stable running locomotion with a simple oscillation of the actuator [19], it requires dynamic control of parameters in order to negotiate with large changes in the environment such as a series of large steps.

In this experiment, we applied a simple machine learning method, the so-called Q-learning algorithm, for optimizing the oscillation frequency of the actuated joint. The system optimizes the motor frequency of every leg step to induce adequate hopping to jump over relatively large steps on the terrain. The sequence of motor frequency is learned through the positive reward proportional to the travelling distance and negative reward in case of fall. Because the learning process requires a number of iterations, we conducted the control optimization in simulation and the learned parameters were transferred to the real-world robotic platform. After a few hundred iterations of the simulation, the system is able to find a sequence of frequency parameters that generates a hopping gait of several leg steps for the locomotion of the given rough terrain (Figure 3(c,d)).

In general, the control architectures of underactuated systems are highly non-linear in a sense that hopping height and forward velocity of this one-legged

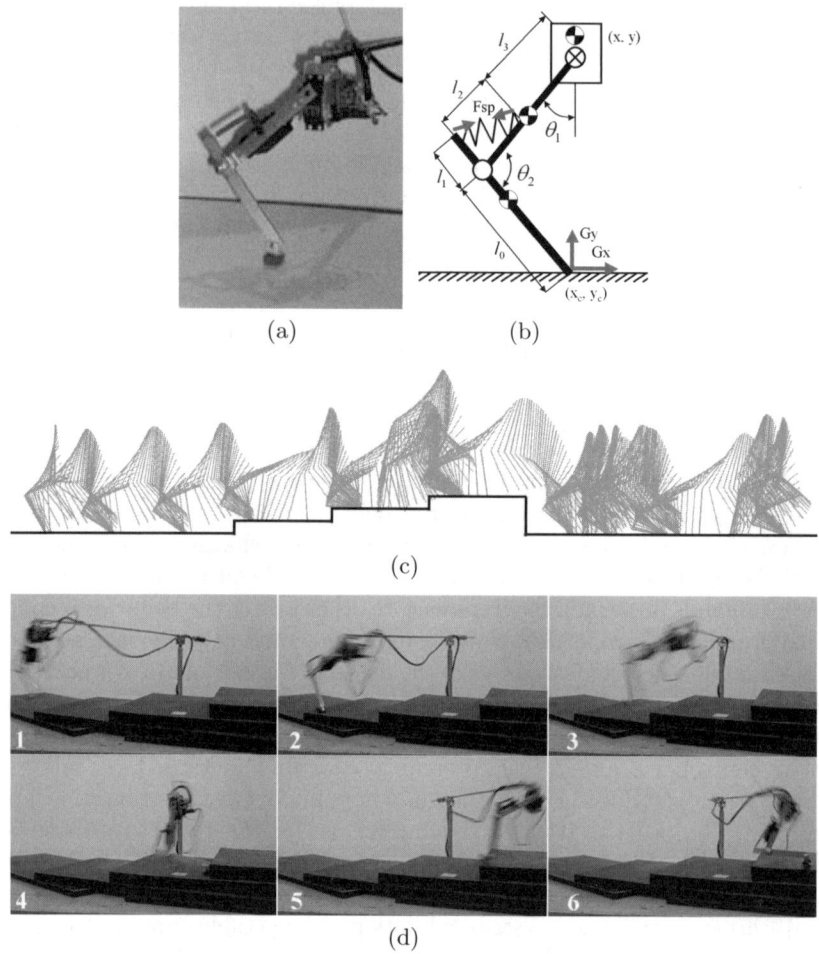

(a)   (b)

(c)

(d)

Fig. 3. One-legged hopping robot traversing rough terrains. (a) Photograph and (b) schematic of the one-legged hopping robot, that consists of one servomotor at the hip joint and two limb segments connected through compliant passive joint. (c) Optimization results of motor control in simulation. The optimized sequence of motor frequency exhibits 12 leg steps successfully travelling through a rough terrain. (d) Time-series photographs of the robot hopping over the steps. The motor control parameter was first optimized in simulation and transferred to the robot for the real-world experiment.

robot, for example, are not fully proportional to the motor oscillation frequency. Therefore, in order to achieve locomotion in complex environment, it is necessary to have adaptive control architectures such as a learning process shown in this case study. However, if the mechanics is properly designed, these adaptive control architectures can be kept quite simple. In fact the optimization process of the one-legged robot searches a sequence of one control parameter only, i.e. the frequency of motor oscillation. Simplicity of control results in a reduced parameter space and less exploration, which leads to considerable speed-up of the learning process.

5 Discussion: Challenges and Perspectives

In the navigation studies in general, means of locomotion and body structures are not explicitly considered, and the research is typically centered around the issues of sensing, modelling of the environment, and planning. However, this article showed that it is essential to investigate physical system-environment interactions in locomotion in order to scale up the performance and complexity of navigation tasks significantly. In this section, we elaborate how the dynamics of underactuated systems is related to a new framework of navigation based on the case studies presented.

One of the most fundamental open problems is still in the level of mechanics. Generally the exploitation of mechanical properties in underactuated systems provides energy efficiency [11,20], recovery of periodic behavioral patterns from disturbances [9,21,22,25], and the increase of behavioral variations derived from body dynamics [26,23,24]. While we still do not fully understand how to design "adaptive mechanics", it is important to note that mechanics is significantly related to motor control and perception, hence navigation and locomotion cannot be independent problems.

Another challenge lies in the adaptive dynamic control architecture, which is a prerequisite for underactuated systems as explained in the case study of the one-legged hopping robot. It is still an active research topic, and a number of different approaches are currently investigated (e.g. [27,28,29,30]). Along this line of research, we expect to understand how underactuated systems will actively explore their body dynamics. By obtaining the capabilities and limits of their own body, they will be able to deal rapidly and precisely with complex environment.

Although we have not explicitly discussed so far how perception processes are related to mechanical properties and motor control, it is a highly important issue in the context of navigation. In fact, the use of body dynamics can be used for better sensing [31]. Because the sensing and the recognition processes are fundamental open problems in navigation, underactuated systems should be investigated further in order to gain a comprehensive understanding of the sensory-motor processes.

Acknowledgement

This work is supported by the Swiss National Science Foundation (Grant No. 200021-109210/1), the Swiss National Science Foundation Fellowship for Prospective Researchers (Grant No. PBZH2-114461), and the German Research Foundation (DFG, SE1042). The authors would like to appreciate Marc Ziegler, Jürgen

Rummel, Gabriel Gomez, Alex Schmitz, and Russ Tedrake for the fruitful discussion and collaboration during the robotic experiments presented in this article.

References

1. Leonard, J.J., Durrant-Whyte, H.F.: Simultaneous map building and localization for an autonomous mobile robot. In: IEEE/RSJ International Workshop on Intelligent Robots and Systems, pp. 1442–1447 (1991)
2. Brooks, R.A.: Intelligence without representation. Artificial Intelligence 47, 139–159 (1991)
3. Arkin, R.C.: Behavior-based robotics. MIT Press, Cambridge (1998)
4. Brooks, R.A.: A robot that walks: Emergent behaviors from a carefully evolved network. Neural Computation 1(2), 253–262 (1989)
5. Braitenberg, V.: Vehicles - Experiments in synthetic psychology. MIT Press, Cambridge (1984)
6. Pfeifer, R., Scheier, C.: Understanding Intelligence. The MIT Press, Cambridge (1999)
7. Iida, F., Pfeifer, R., Steels, L., Kuniyoshi, Y. (eds.): Embodied Artificial Intelligence. LNCS (LNAI), vol. 3139. Springer, Heidelberg (2004)
8. Pfeifer, R., Iida, F., Bongard, J.: New robotics: Design principles for intelligent systems. Artificial Life 11(1-2), 99–120 (2005)
9. McGeer, T.: Passive Dynamic Walking. The International Journal of Robotics Research 9(2), 62–82 (1990)
10. Collins, S.H., Wisse, M., Ruina, A.: A three-dimentional passive-dynamic walking robot with two legs and knees. International Journal of Robotics Research 20, 607–615 (2001)
11. Collins, S., Ruina, A., Tedrake, R., Wisse, M.: Efficient bipedal robots based on passive dynamic walkers. Science Magazine 307, 1082–1085 (2005)
12. Ziegler, M., Iida, F., Pfeifer, R.: "Cheap" underwater locomotion: Roles of morphological properties and behavioural diversity. In: Proceedings of Climbing and Walking Robots (2006)
13. McMahon, T.A., Cheng, G.C.: The mechanics of running: How does stiffness couple with speed? Journal of Biomechanics 23(suppl. 1), 65–78 (1990)
14. Blickhan, R.: The spring-mass model for running and hopping. Journal of Biomechanics 22, 1217–1227 (1989)
15. Seyfarth, A., Geyer, H., Guenther, M., Blickhan, R.: A movement criterion for running. Journal of Biomechanics 35, 649–655 (2002)
16. Iida, F., Minekawa, Y., Rummel, J., Seyfarth, A.: Toward a human-like biped robot with compliant legs. In: Arai, T., et al. (eds.) Intelligent Autonomous Systems - 9, pp. 820–827. IOS Press, Amsterdam (2006)
17. Iida, F., Rummel, J., Seyfarth, A.: Bipedal walking and running with compliant legs. In: International Conference on Robotics and Automation (ICRA 2007), pp. 3970–3975 (2007)
18. Iida, F., Tedrake, R.: Motor control optimization of compliant one-legged locomotion in rough terrain. In: Proc. of the IEEE/RSJ International Conference on Intelligent Robots and Systems (IROS 2007) (in press, 2007)
19. Rummel, J., Iida, F., Seyfarth, A.: One-legged locomotion with a compliant passive joint. In: Arai, T., et al. (eds.) Intelligent Autonomous Systems – 9, pp. 566–573. IOS Press, Amsterdam (2006)

20. Buehler, M., Battaglia, R., Cocosco, A., Hawker, G., Sarkis, J., Yamazaki, K.: Scout: A simple quadruped that walks, climbs and runs. In: Proc. Int. Conf on Robotics and Automation, pp. 1707–1712 (1998)
21. Kubow, T.M., Full, R.J.: The role of the mechanical system in control: a hypothesis of self-stabilization in hexapedal runners. Phil. Trans. R. Soc. Lond. B 354, 849–861 (1999)
22. Raibert, H.M.: Legged Robots That Balance. MIT Press, Cambridge (1986)
23. Iida, F., Pfeifer, R.: "Cheap" rapid locomotion of a quadruped robot: Self-stabilization of bounding gait. In: Groen, F., et al. (eds.) Intelligent Autonomous Systems 8, vol. 35, IOS Press, Amsterdam (2004)
24. Iida, F., Gomez, G.J., Pfeifer, R.: Exploiting body dynamics for controlling a running quadruped robot. In: ICAR 2005, July 18th-20th, Seattle, U.S.A. pp. 229–235 (2005)
25. Cham, J.G., Bailey, S.A., Clark, J.E., Full, R.J., Cutkosky, M.R.: Fast and robust: hexapedal robots via shape deposition manufacturing. The International Journal of Robotics Research 21(10), 869–882 (2002)
26. Raibert, M.H.: Trotting, pacing and bounding by a quadruped robot. Journal of Biomechanics 23(suppl. 1), 79–98 (1990)
27. Taga, G., Yamaguchi, Y., Shimizu, H.: Self-organized control of bipedal locomotion by neural oscillators in unpredictable environment. Biological Cybernetics 65, 147–159 (1991)
28. Tedrake, R., Zhang, T.W., Seung, H.S.: Stochastic policy gradient reinforcement learning on a simple 3D biped. In: Proc. of the 10th Int. Conf. on Intelligent Robots and Systems, pp. 2849–2854 (2004)
29. Geng, T., Porr, B., Wörgötter, F.: A reflexive neural network for dynamic biped walking control. Neural Computation 18(5), 1156–1196 (2006)
30. Buchli, J., Iida, F., Ijspeert, A.J.: Finding resonance: Adaptive frequency oscillators for dynamic legged locomotion. In: Proc. of the IEEE/RSJ International Conference on Intelligent Robots and Systems (IROS 2006), pp. 3903–3909 (2006)
31. Iida, F., Pfeifer, R.: Sensing through body dynamics. Robotics and Autonomous Systems 54(8), 631–640 (2006)

On the Task Distribution Between Control and Mechanical Systems

A Case Study with an Amoeboid Modular Robot

Akio Ishiguro and Masahiro Shimizu

Dept. of Electrical and Communication Engineering, Tohoku University,
6-6-05 Aoba, Aramaki, Aoba-ku, Sendai 980-8579, Japan

Abstract. This paper introduces our robotic case study which is intended to intensively investigate the neural-body coupling, *i.e.*, how the task distribution between control and mechanical systems should be achieved, so as to emerge useful functionalities. One of the significant features of this case study is that we have employed a collective behavioral approach. More specifically, we have focused on an "embodied" coupled nonlinear oscillator system by which we have generated one of the most primitive yet flexible locomotion, *i.e.*, amoeboid locomotion, in the hope that this primitiveness allows us to investigate the neural-body coupling effectively. Experiments we have conducted strongly support that there exists an "ecologically-balanced" task distribution, under which significant abilities such as real-time adaptivity emerge.

1 Introduction

The behavior of a robotic agent emerges through the dynamics stemming from the tight interplay between the control system, mechanical system, and environment. Despite the existence of tight interdependency between control and mechanical systems, traditional robotics have often ignored this and have focused on either mechanical designs or control architectures in isolation. Generally speaking, system enhancement has been achieved normally by increasing the complexity of its control system. This, however, causes serious problems, particularly in terms of adaptivity and energy efficiency.

Under these circumstances, recently the importance of the following suggestions has been widely recognized: (1) there should be an "ecologically-balanced" task distribution between control and mechanical systems; and (2) under which one can expect that quite interesting phenomena, *e.g.*, real-time adaptivity and high energy efficiency, will emerge[1]-[3]. However, there are still a number of issues that remained to be understood about how such task distribution between control and mechanical systems can be achieved[4][7].

Now a question arises. How can we investigate the validity of the suggestions above effectively? One may say that one of the promising ways is to focus on a primitive living system, and to mimic its behavior in a synthetic manner, *i.e.*, building a robotic agent. To do so, we have employed *slime mold* as a model living system[5][6], and have modeled this as an "embodied" coupled nonlinear

M. Lungarella et al. (Eds.): 50 Years of AI, Festschrift, LNAI 4850, pp. 144–153, 2007.

oscillator systems. In this paper, we introduce our robotic case study dealing with a two-dimensional modular robot called "Slimebot" that exhibits amoeboid locomotion by taking full advantages of the interplay between its control and mechanical systems. Owing to this approach, we show that there exists an "ecologically-balanced" coupling, under which significant abilities such as real-time adaptivity effectively emerges.

The rest of this paper is structured as follows. The following section briefly explains how we have designed our modular robot Slimebot and illustrates some of the highlight data taken from the experiments conducted. Section 3 then illustrates discussion based on the results obtained from this case study, followed by the conclusion.

2   A Robotic Case Study: *Slimebot*

In this section, we describe the mechanical structure and the distributed control algorithm of Slimebot.

2.1   The Method

The mechanical structure. We consider a two-dimensional Slimebot whose task is to move toward a goal light. Slimebot consists of many identical modules, each of which has a mechanical structure like the one shown in Fig. 1 (a). A schematic of the entire system is also illustrated in Fig. 1 (b). Each module is equipped with six independently-driven telescopic arms and a ground friction control mechanism (explained later). Each module is also equipped with the following two types of sensors: (1) an omnidirectional light-sensitive sensor for detecting the goal light; and (2) a sensor which informs the module whether it is positioned as an *inner* module or a *surface* module in the entire system. For attachment to other modules, the circumference of the module is covered by a "functional material". More specifically, we use a *genderless Velcro strap*: when two halves of Velcro come into contact, they easily stick together; however when a force greater than the Velcro's yield strength is applied, the halves automatically separate. We expect that by exploiting the properties of Velcro, a spontaneous inter-module connection control mechanism is realized which not only reduces the computational cost required for the connection control, but also allows to harness emergence to achieve more adaptivity. The characteristics of the inter-module connections are mainly determined by the yield stress of the Velcro straps employed: connection between the modules is established spontaneously when the arms of each module make contact; disconnection occurs if the disconnection stress exceeds the Velcro's yield stress. We also assume that local communication between the connected modules is possible. Such communication will be used to create a phase gradient inside the modular robot (discussed below). In this study, each module is moved by the telescopic actions of the arms and by the ground friction. Note that the individual modules do not have any mobility but can move only by "cooperating" with other modules.

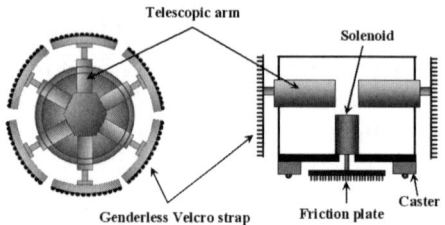

(a)Mechanical structure of each module. (Left) top view. (Right) cross-section of side view.

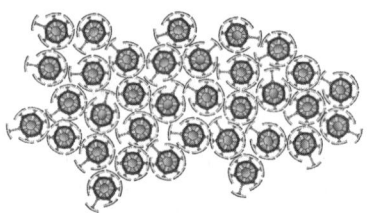

(b) Entire system (top view)

Fig. 1. Schematic of Slimebot

2.2   The Control Algorithm

In this section, we discuss how the mechanical structure described above can generate stable and continuous locomotive patterns without losing the coherence of the entire system. To this end, each module is endowed with a nonlinear oscillator. Through mutual entrainment (frequency locking) between the oscillators, rhythmic and coherent locomotion is produced. In what follows, we give a detailed explanation of this algorithm.

Basic operation. At any time, each module in the Slimebot can take one of two mutually exclusive modes : *active mode* and *passive mode*. As shown in Fig. 2, a module in the active mode actively contracts/extends the connected arms, and simultaneously reduces the ground friction. In contrast, a module in the passive mode increases the ground friction, and returns its arms to their original length. Note that a module in the passive mode does not move itself but serves as a "supporting point" for efficient movement of the modules in the active mode.

Phase gradient through the mutual entrainment. In order to generate rhythmic and coherent locomotion, the mode alternation in each module should be controlled appropriately. Of course, this control should be done in a "decentralized" manner, and its algorithm should not depend on the number of the modules and the morphology of the Slimebot. To do so, we have focused on the "phase gradient" created through the mutual entrainment among locally-interacting nonlinear oscillators in the Slimebot, exploiting this as a key information for the mode alternation. Therefore, the configuration of the resulting

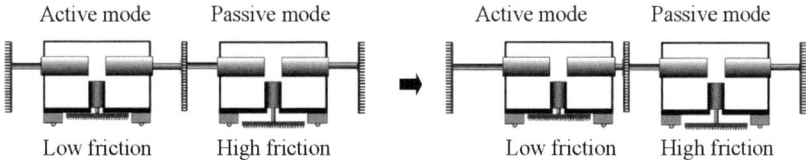

Active mode   Passive mode   Active mode   Passive mode

Low friction   High friction   Low friction   High friction

Fig. 2. Schemtic of the active mode and the passive mode. Side view of the connected modules is shown for clarity.

phase gradient is extremely important. In the following, we will explain this in more detail. As a model of a nonlinear oscillator, the *van der Pol oscillator* (hereinafter VDP oscillator) was employed, since this oscillator model has been well-analyzed and widely used for its significant entrainment property. The equation of VDP oscillator implemented on module i is given by

$$\alpha_i \ddot{x}_i - \beta_i(1 - x_i^2)\dot{x}_i + x_i = 0, \tag{1}$$

where the parameter α_i specifies the frequency of the oscillation. β_i corresponds to the convergence rate to the limit cycle.

The local communication among the physically connected modules is done by the local interaction among the VDP oscillators of these modules, which is expressed as:

$$x_i = x_i^{\text{tmp}} + \varepsilon \left\{ \frac{1}{N_i(t)} \sum_{j=1}^{N_i(t)} x_j^{\text{tmp}} - x_i^{\text{tmp}} \right\}, \tag{2}$$

where x_i^{tmp} and $N_i(t)$ represent the state just before the local interaction, and the number of modules neighboring module i at time t, respectively. x_j^{tmp} is the state of the VDP oscillator implemented into neighboring module j, which is physically connected with module i. This is the information obtained through the local communication. The parameter ε specifies the strength of the interaction. Note that this local interaction acts like a *diffusion*.

When VDP oscillators interact according to equation (2), significant phase distribution can be created effectively by varying the value of α_i in equation (1) for some of the oscillators. In order to create an equiphase surface effective for generating locomotion, we have introduced a simple sensory feedback mechanism. More specifically, we set the value of α_i as:

$$\alpha_i = \begin{cases} 0.7 \text{ if the module detects the goal light.} \\ 1.3 \text{ if the module is positioned as a } surface \ module. \\ 1.0 \text{ if the module is positioned as an } inner \ module. \end{cases} \tag{3}$$

Note that the value of α_i is increased when the module is positioned as a surface module in the entire system. As a result, the frequency of the VDP oscillators in

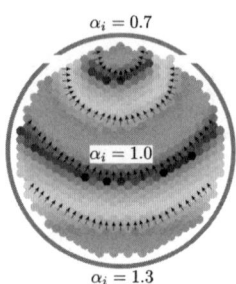

Fig. 3. Phase distribution created through the mutual entrainment among the VDP oscillators in a circular arrangement. The gray scale denotes the value of the phase at the corresponding point. Each arrow represents the direction of the gradient vector at the corresponding point. The goal light is given from the top of the figure.

the outer modules will be relatively decreased compared to the ones in the inner modules. This allows us to create the phase gradient inside the modular robot, which can be effectively exploited to endow the entire system with a *cohesive force*, similar to the effect of surface tension. Figure 3 shows the phase distribution created when the modules are placed to form a disk-like shape. The top and bottom of the figure corresponds to the front and rear of the modular robot, respectively. In the figure, the arrows, each of which represents the direction of the gradient vector at the corresponding point, are also depicted for clarity.

Generating locomotion. Based on the above arrangements, here we will explain how we have designed the control algorithm. More specifically, we will show how the mode alternation and the arm extension/contraction in each module is controlled by exploiting the phase gradient created from the aforementioned mutual entrainment among the locally-interacting VDP oscillators.

The mode alternation in each module is simply done with respect to the phase of its VDP oscillator, which is expressed as:

$$\begin{cases} 0 \leq \theta_i(t) < \pi \ : \ \text{active mode} \\ \pi \leq \theta_i(t) < 2\pi : \ \text{passive mode} \end{cases} \tag{4}$$

where, $\theta_i(t)$ denotes the phase of oscillation of the VDP oscillator in module i at time t, which is written by

$$\theta_i(t) = \arctan \frac{\dot{x}_i(t)}{x_i(t)} \quad (0 \leq \theta_i(t) < 2\pi). \tag{5}$$

On the other hand, the degree of the extension/contraction of each arm mounted on module i in the active mode is individually determined with respect to the phase difference with its corresponding neighboring module. This is given by

$$F_i^m(t) = -k\{\theta_j(t) - \theta_i(t)\}, \tag{6}$$

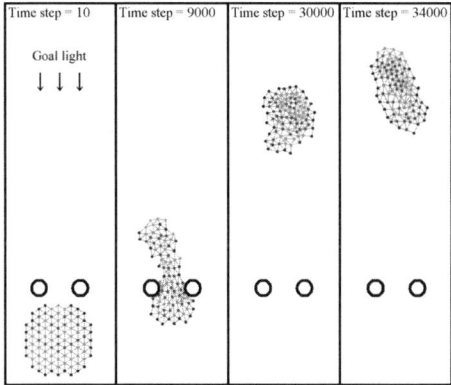

Fig. 4. Representative data of the transition of the morphology (see from left to right)

where, $F_i^m(t)$ is the force applied for the extension/contraction of the m-th arm of module i at time t. k is the coefficient. $\theta_j(t)$ represents the phase of neighboring module j physically connected to the m-th arm of module i.

Due to the algorithm mentioned above, the degree of arm extension/contraction of each module will become most significant along the phase gradient (see Fig. 3), and the timings of the mode alternation are propagated from the modules detecting the goal light to the surface ones as traveling waves. As a result, all the modules are encouraged to move toward the goal light while maintaining the coherence of the entire system.

2.3   Experiments

Simulation results. In order to confirm the validity of the proposed method, simulations were performed under the condition where $\beta_i = 1.0$; $\epsilon_i = 0.01$; all the modules were placed initially so as to form a disk-like shape. Figure 4 shows a representative result obtained in the case where the number of modules was set to 100. The thick circles in the figures denote obstacles. The goal light is given from the top of each figure, and thus the Slimebot is encouraged to move upward. These snapshots are in the order of the time evolution (see from left to right). As illustrated in the figure, the Slimebot can successfully negotiate the environmental constraints without losing the coherence. Note that around the time step of 30000 in Fig. 4., we temporarily turned off the goal light. As we see from the figure, the Slimebot starts to form a disk-like shape. This is due to the cohesive force which serves to maintain the coherence of the entire system.

Preliminary experiments with a real physical Slimebot. Fig. 5 shows the latest version of the real physical module. Each module is controlled by a control circuit with a 16MHz H8/300H CPU, which was designed specially for this purpose. All the electric power required in the module is supplied from a Lithium-Polymer battery (11.1V). Fig. 5 (b) illustrates the mechanical structure of the module developed. As shown in the figure, both the telescopic arms

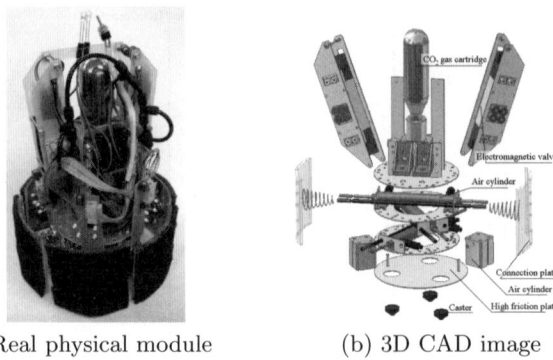

(a) Real physical module   (b) 3D CAD image

Fig. 5. A developed module of the Slimebot

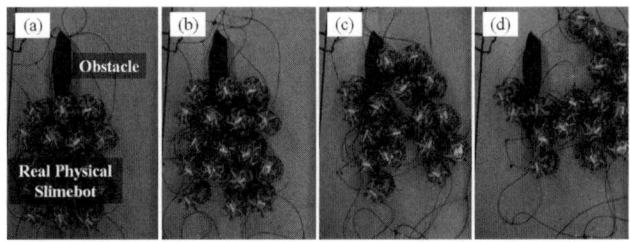

Fig. 6. Adaptive reconfiguration with 17 modules. From left to right: Sequence of snapshots of a typical example of a spontaneous inter-module connection control.

and the ground friction control mechanism are realized by implementing *pneumatic air cylinders*. The power necessary for these air cylinder is supplied from a gas cartridge (CO_2 gas, 0.3MPa) implemented onto the top of the module. Owing to this design, each module can operate in a fully self-contained manner approximately for 30 minutes without suffering from problems such as a cable entanglement.

Due to the lack of space, we briefly introduce a preliminary experiment with the use of real physical Slimebot. Fig. 6 depicts a representative experimental result with 17 modules. Fig. 7 illustrates a simulation result in which almost the same condition was applied as the one in Fig. 6. As can be seen from these figures, interestingly, good qualitative agreement is observed.

3   Discussion: Lessons from This Case Study

In this section we briefly investigate "the relationship between control and mechanical systems as it should be", or we may say "the well-balanced coupling between control and mechanical systems". For the ease of intuitive understanding, we consider this by using Fig. 8. The figure metaphorically illustrates the

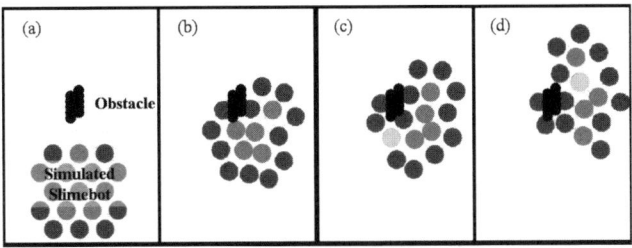

Fig. 7. Simulation result conducted for comparison with the experiment indicated in Fig. 6. See from (a) to (d)

possible task distribution between control and mechanical systems in the generation of behavior.

In Fig. 8, we will immediately notice that the current robots are driven by such extremely different approaches: the meeting point between control and mechanical systems lies either around the left or right extremity; almost nothing exists in between (*i.e.* around the region C).

Now questions arise, which can be summarized as followings: the first is closely related to the "well-balanced coupling" or the "relationship as it should be" between control and mechanical systems. Should robots be designed in such a way that their control and mechanical systems are coupled around the region C?; the second and the last concerns "emergent phenomena". Should we see the region C simply from the viewpoint of trade-off? If not, what sort of interesting properties are expected to be observed in this region?

In what follows, we show simple experiments in order to investigate these questions. Fig. 9 illustrates the time evolution of the morphology of Slimebot under different Velcro strength. Note that the Velcro strength acts as a primary coupling parameter between its control and mechanical systems. This is one of the significant features of Slimebot. As illustrated in the figure, Slimebot having an appropriate Velcro strength exhibits adaptive locomotion (see

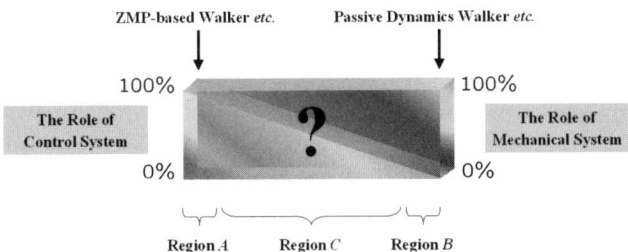

Fig. 8. A graphical representation of possible task distribution between control and mechanical systems in generating behavior. Most of the current robots are driven under the task distribution either around the left or right extremity.

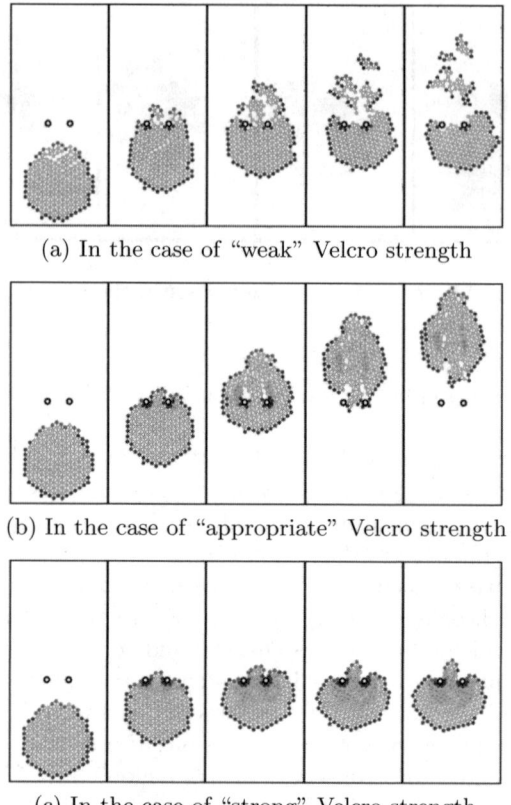

(a) In the case of "weak" Velcro strength

(b) In the case of "appropriate" Velcro strength

(c) In the case of "strong" Velcro strength

Fig. 9. The time evolution of the morphology of Slimebot under different Velcro strength (see from left to right). The thick circles in the figures denote obstacles.

Fig. 9(b)), whilst the ones with a weak and a strong Velcro strength cannot negotiate the environment. This suggests that there exists a well-balanced neural-body coupling, under which significant abilities —in this case real-time adaptivity— emerge.

4   Conclusion

This paper has investigated how the task distribution between control and mechanical systems should be achieved so as to emerge useful functionalities such as adaptivity, taking a real-time morphology control of a two-dimensional modular robot called Slimebot as a practical example. In our case study, aiming at the well-balanced task distribution between control and mechanical systems, we focused on a functional material (*i.e.* genderless Velcro strap) and the phase gradient created through the mutual entrainment between locally interacting nonlinear oscillators (*i.e.* VDP oscillators), the former of which was utilized as a spontaneous inter-module connection control mechanism and the latter of which as a core mechanism for generating locomotion.

Owing to focusing on amoeboid locomotion —one of the most primitive types of locomotion—, the results obtained strongly support the validity of our working hypothesis, stating that well-balanced task distribution between control and mechanical systems plays an essential role to elicit interesting emergent phenomena, which can be exploited to increase adaptivity, scalability, fault tolerance and so on. This research is a first step to shed light on this point.

Acknowledgments

This work has been partially supported by a Grant-in-Aid for Scientific Research on Priority Areas "Emergence of Adaptive Motor Function through Interaction between Body, Brain and Environment" from the Japanese Ministry of Education, Culture, Sports, Science and Technology. The authors would like to thank Prof. Toshihiro Kawakatsu, Tohoku University, for frequent, stimulating, and helpful suggestions, and Prof. Rolf Pfeifer, University of Zurich, for his continuous encouragement.

References

McGeer, T.: Passive dynamic walking. The International Journal of Robotics Research 9(2), 61–82 (1990)

Pfeifer, R., Iida, F.: Morphological computation: connecting body, brain and environment. Japanese Scientific Monthly 58(2), 48–54 (2005)

Matsushita, K., Lungarella, M., Paul, C., Yokoi, H.: Locomoting with less computation but more morphology. In: Proc. of the 2005 IEEE International Conference on Robotics and Automation, pp. 2020–2025 (2005)

Ishiguro, A., Kawakatsu, T.: How should control and body systems be coupled? — a robotic case study. In: Iida, F., Pfeifer, R., Steels, L., Kuniyoshi, Y. (eds.) Embodied Artificial Intelligence. LNCS (LNAI), vol. 3139, pp. 107–118. Springer, Heidelberg (2004)

Nakagaki, T., Yamada, H., Tóth, Á.: Maze-solving by an Amoeboid Organism. Nature 407, 470 (2000)

Takamatsu, A., Tanaka, R., Yamada, H., Nakagaki, T., Fujii, T., Endo, I.: Spatiotemporal Symmetry in Rings of Coupled Biological Oscillators of Physarum Plasmodial Slime Mold. Physical Review Letters 87(7), 78102 (2001)

Ishiguro, A., Shimizu, M., Kawakatsu, T.: Don't try to control everything!: an emergent morphology control of a modular robot. In: Proc. of the 2004 IEEE International Conference on Robotics and Automation, pp. 981–985 (2004)

Bacteria Integrated Swimming Microrobots

Bahareh Behkam and Metin Sitti

NanoRobotics Laboratory, Department of Mechanical Engineering,
Carnegie Mellon University, Pittsburgh, PA 15213, USA
{behkam,sitti}@cmu.edu

Abstract. A new approach of integrating biological microorganisms such as bacteria to an inorganic robot body for propulsion in low velocity or stagnant flow field is proposed in this paper with the ultimate goal of fabricating a few hundreds of micrometer size swimming robots. To show the feasibility of this approach, *Serratia marcescens* bacteria are attached to microscale objects such as 10 micron polystyrene beads by blotting them in a bacteria swarm plate. Randomly attached bacteria are shown to propel the beads at an average speed of approximately 15 μm/sec stochastically. Using chemical stimuli, bacteria flagellar propulsion is halted by introducing copper ions into the motility medium of the beads, while ethylenediaminetetraacetic acid is used to resume their motion. Thus, repeatable on/off motion control of the bacteria integrated mobile beads was shown. On-board chemical motion control, steering, wireless communication, sensing, and position detection are few of the future challenges for this work. Small or large numbers of these microrobots can potentially enable hardware platforms for self-organization, swarm intelligence, distributed control, and reconfigurable systems in the future.

Keywords: Microrobotics, bacteria flagellar propulsion, swimming microrobots, miniature mobile robots, distributed systems.

1 Introduction

Miniature mobile robots have unique abilities such as accessing small spaces and the capability of massively parallel operation with large numbers of agents with distributed control. Due to these characteristics, they are indispensable for health-care applications such as minimally invasive diagnosis and treatment inside or outside of the human body, mobile sensor network applications for environmental monitoring and search and rescue, inspecting and repairing few millimeter diameter pipes in nuclear plants and space shuttle surfaces in space, and entertainment and educational applications. Many miniature robots with various locomotion capabilities such as walking, flying, climbing, crawling, walking on water, and swimming have been proposed in the literature for some of these applications [1]. However, currently the overall size of the miniature robots is limited to centimeter or millimeter scale mainly due to the miniaturization limits of the on-board actuators and power sources. Scaling these robots down to micrometer scale with novel actuation and powering methods is one of the grand challenges of micro/nano-robotics, which is a newly emerging field of robotics.

M. Lungarella et al. (Eds.): 50 Years of AI, Festschrift, LNAI 4850, pp. 154–163, 2007.

This work specifically focuses on actuation (propulsion) of microscale swimming robots in stagnant or low velocity unstructured liquid environments. For these robots, viscous forces dominate over inertia (i.e., the Reynolds number $\ll$ 1). Since inertia plays no role, it is well-known that the propulsion methods based on reciprocatory motion, such as fin flapping of the fish, become ineffective [2]. Thus, an asymmetric motion pattern, such as dissimilar power/recovery stroke of an elastic flap or rotation of a helical structure, is required for effective propulsion within the very small Reynolds number domain. Potential target applications of such swimming microrobots include: (1) *In situ* and minimally invasive screening of diseases and targeted local drug delivery inside the stagnant liquid environments of the human body such as the urinary tract, eyeball cavity, ear, and cerebrospinal fluid [3-5]; (2) Environmental monitoring of liquid environments for toxic or pathogenic biochemical agents; (3) Inspection and maintenance of liquid filled pipes in spacecrafts and nuclear plants with small diameters for potential cracks and leaks. These robots will have the unique advantage of being able to travel to currently inaccessible or hard to access small spaces, and will be massively parallel by fabricating and controlling large numbers of them in the future.

To solve the actuation grand challenge for microscale mobile robots, various approaches have been recently proposed in literature. In the first approach, off-board actuation and powering has enabled the microrobot. Donald *et al.* [6] developed a non-tethered 60x250x10 μm silicon microrobot, which can walk on a flat substrate using the scratch drive electrostatic actuation principle. Here, the power lines on the substrate supply the electrical energy. This actuation and powering method is limited to motion on a structured surface and cannot be applied to microrobots in liquid and unstructured environments. Similar approach has been also implemented using external magnetic actuation. Yesin *et al.* [7] used Helmotz-coil based three-dimensional (3-D) magnetic field gradients to propel and steer a ferroelectric microfabricated swimmer of the order of 100s of micrometers inside a stagnant liquid environment for potential applications inside the human eye for diagnosis and treatment of diseases. Furthermore, magnetic field gradients were also used to undulate a tail consisting of colloidal magnetic particles linked by DNA and attached to a red blood cell as another microscale swimming propulsion demonstration [8]. These methods require high magnetic field gradients for propulsion of a microscale robot, which could potentially heat and damage biological tissues. Moreover, expensive and bulky external magnetic instrumentation, such magnetic resonance imaging (MRI), is required to encompass the microrobot's operation environment. Therefore, this method of propulsion can be used only for microrobots that operate in confined spaces such as inside the human body, and it is challenging to control the robot locomotion and position due to nonlinearities in the magnetic forces depending on the distance.

In a rather different approach, biological motors that are isolated from microorganisms are utilized as novel micro/-nanoscale actuators. Mantemagno *et al.* [9] used F-ATPase rotary biomotors to rotate a chemically attached 100s of nanometer size nickel rods at around 8 Hz using adenosine triphosphate (ATP) as the source of energy in liquid environment. This method requires the integration and attachment of the synthetic propeller with the biomotor rotor using chemical

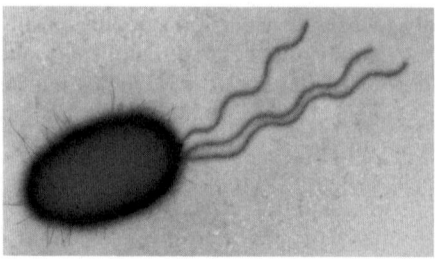

Fig. 1. Transmission electron microscope (TEM) image of an *E. coli* bacterium (x3515). Three flagella are shown. Diameter of bacteria and each flagellum are about 500 nm and 20 nm, respectively (Image courtesy of Dennis Kunkel Microscopy).

assembly, which is a very challenging and low-yield process. Moreover, a continuous supply of ATP is required to move the biomotor, which could be a limitation for some liquid environments and applications.

2 Approach

This work investigates bacteria assisted propulsion for swimming microrobots as a novel microrobotic actuation approach. The authors propose to use peritrichous bacteria for controlled propulsion of a swimming microrobot robustly and efficiently. An inorganic microrobot body is propelled by the helical flagella – only about 20 nanometers in diameter (Fig. 1) – of the bacteria attached to it. Fig. 2 shows the conceptual drawing of a *hybrid (biotic/abiotic)* swimming microrobot propelled by the bacteria attached to one of the flat ends of a polymeric micro-disk attached to the base of the microfabricated body of the robot. Here, a large number of couple of micrometer long bacteria are attached to a functionalized polymer surface by their bodies rather than their flagella, so that the flagella can rotate freely and propel the robot body. Using chemical stimuli, bacterial flagellar motors are turned on or off when desired for on/off motion control. The actuation and power source are harvested from the bacteria while providing them with the required chemical energy source and environmental conditions.

Key advantages of using whole bacteria as propulsive elements are: (1) No purification and reconstitution is necessary; (2) Simple nutrients such as glucose can be provided for the microorganisms and ion gradients are generated by the cell; (3) Bacteria are robust machines that can be easily integrated with the robot; (4) Sensors are already present in the cell. Thus, bacteria can be precisely controlled by stimulating the receptors embedded in their cell membrane. Therefore, the bacteria integrated microrobot is *robust and adaptive to the changes in the environment*, as bacteria are. Microorganisms of various sizes with different motility organelles have been used for other applications [10]. This new method could be a technologically viable actuation method for miniaturizing swimming and other robots down to micrometer sizes. This method is inspired by a recent scientific work which studies the flow field generated in the vicinity of bacterial carpets and demonstrates random motion of mobile micro-beads propelled by bacteria [11].

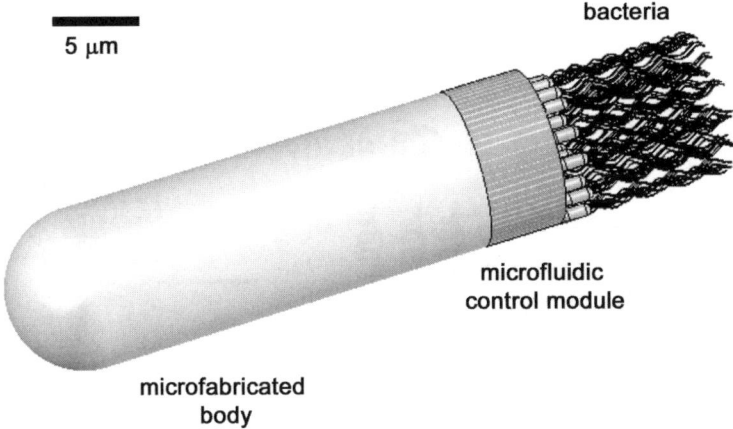

Fig. 2. Conceptual drawing of the bacteria integrated swimming microrobot: the microrobot is propelled by the attached array of bacteria, and its motion is controlled by turning the bacterial flagellar motors on/off using chemical stimuli. The robot body diameter would be of the order of 10s or few 100s of microns.

In this work, we use the peritrichous bacteria *Serratia marcescens* (*S. marcescens*) for propelling the inorganic body of the swimming microrobot. The flagella of these cells are randomly distributed over the cell surface, and each flagellar motor rotates independently of the others. Hydrodynamic interactions among flagella cause them to coordinate, coalesce and bundle behind the cell during swimming. The flagellum is a propulsive organelle that includes a reversible rotary motor embedded in the cell wall, and a filament that extends into the external medium [12]. The filament is a long (around 10 μm) and thin (around 20 nm) cylindrical helix (2.5 μm pitch and 0.5 μm diameter) that rotates at a speed of approximately 100 Hz [13]. Wild type bacteria exhibit random 'run' and 'tumble' behavior, which results random walk of the bacteria. This translates into efficiency reduction and random change of direction of the propulsion for robotic applications. Therefore, genetically engineered bacteria that are commercially available with no tumbling could be preferable. Moreover, for possible biomedical applications in the future, non-pathogenic bacteria would be selected.

3   Integration of Bacteria with the Microrobot Body

The first challenge is the controlled attachment of an array of bacteria to the polymeric microrobot body part. Here, the critical issue is selecting a strain of bacteria which is highly motile and capable of strong and stable adhesion to the robot body. Adhesion density can be controlled by micropatterning the region that bacteria are expected to attach to.

Selection of Bacteria: We choose *S. marcescens* for this work since they are fast (30-47 μm/sec), easy to culture, and can be genetically engineered to have desired motility, and sensing characteristics. These bacteria are grown in Luria broth

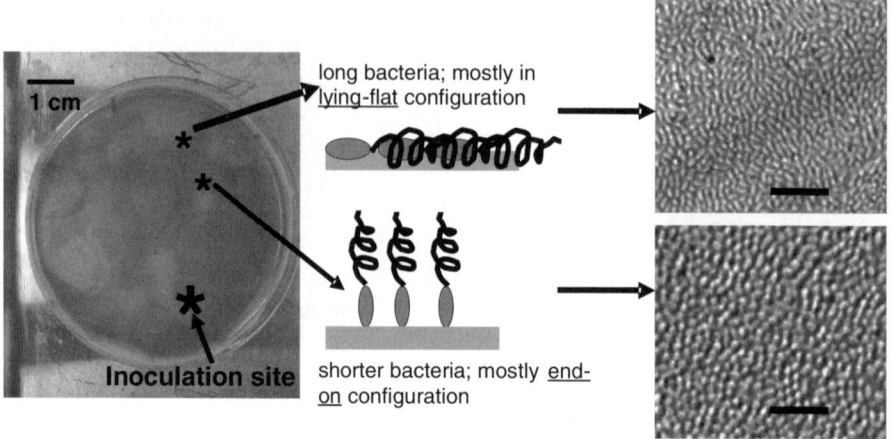

Fig. 3. Lying-flat and end-on attachment of bacteria to a polymer surface by 'blotting': bacteria from behind the leading edge of the swarm on the bacteria culture plate (leftmost photo) adhere in end-on configuration while the ones at the leading edge of the swarm attach in lying-flat configuration. The two optical microscope images on the right show attached bacteria on a flat PDMS substrate which is blotted at the leading edge of the culture (top image) or behind it (bottom image) resulting in lying-flat and end-on configurations, respectively (scale bar is 10 microns).

(L-broth) to saturation. A 1.8 µl aliquot of a 10^{-6} dilution of the saturated L-broth culture is used to inoculate a 9 cm swarm plate (L-broth containing 0.6% Difco Bacto-agar and 5 g/l glucose) off-center. The Petri plate is incubated for 20 hours at $30^{\circ}C$. This results in a swarming colony approximately 7 cm in diameter. The swarmer bacteria are larger and have more flagella than the bacteria grown in other environments.

Bacteria Attachment in End-On Configuration: Blotting is used to attach bacteria to the polymer micro-part using self-organization in massively parallel fashion. By choosing the blotting location of the polymer micro-disks on the swarm plate, it is observed that bacteria attach in end-on or lying-flat configurations [14]. This observation is also confirmed by Darnton *et al.* [11]. Figure 3 shows the bacteria from the leading edge of the swarm attached to a flat PDMS substrate. These bacteria are in lying-flat configuration. In the same figure, the bacteria from farther behind the leading edge of the swarm attach to the PDMS sheet in end-on configuration.

Bacterial Adhesion: Mechanism of adhesion of bacteria to surfaces is not completely understood and is an active area of research. It is speculated that bacterial adhesion occurs in two steps: (1) Reversible adhesion which occurs within few seconds and is mainly due to electrostatic and van der waals forces, and hydrophobic and acid/base interactions; (2) Irreversible attachment happens after the reversible attachment step and is mainly due to surface conformational changes and formation of protein-ligand bonds and production of extracellular polymer substance [15].

Hydrophobicity of the bacteria surface and the substrate, geometry and material properties of the substrate, and chemical composition of the extracellular polymer substance of the bacteria are among the important factors that affect the stability and strength of bacterial adhesion. It should be noted that bacteria properties pertaining to adhesion can vary significantly not only across but also within a genus of bacteria.

After rigorous testing, we have shown that *S. marcescens* ATCC 274 strongly attach to hydrophobic substrates such as PDMS or polystyrene during the blotting process shown in Fig. 3 using mainly hydrophobic and polymeric interaction at short time scales. *S. marcescens* adhere to hydrophilic surfaces such as air-plasma treated PDMS and gold due to polymeric interactions dominantly at a longer time scale [16].

4   Bacteria Attached Micro-bead Propulsion Experiments

To demonstrate the feasibility of using bacterial flagella as actuators for a microscale robot, *10 μm* diameter polystyrene (PS) beads are propelled by several *S. marcescens* bacteria attached to them. Random numbers of bacteria are attached to the micro-bead by the blotting method described in Fig. 3. PS beads are then pipetted into the motility medium and their random displacements are measured and compared with the diffusion length for *10 μm* particles to prove that the beads are propelled by bacteria and their displacement is not due to Brownian motion.

Micro-beads suspended in DI water were added into 1 ml of motility medium (0.01 M potassium phosphate, 0.067 M sodium chloride, 10^{-4} M ethylenediaminetetraacetic acid (EDTA), 0.01 M glucose, and 0.002% Tween-20, pH 7.0) [17]. The solution was vortexed and subsequently centrifuged at 800g. The beads were then concentrated five-fold. A 10 μl aliquot of the final suspension was pipetted into the leading edge of the swarm plate. After 5 minutes, the region was pipetted into 1 ml of motility medium. During these 5 minutes, beads randomly interact with the bacteria swimming on the surface of the swarm plate. Some of these bacteria adhered to the beads. A 100 μl sample of the final suspension was placed in an imaging enclosure. To construct the enclosure, an approximately 500 μm thick PDMS ring was placed on a microscope glass slide. Once the sample is deposited in the chamber, it was covered with a cover slip to prevent evaporation.

The motion of the PS beads was observed with a 60x oil immersion phase objective using an inverted optical microscope (Zeiss Axiovert 200). Figure 4 depicts a PS bead at *t=0* and the same bead at *t=6 s*. The total displacement of the bead was measured to be approximately *90 μm*. On the other hand, the diffusion length L_d for a *10 μm* particle is computed to be around *0.9 μm* from $L_d=(4Dt)^{0.5}$, where $D=k_BT/(6\pi\mu R)$ is the diffusion coefficient, *t=6* sec is the diffusion time, k_B is the Boltzman's constant, T is the absolute temperature of the solution, μ is the dynamic viscosity of water, and R is the radius of the particle. This value is about 100 times smaller than the observed displacement of the bead and this confirms that the bead is actually propelled by the attached bacteria.

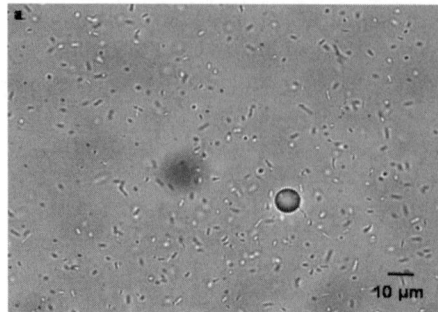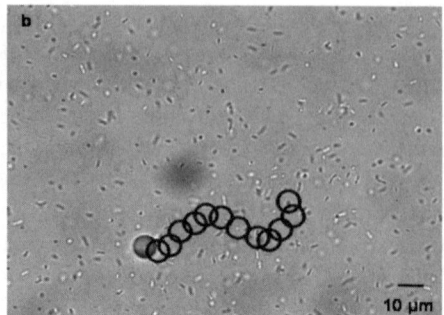

Fig. 4. Phase-contrast inverted optical microscope images of a mobile 10 μm diameter polystyrene micro-bead with several *S. marcescens* bacteria attached to it at (a) t=0 sec and (b) t=6 sec. Micro-bead's path is shown with rings. Every ring represents a 0.5 second interval [18].

5   Chemical Stimulus Based On/Off Motion Control

For numerous applications, it is imperative for a hybrid swimming microrobot to possess the ability to stop moving and subsequently resume its motion repeatedly on demand. External chemical or optical stimuli can be used to modulate the speed of some bacteria such as *S. marcescens*. Bacteria's response to selected chemicals and light of certain wavelengths are respectively known as chemotaxis and phototaxis. Chemo- and phototaxis response of bacteria can be used to control their direction of motion.

In 1973, Adler [17] reported that the absence of a chelating agent in bacterial suspensions will lead to paralysis of the bacteria. A chelating agent is a substance whose molecules can form several bonds to a single metal ion. The reason for the observed phenomenon is that heavy metal ions naturally present in water bond to the flagellar motors of the bacteria and prevent their motion.

In this study, the phenomenon mentioned above, is taken advantage of and the bacteria are purposefully paralyzed only temporarily and in a reversible fashion. To do so, $CuSO_4$ as a source of Cu^{+2} heavy metal ions is added to the bacterial suspension, halting the flagellar motors. Subsequently, EDTA is added to chelate the copper ions and resume the motion of the bacterial flagellar motors.

On/off motion control of the mobile PS beads, shown in Fig. 4, is achieved by stopping and resuming the motion of the flagellar motor of the attached bacteria by using Cu^{+2} and EDTA, respectively [18]. A mobile PS bead is stopped when a small volume of 5×10^{-3} M $CuSO_4$ solution is introduced into the motility environment. The copper ions diffuse in the solution and bond to the rotor of the flagellar motors. Next, by adding a small volume of 7.5×10^{-3} M EDTA to the solution, the copper ions are chelated and the rotors become free and resume their motion and the bead starts to move again. This on/off motion control can be potentially repeated indefinitely. However, using the current off-board method of adding the chemicals used for controlling the motion to the test sample, the total volume increases after every on/off control cycle which can pose a limitation. This can be resolved by implementation of the on-board control module which allows for more localized release of the chemicals in significantly smaller volumes.

6 Future Challenges

There still exist many challenges to be resolved towards further developing of the hybrid swimming microrobots. Firstly, an on-board control module for on/off propulsion control needs to be developed. The current chemical stimulation method is off-board and the $CuSO_4$ or EDTA solutions are manually introduced into the motility medium. This is perhaps sufficient for preliminary motion control experiments of the hybrid microrobots. However, with the current off-board method, the diffusion time of the $CuSO_4$ and EDTA solutions to reach to the bacteria flagella could be very long for larger volumes of liquid, and independent on/off control of large number of microrobots is not possible. Therefore, to reduce the diffusion time of the chemicals and enable independent on/off actuation of large number of microrobots in the future, an on-board propulsion control module will be developed. Such a module could consist of two liquid reservoirs for $CuSO_4$ and EDTA solutions, microvalves for releasing these solutions in a volume-controlled manner, and an off-board switch for actuating the microvalves.

Steering of the microrobot in 3-D will be possible by fabricating the bacterial array at the back of the robot in four separate segments with independent on/off control [3]. By turning each segment selectively on or off, the microrobot could stochastically go forward or turn left, right, up, or down. This concept is very similar to steering a submarine.

In addition to on-board actuation, there are many other future challenges for developing a microscale swimming mobile robot: position detection of the robot and on-board (biochemical, pH, temperature, pressure, flow, etc.) sensors, wireless communication, computing, and control.

7 Conclusions

This paper proposes a new approach for microscale swimming robot actuation by integrating biological microorganisms such as bacteria to an inorganic robot body to propel it in stagnant/low velocity flow fields. To show the feasibility of this approach, *S. marcescens* bacteria are attached to microscale objects such as $10\,\mu m$ polystyrene micro-beads by blotting them in a bacteria swarm plate. Randomly attached bacteria are shown to propel the beads at approximately $15\,\mu m/sec$ average speed, stochastically. Using chemical stimuli, on/off motion control of mobile micro-beads is achieved.

Hundreds of thousands of these hybrid swimming microrobots could be used as inexpensive agents in distributed systems and swarm robotic system platforms when applicable. They could revolutionize health-care and environmental monitoring applications in the near future. These microrobots could also be utilized for artificial intelligence applications. First, since these microrobots have very limited computing power, sensing, and actuation and they behave stochastically, their computation based intelligence would be minor. Instead these microrobots would have swarm intelligence, self-organization, and emerging behavior by the coordination of large numbers of these microrobots using distributed and stochastic control methods in the future.

Acknowledgments. This work is funded by National Science Foundation (IIS-0713354). The authors thank Philip Leduc for allowing the use of Cellular Biomechanics Laboratory facilities and the members of NanoRobotics Laboratory for their helpful comments and suggestions.

References

1. Sitti, M.: Microscale and nanoscale robotics systems - Characteristics, state of the art, and grand challenges. IEEE Robotics and Automation Magazine 53–60 (2007)
2. Purcell, E.M.: Life at low Reynolds number. American Journal of Physics 3–11 (1977)
3. Edd, J., Payen, S., Rubinsky, B., Stoller, M.L., Sitti, M.: Biomimetic Propulsion for a Swimming Surgical Microrobot. In: Proc. of the IEEE Int. Conf. on Intelligent Robots and Systems pp. 2583–2588 (2003)
4. Kosa, G., Shoham, M., Zaaroor, M.: Propulsion Method for Swimming Microrobots. IEEE Tran. Robotics, 137–150 (2007)
5. Behkam, B., Sitti, M.: Design Methodology for Biomimetic Propulsion of Miniature Swimming Robots. ASME Journal of Dynamic Systems, Measurement, and Control, 36–43 (2006)
6. Donald, B.R., Levey, C.G., McGray, C.D., Paprotny, I., Russ, D.: An Untethered, Electrostatic, Globally Controllable MEMS Microrobot. Journal of Microelectromechanical Systems, 1–15 (2006)
7. Yesin, K.B., Vollmers, K., Nelson, B.J.: Modeling and Control of Untethered Biomicrorobots in a Fluidic Environment Using Electromagnetic Fields. The International Journal of Robotics Research, 527–536 (2006)
8. Dreyfus, R., Baudry, J., Roper, M.L., Fermigier, M., Stone, H.A., Bibette, J.: Microscopic artificial swimmers. Nature, 862–865 (2005)
9. Soong, R.K., Bachand, G.D., Neves, H.P., Olkhovets, A.G., Craighead, H.G., Montemagno, C.D.: Powering an Inorganic Nanodevice with a Biomolecular Motor. Science, 1555–1558 (2000)
10. Weibel, D.B., Garstecki, P., Ryan, D., DiLuzio, W.R., Mayer, M., Seto, J.E., Whitesides, G.M.: Microoxen: Microorganisms to move microscale loads. PNAS 11963–11967 (2005)
11. Darnton, N., Turner, L., Breuer, K., Berg, H.: Moving Fluid with Bacterial Carpets. Biophysical Journal, 1863–1870 (2004)
12. Berg, H.: The Rotary Motor of Bacterial Flagella. Annual Review of Biochemistry, 19–54 (2003)
13. Leifson, E.: Atlas of Bacterial Flagellation. Academic Press, New York and London (1960)
14. Behkam, B., Sitti, M.: Toward Hybrid Swimming Microrobots: Propulsion by an Array of Bacteria. In: Proc. of the IEEE Int. Conf. of Engineering in Medicine and Biology pp. 2421–2424 (2006)
15. Jones, J., Feick, J., Imoudu, D., Chukwumah, N., Vigeant, M., Velegol, D.: Oriented adhesion of E. coli to polystyrene particles. Applied & Environmental Microbiology, 6515–6519 (2003)

16. Behkam, B., Sitti, M.: Mechanism of Adhesion of Serratia marcescens ATCC 274 to hydrophobic and hydrophilic substrates (In preparation)
17. Adler, J.: A method for measuring chemotaxis and use of the method to determine optimum conditions for chemotaxis by Escherichia coli. Journal of General Microbiology, 77–91 (1973)
18. Behkam, B., Sitti, M.: Bacterial Flagella-Based Propulsion and On/Off Motion Control of Microscale Objects. Applied Physics Letters 23902–23904 (2007)

Adaptive Multi-modal Sensors

Kyle I. Harrington[1] and Hava T. Siegelmann[2]

[1] School of Cognitive Science, Hampshire College, Amherst, MA 01002 USA
kyle@kephale.com
[2] Department of Computer Science, University of Massachusetts, Amherst,
MA 01003 USA

Abstract. Compressing real-time input through bandwidth constrained connections has been studied within robotics, wireless sensor networks, and image processing. When there are bandwidth constraints on real-time input the amount of information to be transferred will always be greater than the amount that can be transferred per unit of time. We propose a system that utilizes a local diffusion process and a reinforcement learning-based memory system to establish a real-time prediction of an entire input space based upon partial observation. The proposed system is optimized for dealing with multi-dimension input spaces, and maintains the ability to react to rare events. Results show the relation of loss to quality and suggest that at higher resolutions gains in quality are possible.

1   Introduction

Sensor systems are often required to transfer spatially related data across bandwidth constrained connections. This data can come from many forms: visual, auditory, electrical, etc. [1,2,3,6,8,9,19,21]. We propose a system that compensates for these constraints by accessing a subset of the available input and performs a real-time spatio-temporal extrapolation for the values of the unknown input. The result of this extrapolation is an expectation of the input space. Once an expectation of the input space is established behaviors can be performed, such as reacting, planning, and learning [12,20]. The result of the system is a smaller input size causing a faster update rate, this increases the potential for reactivity, the relevancy of plans, and pertinence of knowledge. The system described in section 2 performs this recreation.

Embodied intelligent systems have sensorimotor loops. These loops allow such systems to observe the environments with which they may interact. Intelligent systems can then learn to exploit sensorimotor relationships within the environment, for example causal relationships. The work of Lungarella and Sporns provides a foundation for understanding how to learn and exploit such relationships [7]. The authors suggest an inherent link between a system's physical representation and information flow within the system. However, in order to exploit such relationships the system must have an internal representation of its physical state.

M. Lungarella et al. (Eds.): 50 Years of AI, Festschrift, LNAI 4850, pp. 164–173, 2007.

An internal representation of physical state and innate knowledge of the sensor and motor systems is not always given; however, they can be learned. Olsson *et al.* describe a system which learns a model of its sensorimotor system with no *a priori* knowledge [10]. The system initially performs exploratory actions to develop a map of the relationships between the sensor and motor systems. An entropy based metric is used for measuring the informational distance between sensors. The system which is described in this paper follows an alternative approach of strictly utilizing spatial and temporal relationships between sensors.

Related work in robotics, wireless sensor networks, and image processing approach the bandwidth constraint problem at different levels. Rixner *et al.* use a bandwidth hierarchy specific to media applications [13]. Webb developed a new set of communications primitives for parallel robotics image processing [19]. Hull, Jamieson, and Balakrishnan used a rule-based approach for real-time bandwidth allocation [4]. The proposed system uses temporal and spatial information to allocate bandwidth in real-time.

A similar approach was investigated by Schneider *et al.* in a power grid control application in which distributed value functions were used. Their system allows nodes to learn a value function which estimates future rewards at every node in the system [14]. In the context of the proposed system this means that each sensor stores an estimate of all activity in the system. The most prominent difference is that diffusion is used as the method for distributing this estimate along the sensor grid.

2   System

The system consists of a lattice of locations where sensors can reside. Each and every sensor is capable of accessing multiple types of input at its location. Examples of these types of input are red, green, and blue sensing modes Sensor activation ranges from $[0, 1]$ for each mode. 0 means that there is no activity in the given mode, and 1 means the mode is maximally active. Each type of input is referred to as a sensing mode. By adaptively selecting each sensor's active sensing modes based upon a set of sensor memory chemicals it is possible to reduce the number of sensors necessary to adequately sense the environment, as well as the reduce the amount of bandwidth used by sensors. These reductions are the result of an iterative chemical diffusion process. These components are further described in the following subsections.

2.1   Sensors

The sensors are homogeneous, in that each sensor is capable of sensing the same number of modes as the others. In order to replicate the observed environment every sensor must observe every mode in the environment, this is equivalent to performing a complete copy of the environment. However, in many cases the environment changes in a temporally related manner. This temporal relation is exploited by sensing fewer than the total available number of modes, and

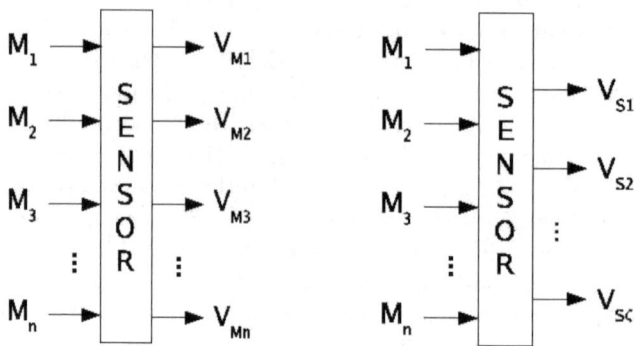

Fig. 1. First, a sensor accessing all n modes at once which requires n observations. Second, a sensor accessing ζ dynamically selected modes which requires ζ observations, where $\zeta < n$.

extrapolating the values of modes based upon learned sensor histories, which we refer to as chemical because they are distributed via diffusion.

2.2   Sensor Chemicals

Sensor activity is stored in chemicals just as value functions are learned by temporal-difference reinforcement learning[15]. For each sensing mode, there are two chemicals that act as sensory memory, C_S, short-term memory and C_L, long-term memory. The parameters γ_S and γ_L are the discount-rate, the rate at which a value fades from memory, and $\gamma_S < \gamma_L$. These chemicals are produced by the rate equations

$$C_S = C_S + \alpha((S_A + \gamma_S C_S) - C_S)$$
$$C_L = C_L + \alpha((S_A + \gamma_L C_L) - C_L)$$

where, S_A is the activation value of the sensor for mode A, and α is the learning rate. In this case *state* refers to the chemical configuration. An example of sensor chemicals within a sensor changing over time can be seen in figure 2. The sensor chemicals also represent short- and long-term activation values. This allows the short-term chemical to be used directly for reconstruction.

In order for the system it must be able to reconstruct an expectation of the environment based upon the current chemical configuration. Reconstruction is a simple local process performed at each sensor for every sensing mode. If the sensor has already observed activity in the given mode, then the activity is already known and that value is used. Otherwise, the value of the short-term chemical is used. This allows the system to automatically maintain expectations of activity in every observable mode.

The diffusion coefficients of the sensor chemicals are numbers in the range of $[0, 1]$. This allows the system to exhibit a continuum of behaviors. When the diffusion coefficient is set to 0 all local history is retained, to 1 all local history is

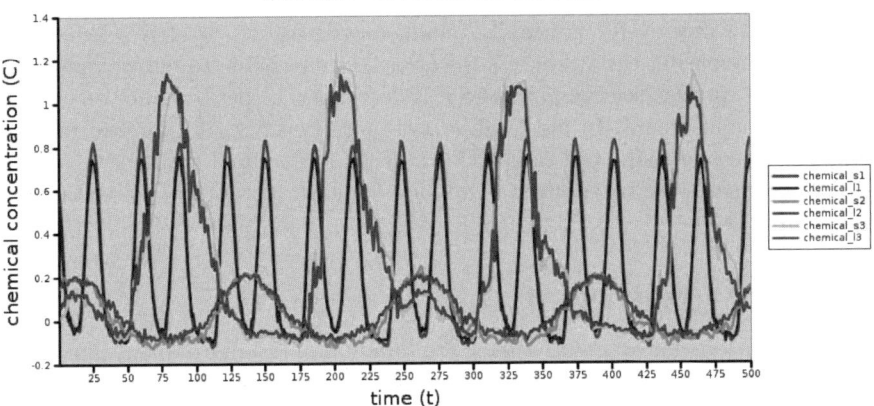

Fig. 2. Sensor chemicals over time for an example sensor. 2 of 3 possible modes were active in this case. The sensor switches its 2 active modes based upon the current concentration of short- and long-term chemicals.

distributed to neighboring sensors causing a global history. The diffusion coefficients allow control to be exercised on the amount of information that is shared amongst neighboring nodes.

2.3   Diffusion

The premise for using diffusion is inspired by Turing's explanation of pattern formation by reaction-diffusion [17]. Two properties appear by diffusion. First, older information can be retained. By diffusing off the edges of the lattice some information is lost. Second, localized diffusion propagates information globally over time, allowing sensors to anticipate unobserved activity. Diffusion occurs according to $C_r^i = C_r^i + D_i \sum_{r \sim r'} (C_{r'}^i - C_r^i)$, where C^i is the amount of sensor chemical i, and D_i is the diffusion coefficient for sensor chemical i. The relation $r \sim r'$ is held for the Von Neumann neighborhood [18] of the target node for diffusion along the lattice, and the relation $r \sim r'$ is from a sensor to the node it occupies for diffusion into the sensor.

2.4   Sensor Mode Selection

A variable $\zeta \in [0, m]$, where m is the total number of modes, controls the number of active modes in each sensor. ζ modes are selected incrementally using ϵ-greedy based on the greatest difference between C_S and C_L for each mode. The ϵ-greedy selection process has two results, the greedy result where the selected mode $max(C_S - C_L)$, or the random result where a random mode is selected. Greedy is always selected, unless $p < \epsilon$, where p is the probability of selecting a random mode [16]. As is mentioned with respect to sensing rare events, the value of ϵ controls the minimum frequency at which events can be detected.

2.5   Sensing Rare Events

The sensing of rare events is heavily reliant on the use of ϵ-greedy selection for modes. By increasing the value of ϵ towards 1 it is possible to detect more rare events by the random selection; however, it decreases the performance on sensing the overall environment. In most cases it is more beneficial to, instead of or in conjunction to increasing the value of ϵ, to increase the number of active modes. This is because ϵ-greedy selection is applied for each mode to be sensed during a given update.

2.6   Observed Environment

The specifications for an environment are minimal. The environment must have a function that returns the value for a region defined by a point and a radius, shown in figure 3. In our case the radius is a Von Neumann neighborhood radius and the value for the region is the sum of activity at each point within that region, Fig. 3 illustrates this. The environment can consist of multiple modalities to be sensed, in this case a function is necessary for each mode. It is also possible for the observed environment to be dynamic with respect to time. The dynamics of the environment should be spatially and temporally related.

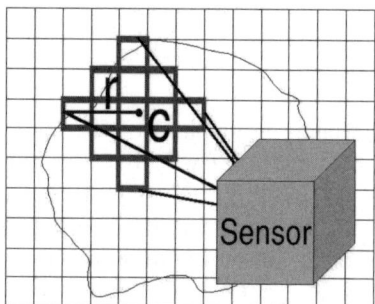

Fig. 3. Sensor with center, c, and radius, r=2. This sensor would return the sum of the values of each highlighted cell.

3   Experiments

3.1   Experiment I: Rare Events Are Detected

The system's ability to detect rare events was evaluated in an environment with a blue background and one small red circle, which moved across the environment infrequently from some same edge 4 times over 1000 iterations. The area of the red circle was 7% of the entire environment. A prevalent tactic for compression is spatial generalization; however, when generalizing based on the spatial relationships of sensory information it is frequent that small rare events may

Fig. 4. Snapshot of experiment I, a blue environment with red circle that moves into the environment periodically. First is the original image. Second are the observations colored for respective their modes, black sensors are observing modes for which there is no activity. Third is the chemical prediction of original image. The environment and the sensor lattice are 100x100, the red circle is of radius 15. $\zeta = 1$, meaning only one color can be sensed by any given sensor.

be filtered out. The use of the ϵ-greedy method allows for small rare events to be fairly easily detected if the value of epsilon is adjusted according to the size of potential rare events. The system strictly favors more recent sensor memory, regardless of its value. This experiment was designed to test whether the system was still capable of responding to new events even after saturating its memory with a single mode.

For the case of rare events the set of parameters that were of the most interest were a mode compression of $\zeta = 1$ given a blue background and a relatively small mobile red event. This is because this parameter setting allows the system to be saturated with activity from one mode, then a rare event from another mode is presented. The detection of the rare event can then be observed. Results for $\zeta = 1$ for 100x100 environments had an average error less than 2.5%. The quality of this detection is exemplified in Fig. 5, which illustrates a 100x100 observed environment with no resolution compression, and a mode compression of 1.0.

3.2   Experiment II: Simultaneous Activity Is Observed for All ζ

This experiment was designed to demonstrate the system's ability to handle simultaneous activity. For this experiments there were 3 modes which were represented by the colors, red, green, and blue. All instances were evaluated for 1000 iterations. The experiment utilizes three trajectories, each represented by a different colored circle. The trajectories are as follows: red, a half-circle from top-left to bottom-left, green, a circle rotating around the environment, and blue, a horizontal hourglass across the environment. The trajectories were selected such that each combination of overlapping circles occurred multiple times. This is to ensure that even with small values of ζ all modes are still sensed. This environment was evaluated in three cases, A, a 10x10 observable environment and 10x10 sensor lattice, B, a 100x100 observable environment and 10x10 sensor lattice, and C, a 100x100 observable environment and 100x100 sensor lattice.

Our results show the utility of this system with variety of dimensions of the sensor grid and environment with respect to simultaneous activity. We evolved a

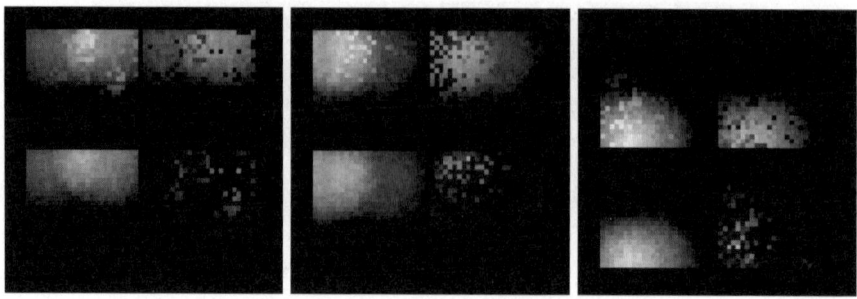

Fig. 5. Image of the system demonstrating the ability to represent 3 modes with 2 active modes. Three circles, one for each mode, follow unique trajectories with multiple intersections. Images of 3 of these intersections are shown, each of which has the following 4 displays Top left, sensors' prediction of the environment. Top right, activity recorded by sensors. Bottom left, original image. Bottom right, the absolute value of difference between the original and predicted image.

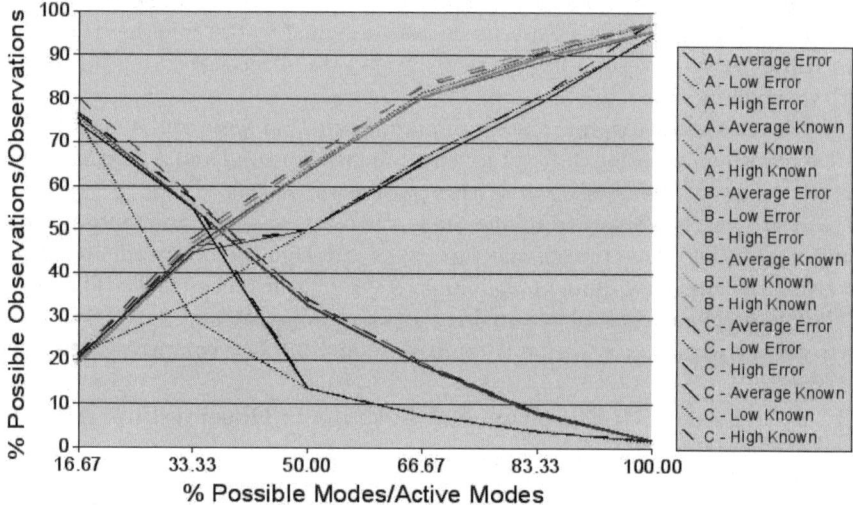

Fig. 6. Results for the three cases in experiment II, A, a 10x10 observable environment and 10x10 sensor lattice, B, a 100x100 observable environment and 10x10 sensor lattice, and C, a 100x100 observable environment and 100x100 sensor lattice. The relationship between the number of active modes and prediction accuracy is shown. For cases A and B the percent error is approximately equal to the amount of the environment that is not observed. For case C the percent error is less than the amount of the environment that is not observed.

population of 50 parameter settings for 1, 000 different parameter settings. Fig.6 illustrates the effect of mode compression, $\frac{c}{m}$ where m is the number of modes, on the quality of the final product. We define this quality in terms of prediction error, which is simply defined as the difference between the original image and the prediction summed over all modes for all sensors. Cases A and B show the system producing results of approximately equal quality to percentage of the environment that is known. Both of these cases consisted of low resolution grids, yet the system still maintained an output that was at least consistent with the amount of compression, if not actually providing enough inference to reduce the error beyond minimal expectation. In both cases the amount of error and the amount of compression sum up to approximately 100%. This means the amount of error is equivalent to the amount of the environment that the system did not observe. The similarity between the cases A and B suggest that the size of the sensor grid limits the quality of the compression. Case C illustrates the difference between compression and error with an image of higher resolution. The compression and error sum to significantly less than 100%.

4 Conclusions

It is important that the results with low resolutions show the compression to maintain a total quality that does not decrease beyond that of unprocessed input with an equivalent bandwidth constraint. This observed quality threshold is suggestive when considered with the results in Fig. 8. By increasing the resolution of the image it possible for the quality to increase above the quality of uncompressed data through an equivalently small bottleneck. This suggests that it is possible for high resolution images to maintain smaller storage and/or network transfer footprints. These benefits are similar to those Kansal et al. obtained by using motion control [5].

We have presented a system which allows real-time input to be scaled through a bandwidth constraint while maintaining a level of quality appropriate to the amount of compression used. The system does not require any overhead bandwidth, instead selects which values from the environment are transferred. Values are recorded as sensor chemicals which are diffused across the sensor grid. When used at higher resolutions, some values of ζ allowed for quality surpassing 100%. Our results suggest that this system is useful for compressing some types of real-time input through bandwidth constraints.

5 Future Work

Future work will investigate an implementation of the previously described system in a 3-dimensional environment. Additionally, discretization of sensor input will be used to further reduce bandwidth usage [11]. Embodied implementations of this system should investigate sensorimotor regularities induced by the addition of a motor system for further optimization [7].

Acknowledgements

We acknowledge funding from NSF grant ECS-0501432.

References

1. Babilonia, F., Mattiab, D., Babilonia, C., Astolfib, L., Salinarie, S., Basiliscoa, A., Rossinic, P.M., Marcianib, M.G., Cincotti, F.: Multimodal integration of EEG, MEG, and fMRI data for the solution of the neuroimage puzzle. Magnetic Resonance Imaging 22, 1471–1476 (2004)
2. Gottesfeld Brown, L.M.: Registration of Multimodal Medical Images - Exploiting Sensor Relationships. PhD thesis, Columbia University (1996)
3. DuFaux, F., Moscheni, F.: Motion estimation techniques for digital TV: A review and a new contribution. Proceedings of IEEE 83(6), 858–876 (1995)
4. Hull, B., Jamieson, K., Balakrishnan, H.: Bandwidth management in wireless sensor networks. Technical report, Massachusetts Institute of Technology (2003)
5. Kansal, A., Kaiser, W., Pottie, G., Srivastava, M., Sukhatme, G.S.: Virtual high-resolution for sensor networks. In: ACM SenSys, ACM Press, New York (2006)
6. Kulkarni, P., Ganesan, D., Shenoy, P., Lu, Q.: Senseye: a multi-tier camera sensor network. In: MULTIMEDIA 2005. Proceedings of the 13th annual ACM international conference on Multimedia, New York, NY, USA, pp. 229–238 (2005)
7. Lungarella, M., Sporns, O.: Mapping information flow in sensorimotor networks. PLoS Computational Biology 10, 1301–1312 (2006)
8. Mario, X.H.: Load balanced, energy-aware communications for mars sensor networks. In: IEEE Aerospace Conference Proceedings (2002)
9. Olsson, L., Nehaniv, C.L., Polani, D.: Sensory channel grouping and structure from uninterpreted sensor data. In: Proceedings of NASA/DoD Conference on Evolvable Hardware (2004)
10. Olsson, L., Nehaniv, C.L., Polani, D.: From unknown sensors and actuators to visually guided movement. In: Proceedings of the 4th International Conference on Development and Learning (2005)
11. Olsson, L., Nehaniv, C.L., Polani, D.: Sensor adaptation and development in robots by entropy maximization of sensory data. In: CIRA 2005. Proceedings of IEEE International Symposium on Computational Intelligence in Robotics and Automation (2005)
12. Prati, A., Vezzani, R., Benini, L., Farella, E., Zappi, P.: An integrated multi-modal sensor network for video surveillance. In: VSSN 2005. Proceedings of the third ACM international workshop on Video surveillance & sensor networks, New York, NY, USA, pp. 95–102 (2005)
13. Rixner, S., Dally, W.J., Kapasi, U.J., Khailany, B., Lopez-Lagunas, A., Mattson, P.R., Owens, J.D.: A bandwidth-efficient architecture for media processing. In: International Symposium on Microarchitecture, pp. 3–13 (1998)
14. Schneider, J., Wong, W.-K., Moore, A., Riedmiller, M.: Distributed value functions. In: Proceedings of the 16th International Conference on Machine Learning, pp. 371–378. Morgan Kaufmann, San Francisco, CA (1999)
15. Sutton, R.S.: Learning to predict by the methods of temporal differences. Machine Learning 3, 9 (1988)
16. Sutton, R.S., Barto, A.G.: Reinforcement Learning: An Introduction. MIT Press, Cambridge (1998)
17. Turing, A.M.: The chemical basis of morphogenesis. Philosophical Transactions of the Royal Society of London. Series B, Biological Sciences 237(641), 37–72 (1952)
18. von Neumann, J.: Theory of Self-Reproducing Automata. University of Illinois Press, Urbana, Illinois (1966)

19. Webb, J.A.: Latency and bandwidth considerations in parallel robotics image processing. In: Proceedings of Supercomputing, pp. 230–239 (1993)
20. Weise, T., Geihs, K.: Genetic programming techniques for sensor networks. In: Proceedings of 5. GI/ITG KuVS Fachgespräch "Drahtlose Sensornetze", pp. 21–25 (2006)
21. Zhao, Y., Garden Richardson, I.E.: Computational complexity management of motion estimation in video encoders. In: DCC, p. 483. IEEE Computer Society, Los Alamitos (2002)

What Can AI Get from Neuroscience?

Steve M. Potter

Laboratory for Neuroengineering
Department of Biomedical Engineering
Georgia Institute of Technology
313 Ferst Dr NW, Atlanta, GA, USA 30332-0535
steve.potter@bme.gatech.edu
http://neuro.gatech.edu

Abstract. The human brain is the best example of intelligence known, with unsurpassed ability for complex, real-time interaction with a dynamic world. AI researchers trying to imitate its remarkable functionality will benefit by learning more about neuroscience, and the differences between Natural and Artificial Intelligence. Steps that will allow AI researchers to pursue a more brain-inspired approach to AI are presented. A new approach that bridges AI and neuroscience is described, Embodied Cultured Networks. Hybrids of living neural tissue and robots, called hybrots, allow detailed investigation of neural network mechanisms that may inform future AI. The field of neuroscience will also benefit tremendously from advances in AI, to deal with their massive knowledge bases and help understand Natural Intelligence.

Keywords: Neurobiology, circular causality, embodied cultured networks, animats, multi-electrode arrays, neuromorphic, closed-loop processing, Ramon y Cajal, hybrot.

1 Introduction

An alien power plant was unearthed in a remote South American jungle. After excavating and dusting it off, the archeologists flip the switch, and it still works! It generates electricity continuously without needing fuel. Wouldn't we want to make more of these power plants? Wouldn't we want to know how this one works? What if the scientists and engineers who design power plants saw photos of the locals using electricity from the alien power plant, and knew it reliably powers their village. Yet they ignore this amazing artifact, and feel it has little relevance to their job. Although this scenario seems implausible, it is analogous to the field of AI today. We have, between our ears, a supremely versatile, efficient, capable, robust and intelligent machine that consumes less than 100 Watts of power. If AI were to become less artificial, more brain-like, it might come closer to accomplishing the feats routinely carried out by *Natural Intelligence* (NI). Imagine an AI that was as adept as humans at speech and text understanding, or reading someone's mood in an instant. Imagine an AI with human-level creativity and problem solving. Imagine a dexterous AI, which could precisely and adaptively manipulate or control physical artifacts such as violins, cars, and balls. Humans, thanks to our complex nervous system, are especially

M. Lungarella et al. (Eds.): 50 Years of AI, Festschrift, LNAI 4850, pp. 174–185, 2007.

good at interacting with the world in real time in non-ideal situations. Yet, little attention in the AI field has been directed toward actual brains. Although many of the brain's operating principles are still mysterious, thousands of neuroscientists are working hard to figure them out.[1]

Unfortunately, the way neuroscientists conduct their research is often very reductionistic [1], building understanding from the bottom up by small increments. A consequence of this fact is that trying to learn, or even keep up with, neuroscience is like trying to drink from a fire hose. General principles that could be applied to AI are hard to find within the overwhelming neuroscience literature.

AI researchers, young and old, might do well to become at least somewhat bilingual. Taking a neuroscience course or reading a neuroscience textbook would be a good start. Excellent textbooks include (among others) Neuroscience [2], Neuroscience: Exploring the Brain [3], and Principles of Neural Science [4]. There are several magazines and journals that allow the hesitant to gradually immerse themselves into neuroscience, one toe at a time. These specialize in conveying general principles or integrating different topics in neuroscience. In approximate order of increasing difficulty, some good ones are: Discover, Science News, Scientific American Mind, Cerebrum, Behavioral and Brain Sciences (BBS), Trends in Neuroscience, Nature Reviews-Neuroscience, and Annual Review of Neuroscience. BBS deserves special mention, because of its unusual format: a 'target article' is written by some luminary, usually about a fairly psychological or philosophical aspect of brains. This is followed by in-depth commentaries and criticisms solicited from a dozen or more other respected thinkers about thinking. These responses provide every side of a complex issue, and often include many of the biological foundations of the cognitive functions being discussed. The responses are followed by a counter-response from the author of the target article. BBS is probably the best scholarly journal that regularly includes and combines contributions from both neuroscientists and AI researchers.[2]

In this networked era, the internet can be a cornucopia, or sometimes, a Pandora's Box for AI researchers who want to learn about real brains. Be wary of web pages expounding brain factoids, unless there is some form of peer review that helps maintain the quality and integrity of the information. Wikipedia is rapidly becoming an extremely helpful tool for getting an introduction to any arcane topic, and has an especially elaborate portal to Neuroscience.[3] Caution: it is not always easy to find the source or reliability of information given there. A more authoritative source on the fields of computational neuroscience and intelligence is Scholarpedia.[4] The Society

---

[1] I will define neuroscience as all scientific subfields that aim to study the nervous system (brain, spinal cord, and nerves), including neurophysiology, neuropathology, neuropharma-cology, neuroendocrinology, neurology, systems neuroscience, neural computation, neuro-anatomy, neural development, and the study of nervous system functions, such as learning, memory, perception, motor control, attention, and many others. Neurobiology is thought of today as the basis of all neuroscience (ignoring some lingering dualism) and the terms are often used interchangably.

[2] BBS Online: http://journals.cambridge.org/action/displayJournal?jid=BBS

[3] http://en.wikipedia.org/wiki/WP:NEURO

[4] http://www.scholarpedia.org

for Neuroscience (SFN) website[5] is an excellent and reliable source of introductory articles about many neuroscience topics. The SFN consists of over 30,000 (mostly American) neuroscientists who meet annually and present their latest research to each other. All of the thousands of abstracts for meetings back to the year 2000 are searchable on the Annual Meeting pull-down. Although not itself a repository of introductory neuroscience material, the Federation of European Neuroscience Societies website[6] is a good jumping-off point for all things Euro-Neuro.

2   What Do We Already Know About NI (Natural Intelligence) That Can Inform AI?

2.1   Brains Are Not Digital Computers

John von Neumann, the father of the architecture of modern digital computers, made a number of thought-provoking and influential analogies in his book, "The Computer and the Brain." [5] The brain-as-digital-computer metaphor has proven quite popular, and often gets carried too far. For example, a neuron's action potential[7] is often referred to by the AI field as a biological implementation of a binary coding scheme. This and other misinterpretations of brain biology need to be purged from our thoughts about how intelligence may be implemented. Even with our rudimentary conception of how it is implemented in brains, there are clear differences between computers and brains, such as:

2.2   Brains Don't Have a CPU

The brain's processor is neither "central" nor a "unit". Its processing capabilities seem to be distributed across the entire volume of the brain. Some localized regions specialize in certain types of processing, but not without substantial interaction with other brain areas [6].

2.3   Memory Mechanisms in the Brain Are Not Physically Separable from Processing Mechanisms

Recent research has shown that similar brain regions are activated in recalling a memory as would be during perceiving [7]. This may be because an important part of perceiving is comparing sensory inputs to remembered concepts. Memories are dynamic, and continually re-shaped by the recall process [8]. A computer architecture that unites the processor, RAM, and hard disk into one and the same substrate might be far more efficient. An architecture that implements memory as a dynamic process rather than a static thing may be more capable of interacting in real time with a dynamic world.

---

[5] http://www.sfn.org
[6] http://fens.mdc-berlin.de
[7] Action potentials are regenerative electrical impulses that neurons evolved to send information across long axons. They involve a fluctuation of the neuron membrane potential of ~0.1 V across a few milliseconds.

2.4   The Brain Is Asynchronous and Continuous

The computer is a rare type of artifact that has well-defined (discontinuous) states [9], thanks to the fact that its computational units are always driven to their binary extremes each tick of the system clock. There are many brain circuits that exhibit oscillations [10], but none keeps the whole brain in lock-step the way a system clock does for a digital computer. The phase of some neural events in relation to a circuit's ongoing oscillation is used to code for specific information [11], and phase is a continuous quantity.

2.5   With NI, the Details of the Substrate Matter

Digital computers have been very carefully designed so that the details of their implementation don't influence their computations. Vacuum tubes, discrete transistors, and VLSI transistors, since they all speak Boolean, can all run the same program and produce the same result. There is a clear, intentional separation between the hardware and the software. All neuroscience research so far suggests this separation does not exist in the brain.

How do the details of its substrate influence the brain's computations? Every molecule that makes up the brain is in continuous motion, as with all liquids. The lipid bilayer that comprises the neuron's membrane is often referred to as a 2-dimensional liquid and is part of the neural wetware. The detailed structure of the proteins that make up brain cells can only be determined when they are crystallized in a test tube, that is, purified and stacked into unnatural, static, repeating structures that form good x-ray diffraction patterns. In their functional form, proteins (and all brain molecules) are jostling around, continuously bombarded by the cytoplasm or cerebrospinal fluid that surrounds them, like children frolicking in a pen full of plastic balls. Small details about neurons' structure, such as the morphing of tiny (micron-sized) synaptic components called dendritic spines [12], or the opening and closing of voltage-sensitive or neurotransmitter-sensitive ion channels, affect their function at every moment. All that movement of molecules and parts of cells is the substrate of NI, facilitating or impeding communication between pairs of brain cells and across functional brain circuits. Why should AI researchers concern themselves with the detailed, molecular aspects of brain function? Because, fully duplicating brain functionality may only be possible using a substrate as complex and continuous as living brain cells and their components are.

That disappointing possibility should not keep us from trying at least to duplicate *some* brain functionality by taking cues from NI. Carver Mead, Rodney Douglas, and other neuromorphic engineers have designed useful analog circuits out of CMOS components that take advantage of more of the physics of doped silicon than just its ability to switch from conducting to non-conducting states [13]. The continuous "inter-spike interval" between action potentials in neurons is believed to encode neural information [14] and also seems to be responsible for some of the brain's learning abilities [15]. Neuromorphic circuits that use this continuous-time pulse-coding scheme [13, 47] may be able to process sensory information faster and more efficiently than could digital circuits.

2.6   NI Thrives on Feedback and Circular Causality

The nervous system is full of feedback at all levels, including the body and the environment in which it lives; it benefits in a quantifiable way from being embodied and situated [16, 17]. Unlike many AI systems, NI is highly interactive with the world. Human-engineered systems are more tractable when they employ assembly-line processing of information, i.e., to take in sense data, then process it, then execute commands or produce a solution. Most sensory input to living systems is a dynamic function of recent or ongoing movement commands, such as directing gaze, walking, or reaching to grasp something. With NI, this active perception and feedback is the norm [17, 18]. Animal behaviors abound with circular causality, new sensory input continuously modulating the behavior, and behavior determining what is sensed [19]. One beautiful example of active perception that humans are especially good at is asking questions. If we don't have enough information to complete a task, and a more knowledgeable person is available, we ask them questions. New AI that incorporates question-asking and active perception can solve problems quickly that would take too long to solve by brute force serial computation [16, 20].

There are few brain circuits that involve unidirectional flow of information from the sensors to the muscles. The vast majority of brain circuits make use of what Gerald Edelman calls reentry [21]. This term refers to complex feedback on many levels, which neuroscientists have only begun to map, let alone understand. Neuroscience research suggests that a better understanding of feedback systems with circular causality would help us design much more flexible, capable, and faster AI systems [9].

2.7   NI Uses LOTS of Sensors

One of the most stunning differences between animals and artificial intelligences is the huge number of sensors animals have. NI mixes different sensory modalities to enable rapid and robust real-time control. Our brains are very good at making the best use of whatever sense data are available. Without much training, blind people can deftly navigate unfamiliar places by paying attention to the echoes of sounds they make, even while mountain-biking off road![8][22] Bach y Rita's vibrotactile display placed a video camera's image onto a blind person's skin, in the form of a few hundred vibrating pixels. By actively aiming the camera, the user could "see" tactile images via their somatosensory system, allowing them to recognize faces and to avoid obstacles [23, 24]. The continuous flow of information into the brain from the sense organs is enormous. To make AI less artificial, we could strive to incorporate as much sensing power as we dare imagine. When AI adopts a design philosophy that embraces, rather than tries to minimize high-bandwidth input, it will be capable of increasingly more rapid and robust real-time control.

2.8   NI Uses LOTS of Cellular Diversity

There are more different types and morphologies of cells in the brain than in any other organ, perhaps than all organs and tissues combined. Many of these were catalogued by neuroscientist, Santiago Ramon y Cajal a century ago (Fig. 1) [25, 26], but more

---

[8] http://www.worldaccessfortheblind.org/

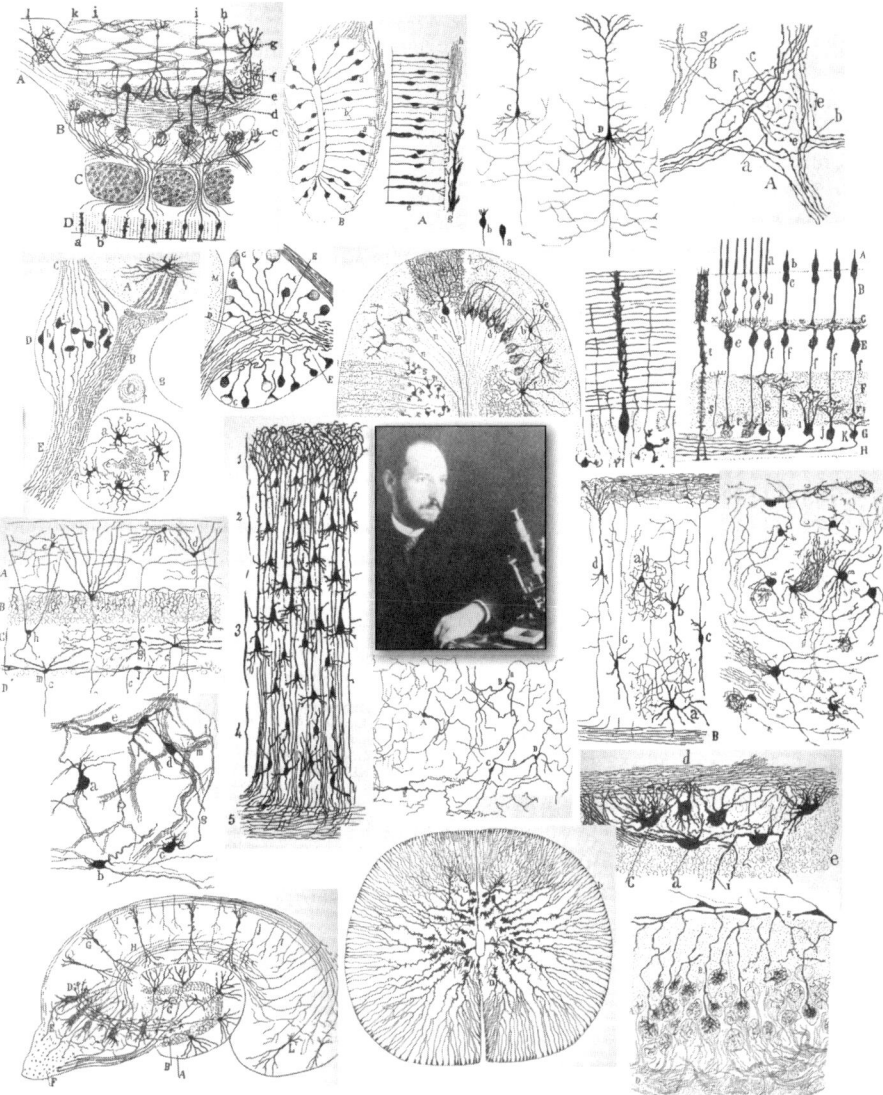

Fig. 1. Neurons and circuits traced a century ago by Spanish neuroanatomist, Santiago Ramon y Cajal (pictured in center). This montage depicts only a few of the many types of neuron morphologies found throughout the nervous system. (Adapted with permission from Swanson & Swanson ©1990 MIT Press).

are still being discovered [27]. Another sign of the brain's complexity is the large amount of genetic information that allows it to develop and function. Both mice and men have ~30,000 genes in their genome, and over 80% of these are active in the brain.[9] All this cellular complexity and diversity may be crucial in creating an intelligent processor that is general-purpose and highly adaptable.

2.9 NI Uses LOTS of Parallelism

The brain's degree of parallelism is not rivaled by any human-made artifact. There are about 100 billion neurons in our brains, each connected to 1,000-10,000 others with over 200,000 km of axons, which we have barely begun to map [28]. The brain's circuits seem to have small-world connectivity [29], i.e., many local connections and relatively few, but crucial, long range connections. The latter integrate the activities of cooperating circuits running simultaneously and asynchronously. Although our knowledge of this elaborate connectivity is rudimentary, some general principles, such as small-world connectivity, could make future AI more capable.

2.10 Delays Are Part of the Computation

It is sometimes mistakenly stated that neurons are slow computational elements, since they fire action potentials at a few hundred Hz at most. The parallelism mentioned above is one way to enable rapid computation with "slow" elements. Modern computers, which are not very parallel if at all, reduce computation and transmission delays in every way possible, from shorter leads, to faster clocks, to backside caches. Any time spent getting information from here to there in a digital computer is viewed as a wasteful impediment to getting on with the computation of the moment. In the brain, delays are not a problem, but an important part of the computation. The subtle timing of action potentials carries information about the dynamics and statistics of the outside world [30]. The relative timing of arrival of two action potentials to the postsynaptic neuron determines whether the strength of their synapse is incrementally increased or decreased [31]. These pulse timings are analog quantities. The brain computes with timing, not Boolean logic [32]. Brain-inspired AI of the future will be massively parallel, have many sensors, and will make good use of delays and the dynamics of interactions between analog signals [33].

3   What Do We *Not* Know About How Brains Work, But Could Learn?

To realize this dream of AI that is closer to NI, there are a number of important questions about how brains work that must be pursued, such as: What is a memory? How do biological networks work? The neurons and glial cells both store and process information in a spatially distributed manner. But we have only a very vague and fuzzy idea of just how they do that. The Blue Brain Project is setting a giant supercomputer (the son of Deep Blue) to the task of simulating just one cortical minicolumn of a few thousand neurons [34]. There is a lot going on at the level of networks that we don't even have the vocabulary to think about yet. Neurobiologists all believe that memories are stored by changes in the physical structure of brain cells such as increases in the number of branches or spines on a neuron's dendritic tree. We

---

[9] See the Allen Institute Brain Atlas, http://www.brainatlas.org/aba/index.shtml

don't all agree about what those changes might be, let alone how the changes are executed when salient sensory input is received. As hinted by Ramon y Cajal's drawings (Fig. 1), neurons have a stunning diversity of morphologies [35]. There is evidence that some aspects of their shape are altered by experience [36-38]. But how that relates to a memory being stored is not known.

4   New Neuroscience Tools

A new type of experimental animal, called the Hybrot, is taking shape in the Laboratory for Neuroengineering at the Georgia Institute of Technology. This is a hybrid robot, an artificial embodiment controlled by a network of living neurons and

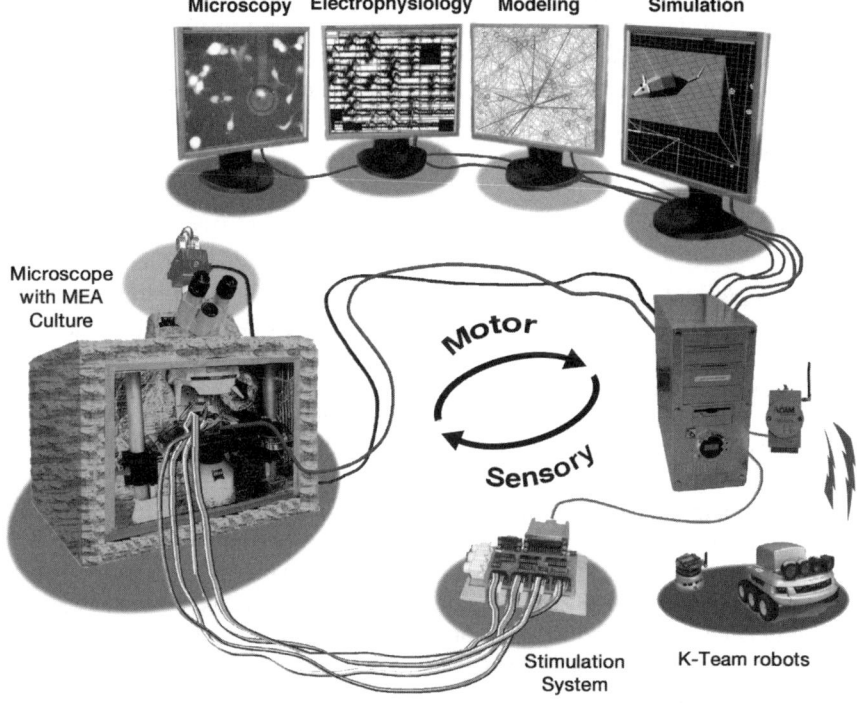

Fig. 2. Hybrots: hybrid neural-robotics for neuroscience research. A living neuronal network is cultured on a multi-electrode array (MEA) where its activity is recorded, processed in real time, and used to control a robotic or simulated embodiment, such as the K-Team Khepera or Koala (pictured at lower right). The robot's input from proximity sensors is converted to electrical stimuli that are fed back to the neuronal network within milliseconds via a custom multi-electrode stimulation system. The hybrot's brain (MEA culture) can be imaged continuously on the microscope while its body behaves and learns. The microscope is enclosed in an incubator (lower left) to maintain the health of the living network. This closed-loop *Embodied Cultured Networks* approach may shed light on the morphological correlates of memory formation, and provide AI researchers with ideas about how to build brain-style AI.

glia cultured on a multi-electrode array (MEA) [39-41]. It will be helpful in studying some of these difficult neuroscience questions. We now have the hardware and software necessary to create a real-time loop whereby neural activity is used to control a robot, and its sensory inputs are fed back to the cultured network as patterns of electrical or chemical stimuli (Fig. 2; [42]).

These embodied cultured networks bring in vitro neuroscience models out of sensory deprivation and into the real world. They form a much needed bridge between neuroscience and AI. An MEA culture is amenable to high-resolution optical imaging [43], while the hybrot is behaving and learning, from milliseconds to months [44]. This allows correlations to be made between neural function and structure, in a living, awake and behaving subject. One of our hybrots, called MEART, was used to create portraits of viewers in a gallery. Its sensory feedback, images of its drawings in progress, affected the next action of the robotic drawing arm, in a closed-loop fashion [45]. This has been used to explore the neural mechanisms of creativity. Whether a network of a few thousand neurons can be creative is still up for debate, but it is vastly more complex than any existing artificial neural network. By studying embodied cultured networks with these new tools, we may learn some new aspects of network dynamics, memory storage, and sensory processing that could be used to make AI less artificial [41].

5   Neuroscience to AI and Back Again

Biologically-inspired artificial neural networks [46], mixed analog/digital circuits [47] and computational neuroscience approaches that attempt to elucidate brain networks [48] (as opposed to cellular properties) are gradually becoming more tightly coupled to experimental neuroscience. The fields of Psychology and Cognitive Science have traditionally made progress using theoretical foundations having little or no basis in neuroscience, due to a lingering Cartesian dualism in the thinking of their practitioners [49]. However, with neuroscience advances in psychopharmacology (e.g., more targeted neuroactive drugs) and functional brain imaging (e.g., functional MRI), Psychology and Cognitive Science advances are becoming increasingly based on and inspired by biological mechanisms. It's time for AI to move in the brainwards direction. This could involve PhD programs that merge AI and neuroscience, journals that seek to unite the two fields, and more conferences (such as the one that spawned this book) at which AI researchers and neuroscientists engage in productive dialogs.

Neuroscientists have not exactly embraced AI either. Both sides need to venture across the divide and learn what the other has to offer. How can neuroscience benefit from AI? As we have seen, brains are far too complicated for us to understand at present. AI can produce new tools for compiling the mass of neuroscience results published, and coming up with connections or general theoretical principles. On a more mundane but important level, we need AI just to help us deal with the massive data sets that modern neuroscience tools produce, such as multi-electrode arrays and real-time imaging. The new field of neuroinformatics has to date been mostly concerned with neural database design and management. Soon, with help from AI, it will incorporate more data mining, knowledge discovery, graphic visualization, segmentation and pattern recognition, and other advances yet to be invented. One can imagine that by increasing the synergy between AI and neuroscience, a bootstrapping process will occur: more neuroscience research will inform better AI, and better AI will give neuroscientists the tools to make more discoveries and interpret them. Where it will lead, who knows, but it will be an exciting ride!

Acknowledgments. I thank Max Lungarella for insightful comments. This work was funded in part by the US National Institutes of Health (National Inst. of Mental Health, National Inst. of Neurological Disorders and Stroke, National Inst. on Drug Abuse), the US National Science Foundation Center for Behavioral Neuroscience, The Whitaker Foundation, The Keck Foundation, and the Coulter Foundation.

References

1. Lazebnik, Y.: Can a biologist fix a radio?—Or, what I learned while studying apoptosis. Cancer Cell 2, 179–182 (2002)
2. Purves, D., Augustine, G.J., Fitzpatrick, D., Hall, W.C., Lamantia, A.S., McNamara, J.O., et al.: Neuroscience, 3rd edn. Sinauer Associates, New York (2004)
3. Bear, M.F., Connors, B., Paradiso, M.: Neuroscience: Exploring the Brain. Lippincott Williams & Wilkins, New York (2006)
4. Kandel, E.R., Schwartz, J.H., Jessel, T.M.: Principles of Neural Science, 4th edn. McGraw-Hill Publishing Co., New York (2000)
5. von Neumann, J.: The Computer and the Brain. Yale University Press, New Haven (1958)
6. Tononi, G., Sporns, O., Edelman, G.M.: A Measure for Brain Complexity: Relating Functional Segregation and Integration in the Nervous System. PNAS 91(11), 5033–5037 (1994)
7. Kosslyn, S.M., Pascual-Leone, A., Felician, O., Camposano, S., Keenan, J.P., Thompson, W.L., et al.: The role of area 17 in visual imagery: convergent evidence from PET and rTMS. Science 284(5411), 167–170 (1999)
8. Alberini, C.M., Milekic, M.H., Tronel, S.: Mechanisms of memory stabilization and de-stabilization. Cell Mol. Life Sci. 63(9), 999–1008 (2006)
9. Bell, A.J.: Levels and loops: the future of artificial intelligence and neuroscience. Philosophical Transactions of the Royal Society of London Series B-Biological Sciences 354(1392), 2013–2020 (1999)
10. Buzsaki, G.: Rhythms of the Brain. Oxford U. Press, Oxford (2006)
11. Wehr, M., Laurent, G.: Odour encoding by temporal sequences of firing in oscillating neural assemblies. Nature 384(6605), 162–166 (1996)
12. Yuste, R., Majewska, A.: On the function of dendritic spines. Neuroscientist 7(5), 387–395 (2001)
13. Liu, S.-C., Kramer, J., Indiveri, G., Delbrück, T., Douglas, R.: Analog VLSI: Circuits and Principles. MIT Press, Cambridge (2002)
14. Reich, D.S., Mechler, F., Purpura, K.P., Victor, L.D.: Interspike intervals, receptive fields, and information encoding in primary visual cortex. Journal of Neuroscience 20(5), 1964–1974 (2000)
15. Froemke, R.C., Dan, Y.: Spike-timing-dependent synaptic modification induced by natural spike trains. Nature 416(6879), 433–438 (2002)
16. Lungarella, M., Pegors, T., Bulwinkle, D., Sporns, O.: Methods for Quantifying the Informational Structure of Sensory and Motor Data. Neuroinformatics 3, 243–262 (2005)

17. Lungarella, M., Sporns, O.: Mapping Information Flow in Sensorimotor Networks. PLOS Computational Biology 2, 1301–1312 (2006)
18. Mehta, S.B., Whitmer, D., Figueroa, R., Williams, B.A., Kleinfeld, D.: Active Spatial Perception in the Vibrissa Scanning Sensorimotor System. PLoS Biol. 5(2), e15 (2007)
19. Chiel, H.J., Beer, R.D.: The brain has a body: adaptive behavior emerges from interactions of nervous system, body and environment. Trends in Neurosciences 20(12), 553–557 (1997)
20. Lipson, H.: Curious and Creative Machines. In: The 50th Anniversary Summit of Artificial Intelligence, Ascona, Switzerland (2006)
21. Sporns, O., Tononi, G., Edelman, G.M.: Connectivity and complexity: the relationship between neuroanatomy and brain dynamics. Neural Networks 13(8-9), 909–922 (2000)
22. Stoffregen, T.A., Pittenger, J.B.: Human echolocation as a basic form of perception and action. Ecological Psychology 7, 181–216 (1995)
23. Bach-y-Rita, P.: Brain Mechanisms in Sensory Substitution. Academic Press, New York (1972)
24. Dennett, D.C.: Consciousness Explained. Little, Brown & Co., Boston (1991)
25. Ramon y Cajal, S.: Histologie du Systeme Nervex de l'Homme et des Vertebres. Maloine, Paris (1911)
26. Ramon y Cajal, S.: Les nouvelles idées sur la structure du système nerveux chez l'homme et chez les vertébrés (L. Azoulay, Trans.). C. Reinwald & Cie, Paris (1894)
27. Kalinichenko, S.G., Okhotin, V.E.: Unipolar brush cells–a new type of excitatory interneuron in the cerebellar cortex and cochlear nuclei of the brainstem. Neurosci. Behav. Physiol. 35(1), 21–36 (2005)
28. Sporns, O., Tononi, G., Kötter, R.: The Human Connectome: A Structural Description of the Human Brain. PLOS Computational Biology 1, 245–251 (2005)
29. Watts, D.J., Strogatz, S.H.: Collective dynamics of 'small-world' networks. Nature 393(6684), 440–442 (1998)
30. Gerstner, W., Kreiter, A.K., Markram, H., Herz, A.V.M.: Neural codes: Firing rates and beyond. Proc. Natl. Acad. Sci. USA 94, 12740–12741 (1997)
31. Bi, G.Q., Poo, M.M.: Synaptic modifications in cultured hippocampal neurons: Dependence on spike timing, synaptic strength, and postsynaptic cell type. Journal of Neuroscience 18, 10464–10472 (1998)
32. Izhikevich, E.M.: Polychronization: Computation with spikes. Neural Computation 18(2), 245–282 (2006)
33. Maass, W., Natschlager, T., Markram, H.: Real-time computing without stable states: a new framework for neural computation based on perturbations. Neural Comput. 14(11), 2531–2560 (2002)
34. Markram, H.: The blue brain project. Nat. Rev. Neurosci. 7(2), 153–160 (2006)
35. Mel, B.W.: Information-processing in dendritic trees. Neural Computation 6, 1031–1085 (1994)
36. Weiler, I.J., Hawrylak, N., Greenough, W.T.: Morphogenesis in Memory Formation - Synaptic and Cellular Mechanisms. Behavioural Brain Research 66(1-2), 1–6 (1995)
37. Majewska, A., Sur, M.: Motility of dendritic spines in visual cortex in vivo: Changes during the critical period and effects of visual deprivation. Proc. Natl. Acad. Sci. USA 100(26), 16024–16029 (2003)
38. Leuner, B., Falduto, J., Shors, T.J.: Associative Memory Formation Increases the Observation of Dendritic Spines in the Hippocampus. J. Neurosci. 23(2), 659–665 (2003)

39. DeMarse, T.B., Wagenaar, D.A., Potter, S.M.: The neurally-controlled artificial animal: A neural-computer interface between cultured neural networks and a robotic body. Society for Neuroscience Abstracts 28, 347.1 (2002)
40. DeMarse, T.B., Wagenaar, D.A., Blau, A.W., Potter, S.M.: The neurally controlled animat: Biological brains acting with simulated bodies. Autonomous Robots 11, 305–310 (2001)
41. Bakkum, D.J., Shkolnik, A.C., Ben-Ary, G., Gamblen, P., DeMarse, T.B., Potter, S.M.: Removing some 'A' from AI: Embodied Cultured Networks. In: Iida, F., Pfeifer, R., Steels, L., Kuniyoshi, Y. (eds.) Embodied Artificial Intelligence. LNCS (LNAI), vol. 3139, pp. 130–145. Springer, Heidelberg (2004)
42. Potter, S.M., Wagenaar, D.A., DeMarse, T.B.: Closing the Loop: Stimulation Feedback Systems for Embodied MEA Cultures. In: Taketani, M., Baudry, M. (eds.) Advances in Network Electrophysiology using Multi-Electrode Arrays, pp. 215–242. Springer, New York (2006)
43. Potter, S.M.: Two-photon microscopy for 4D imaging of living neurons. In: Yuste, R., Konnerth, A. (eds.) Imaging in Neuroscience and Development: A Laboratory Manual, pp. 59–70. Cold Spring Harbor Laboratory Press (2005)
44. Potter, S.M., DeMarse, T.B.: A new approach to neural cell culture for long-term studies. J. Neurosci. Methods 110, 17–24 (2001)
45. Bakkum, D.J., Chao, Z.C., Gamblen, P., Ben-Ary, G., Potter, S.M.: Embodying Cultured Networks with a Robotic Drawing Arm. In: The 29th IEEE EMBS Annual International Conference (2007)
46. Granger, R.: Engines of the brain: the computational instruction set of human cognition. AI Magazine 27, 15 (2006)
47. Linares-Barranco, A., Jiminez-Moreno, G., Linares-Barranco, B., Civit-Balcells, A.: On Algorithmic Rate-Coded AER Generation. IEEE Transactions on Neural Networks 17, 771–788 (2006)
48. Seth, A.K., McKinstry, J.L., Edelman, G.M., Krichmar, J.L.: Visual binding through reentrant connectivity and dynamic synchronization in a brain-based device. Cerebral Cortex 14, 1185–1199 (2004)
49. Damasio, A.R.: Descartes' Error: Emotion, Reason, and the Human Brain. Gosset/Putnam Press, New York (1994)

Dynamical Systems in the Sensorimotor Loop: On the Interrelation Between Internal and External Mechanisms of Evolved Robot Behavior

Martin Hülse[1], Steffen Wischmann[2], Poramate Manoonpong[2], Arndt von Twickel[3], and Frank Pasemann[4]

[1] University of Wales, Aberystwyth, UK
msh@aber.ac.uk
[2] BCCN, University of Göttingen, Germany
[3] University of Osnabrück, Germany
[4] Fraunhofer IAIS, Sankt Augustin, Germany

Abstract. This case study demonstrates how the synthesis and the analysis of minimal recurrent neural robot control provide insights into the exploration of embodiment. By using structural evolution, minimal recurrent neural networks of general type were evolved for behavior control. The small size of the neural structures facilitates thorough investigations of behavior relevant neural dynamics and how they relate to interactions of robots within the sensorimotor loop. We argue that a clarification of dynamical neural control mechanisms in a reasonable depth allows quantitative statements about the effects of the sensorimotor loop and suggests general qualitative implications about the embodiment of autonomous robots and biological systems as well.

1 Introduction

The framework of embodied artificial intelligence has impressively demonstrated that problems in behavior control of autonomous robots, which seem to be rather complicated if approached from a mere computational perspective, but turn out to be surprisingly simple when characteristics of the sensorimotor loop are taken into account appropriately[1]. The challenge is that we usually do not know a priori what "appropriate" means, because the sensorimotor loop involves all the physical properties of the robot (inertia, friction, resonances, shape, etc.) as well as its interaction with the world. Therefore, Evolutionary Robotics (ER) is proposed as a promising testbed for studying the power of embodiment [2,3]. Artificial evolution provides the exploration of hitherto unknown and efficient solutions by reducing prejudices and predispositions made by a human designer [4].

As an example, Nolfi [5] describes the emergence of modularity in evolved neural control, which does not correspond to task decomposition as a human observer would assume. Based on such networks, Ziemke [6] emphasizes the relevance of recurrent neural networks (RNNs) in the context of multi-functional and context-sensitive behavior control. In contrast, Suzuki et al. [7] demonstrate

M. Lungarella et al. (Eds.): 50 Years of AI, Festschrift, LNAI 4850, pp. 186–195, 2007.

how a simple feed-forward structure in conjunction with robot-environment interactions realizes robust and adaptive behavior control through complex visual sensorimotor mappings.

Within the realm of feed-forward and recurrent neural control quantitative statements about the properties of the sensorimotor loop are needed [8]. It should be clarified where recurrent neural control structures are necessary and where simple feed-forward mappings are sufficient if the body, the dynamics of the environment, and the action-perception processes of a robot are taken into account. The difficulty in deriving qualitative statements from the effects of the sensorimotor loop is twofold. On the one hand, it is impossible to find a formal description of the sensorimotor loop including all relevant aspects. On the other hand, if RNNs are used for complex behavior control, usually only the parameters, but not the structure, of a predefined neural network are optimized by evolution. In the majority of the cases, the resulting control structures are high dimensional systems. But high dimensionality makes it practically infeasible to clarify whether complex behavior is basically generated by the control structure or results from robot-environment interactions.

While the first point let us conclude that qualitative statements about the impact of the sensorimotor loop on robot control can be made only indirectly, that is, based on a reasonable understanding of the evolved neurodynamics. The second aspect, namely high dimensionality, seems to counter it. The objective of this paper is to introduce a strategy in ER supporting this approach termed as *synthesis and analysis of minimal recurrent neural robot control*. Further on, we will give representative examples where an application of this strategy has provided us with enlightening examples demonstrating the importance of the sensorimotor loop on behavior control for autonomous mobile robots. The experiments will show how robot-environment interactions give rise to integrated and induced oscillations, the use of transient effects, and the emergence of rhythms and behavior coordination. All these examples show how complex behavior relevant dynamics provided by RNNs are modulated by the sensorimotor loop.

2   Synthesis and Analysis of Minimal Recurrent Neural Controllers

We are using a standard additive neuron model with sigmoidal transfer function $f(x)$ and discrete time dynamics: $a_i(t + 1) = \Theta_i + \sum_{j=0}^{n} w_{ij} \cdot f(a_j(t))$, $i = 1, \ldots, n$, where a_i is the activation of neuron i, w_{ij} the weight of the synapse projecting from neuron j, and Θ_i the bias term. Already small recurrent networks of this type can generate complex dynamics [9]. That's why we apply an evolutionary algorithm called ENS^3 (*evolution of neural systems by stochastic synthesis*) to evolve neural connectivity structures (hidden neurons and synapses) *and* optimize their corresponding parameters (weight and bias terms) at the same time. By modifying certain stochastic variation operators, such as the insertion and deletion probability for structural elements, during evolution we are able to

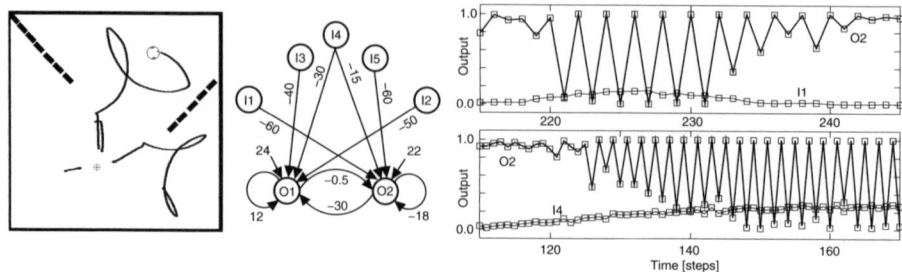

Fig. 1. A reactive light seeking controller ($f(x) = \sigma(x) = 1/(1 + e^{-x})$) utilizing switch-able period-2 oscillations for speed control (see text for details)

enforce the development of minimal neural structures (with respect to the num-ber of hidden neurons and synapses) [10].

To understand the origins of behaviorally relevant dynamics, it is important to clarify the contribution of *minimal* recurrent neural networks. In some occasions, it is possible to directly derive behavior relevant dynamical properties from the structure and parameters of the RNN. But mostly, it is almost impossible to also include the dynamics of robot-environment interactions in order to explain the observed behavior in detail.

In the following we do not provide further details with respect to the para-meter settings of the evolutionary processes. Our focus lies exclusively on the dynamical properties of specific control structures, chosen by us as the best examples, demonstrating the essential mechanisms of frequently observed phe-nomena.

Behavior control by frequency modulation. It is well known from analytical inves-tigations that over-critical negative self-connections of single neurons can gen-erate switchable period-2 oscillations [11]. Analyses of the RNN in Fig. 1 have shown that the behavior control is actually provided by such switchable oscilla-tors. This controller solves a light seeking task.

The diagrams in Fig. 1 (right) show two examples where a period-2 oscillation is modulating the behavior of a Khepera robot. The upper diagram shows the on-off switch of period-2 oscillations of output neuron $O2$. The switching is determined by the left proximity sensor (given by $I1$), that is, a switch-on causes a turn to the right. A period-2 oscillation is also used in front of a light source to generate a stop (Fig. 1, right bottom). In this situation the oscillation is determined by the increased activation of the frontal light sensor ($I4$). In contrast to the turning, both output neurons are synchronously oscillating with period-2 (not shown).

This controller demonstrates, that oscillating output signals can be used for behavior control since the body of the robot operates as an integrator. According to the inertia of the robot's body, effective motor actions result from the mean network output signals. If the robot is standing in front of a light source, both outputs are permanently oscillating between 1 and 0. Hence, the mean over time

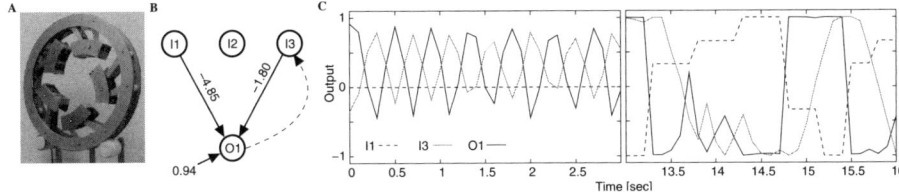

Fig. 2. A: The robot *micro.eve*. B: RNN of one arm $(f(x) = tanh(x))$. C: Neuron output (see text for details).

is 0.5. Due to the applied post-processing an effective motor signal of 0.5 represents a motor speed of zero. Such effects are not superficial results of artificial evolution. Morris and Hooper [12] demonstrated that in biological systems slow muscle contractions are coded by the average amplitudes of rhythmic neural activities.

Induced oscillations. Fig. 2 shows an example where motor oscillations are induced through the environmental loop. The ring-shaped robot *micro.eve* (Fig. 2A) is placed on two passive rollers on which it can rotate around its body center by moving the five independent arms in order to translate the overall center of mass in a coordinated way. For further details about the robot and different control strategies see [13].

Here, the presented RNN (Fig. 2B) is one out of five completely autonomous networks which independently control one of the five arms. Because of information provided by the hall sensor $I1$ and the gyroscope $I2$ (both part of the ring), every RNN gets information about the movement of the common body. Therefore, single controllers can "sense" the resulting effects of the other controllers' behavior. $I3$ gives information about the current motor position of the controlled arm. The output neuron signal $O1$ represents the motor command for the servo motor. As one can see in Fig. 2C the signal is oscillating since the hall sensory input remains zero at the beginning where the robot has to initiate its own rotational movement. These oscillations can not be deduced from the dynamical properties of the network because there are no recurrent connections which can provoke oscillations. Instead, they are caused by the loop through the environment (dashed line in Fig. 2B), which can be described as a reflex oscillator.

The output of $O1$ is sent to the servo motor, and due to the motor's inertia and friction the desired position is reached with a certain delay. The current position of the motor is fed back to the network through $I3$ which has a strong negative connection to $O1$. Therefore, $O1$ produces signals inverse to the current motor position, and this causes the observed oscillations. These oscillations are of utmost importance for initiating a rotation of the ring at all [13]. As soon as the ring starts to rotate, the hall sensor becomes active and due to the much stronger connection from this sensory input ($I1$) the aforementioned oscillations are suppressed depending on the signal strength of $I1$ (see Fig. 2C).

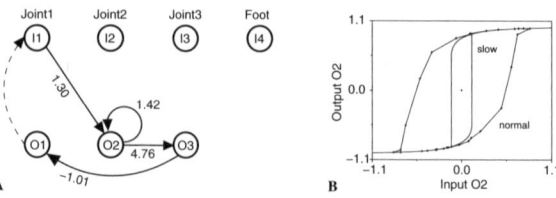

Fig. 3. A: Neural single-leg controller ($f(x) = tanh(x)$). B: Hysteresis of neuron $O2$ (see text for details).

Summarizing these two examples, we can say that in both cases fast oscillations provide important behavior relevant dynamics. However, they differ in their origin. In the first case the oscillations result from the neurodynamics of the RNN. In the second experiment these oscillations emerge from the ongoing robot-environment interactions.

Neural hysteresis in reflex-walking-control. In this section the role of a hysteresis element in a simple neural reflex-oscillator for single-leg (3DOF) control of walking machines is demonstrated. The controller shown in Fig. 3A is one of the simplest and yet one of the most effective controllers found during evolution experiments for the task of forward walking [14]. How does this structure, using only one sensory neuron (neuron $I1$, encoding angular position of joint 1), three motor neurons (neurons $O1, 2, 3$, specifying the desired angles to the servo motors), and four synapses, produce a coordinated walking pattern of a 3DOF leg?

All neurons used for control are connected in a loop ($I1 - O2 - O3 - O1 - I1$) which passes through the environment from neuron $O1$ to neuron $I1$ (dashed line). This sensorimotor loop results in a nonlinear transformation which can be approximated as a negative feedback with a time delay, resulting in a slow oscillatory movement (compare to the aforementioned description of a reflex-oscillator for the *micro.eve* robot).

During the oscillatory movement it was found that the motor-neurons approximately act as bistable elements. The bistability can be explained by the property of neuron $O2$. Neuron $O2$ plays a major role in the controller network. It is the first neuron in a chain which directly couples all motor neurons. The following motor neurons therefore have either the same phase or a phase shifted by 180 degrees (neuron $O3$ in phase, neuron $O1$ in antiphase) when compared to neuron $O2$. Neuron $O2$ has a self-connection larger than 1.0 which makes it a hysteresis element [11]. In Fig. 3B the output of neuron $O2$ is plotted against its input under actual walking conditions (outer curve). The plot shows two effects of the hysteresis element: First, the bistability may be explained by two stable fixed points of the hysteresis domain ($\approx \{-1, 1\}$). Second, the hysteresis element may be approximated as a time delay which adds to that of the environmental loop, therefore contributing to the slow and smooth oscillating walking movement. Finally, it may be noted that the transient is modulated by the frequency

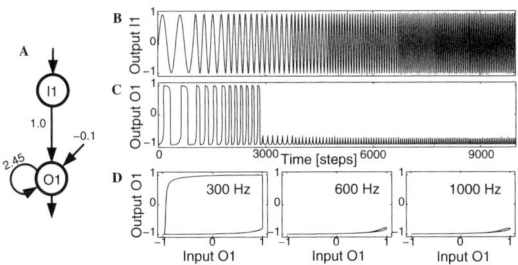

Fig. 4. A: A RNN ($f(x) = tanh(x)$) realizing low-pass filtering at $\approx$ 300 Hz. B,C: Input signal at increasing frequency (from 100 Hz to 1 kHz, 48 kHz sampling rate) and the corresponding output signal. D: The hysteresis effects between input and output signals at certain frequencies.

of the input signal. Under extremely slow (theoretical) walking conditions (inner curve of neuron $O2$ input/output plot) the transient approaches the hysteresis of the system, and therefore becomes much narrower than during actual walking conditions.

Neural processing of auditory signals. Inspired by evolved robot control we deduced a neural structure that realizes a simple hysteresis element (called dynamical neural Schmitt trigger, Fig. 4A). The structure has three parameters that define the width and the shift of the hysteresis domain [15]. For applications it is usually assumed that input signals vary only slowly with respect to the network update frequency. But how do the dynamical properties of the neural Schmitt trigger change when the input values change on arbitrary time scales?

Fig. 4 shows an example where a time series of a continuously increasing frequency is fed into a RNN. At a certain frequency the output remains in the lower saturation domain of the output neuron. Hence, one may argue that hysteresis elements behave as a low-pass filter [16]. We have successfully adapted such a structure to filter background noise of a walking machine and even to recognize low-frequency sounds (i.e., 200 Hz) to perform a sound tropism [16]. These applications demonstrate how a sensory driven dynamical system becomes sensitive to the frequency of the input signal.

The effect of filtering high-frequency signals itself can be explained by the shift of the hysteresis domain and the transients of the system. The self-connection determines how fast (i.e., needed number of time steps) the neuron activation ends near the fixed point. For the isolated system we have stable fixed points in the lower and upper saturation domain (i.e $\approx \{-1, 1\}$). When the input signal is continuously changing, the fixed points vary only slightly and if the amplitude is large enough, one observes the characteristic jumps at the end of the hysteresis domain. However, when a high frequency input signal is applied, because of the slowness of transient dynamics these fixed points are never approached and at a certain frequency orbits may stay near one or the other fixed point. Due to the slowness of the transient dynamics these fixed points are never reached and

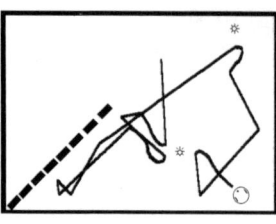 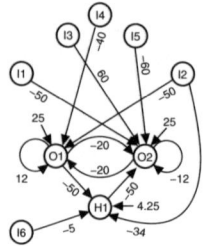 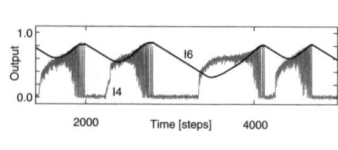

Fig. 5. The recurrent neural network producing a motivational driven robot behavior and the resulting behavior in simulation. The diagram shows the level of energy ($I6$, black) and activation of the frontal light sensor ($I4$, grey) during the interaction ($f(x) = \sigma(x) = 1/(1 + e^{-x})$).

at a certain frequency the orbits may stay near one or the other fixed point if a high frequency input signal is applied. Our presented system has a cut-frequency of $\approx 300Hz$ (compare Fig. 4C). Here, the upper saturation domain will never be reached as a consequence of these slow transients and the bias term. Thus, high-frequency oscillations are suppressed, and therefore, the system acts as a low-pass filter.

Reflex-walking-control and the low-pass filter are both based on bistable elements. The specific control signals, however, are determined by the frequency of the input signals modulating the transients of these hysteresis elements. Both examples, therefore, indicate how one and the same element can act in different ways due to its modulation by the sensorimotor loop.

Rhythmic behavior switching. For the study of behavior switches provided by complex neural dynamics we evolved an RNN to develop a motivational driven robot behavior. We call a robot behavior motivational driven if the neural control is not only determined by current sensor states of external stimuli, but also by an internal level of energy.

As a first simple example for such a motivational driven behavior we used again the Khepera robot and extended a reactive light seeking module (by structure evolution) to a control structure which maintains a certain level of energy while the robot accomplishes an exploration behavior. A resulting network is shown in Fig. 5. As one can see, the already introduced input-output-structure of the reactive light seeking module (see Fig. 1) is extended by one input neuron $I6$. This neuron indicates the current level of the simulated energy reservoir, which is defined as follows: $I6(t + 1) := I6(t) + c_1 \cdot I4(t) - c_2$, $\quad c_1, c_2 > 0$. The constant loss of energy can only be compensated by standing in front of a light source (i.e., by high activations of the frontal light sensor $I4$).

The resulting robot behavior in simulation is also shown in Fig. 5. One can see that the robot switches between exploring the environment and standing close to a light source. The diagram in Fig. 5 indicates that the behavior switches are determined by the level of energy. At a certain intensity of $I6$ (≈ 0.8) the robot is leaving the light source. Further on, the output $I6$ is characterized by slow

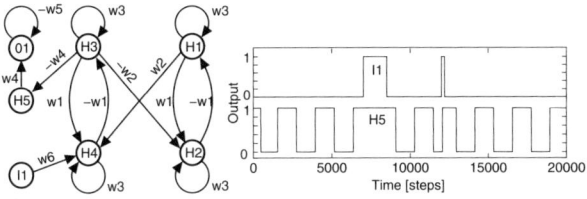

Fig. 6. An internal rhythm generator ($f(x) = \sigma(x) = 1/(1 + e^{-x})$), and how it can be influenced by sensory stimuli

oscillations. But, these slow oscillations are determined by the properties of the energy reservoir (i.e., $c_{1,2}$ in the equation above) *and* by the robot-environment interaction. Note, that this issue leads us to a cyclic causality: on the one hand, $I6$ is determining the behavior switches, and, on the other hand, $I6$ is determined by the robot-environment interaction. The slow oscillations emerge from the sensorimotor loop.

Synchronized rhythms. Fig. 6 (left) shows an implementation of a neural rhythm generator. It is based on a two neuron loop, called SO(2)-network [17]. These networks with a special weight matrix generate quasi-periodic oscillations with a sine-shaped wave form. The period of these oscillations depends only on one parameter in the weight matrix. The coupling of two identical SO(2)-networks can realize stable oscillations with very large periods as demonstrated in [18]. There, a concrete implementation of such a rhythm generator is used to coordinate competing behaviors in groups of up to 150 robots. Each robot is equipped with its own internal rhythm, that is, each robot has a slightly different frequency, which is reminiscent of circadian rhythms found in animals [19]. This rhythm determines whether the robot searches and collects food in the environment or returns to a home area where the collected energy is transfered to the common nest of the group. To maximize the energy level of the nest, it turned out that a coordination of the single behaviors is of great advantage, because the interferences resulting from the interactions of up to 150 robots in a shared environment lead to tremendous mutual obstructions [18].

To achieve a coordinated foraging and homing behavior within the whole group, individual rhythms have to become synchronized. In doing so, a robot needs the ability to communicate its internal state to other robots. One output neuron ($O1$ in Fig. 6, left) triggers a sound signal when it reaches a certain threshold. This neuron is coupled to the pattern generator in a way that sound signaling occurs during the switch from zero to one of the output of neuron $H5$ which amplifies the sine-shaped oscillations of the rhythm generator (Fig. 6, right). This signal can be perceived by nearby robots through the sensory input neuron $I1$. In turn, this perception provokes a phase reset as it can be seen for $H5$ in Fig. 6 (right). This mechanism allows behavior synchronization within a large robot group through minimal local communication. The resulting synchronized collective behavior is a result of local robot-robot and robot-environment interactions.

The impact of slow varying inner rhythms for behavior control has in fact already been demonstrated for robotic applications (see [18]). However, the last two experiments provide minimal examples for the emergence as well as the synchronization of slowly oscillating rhythms within the sensorimotor loop, and for both cases the essential elements of the interplay between internal neural dynamics and external world can be clearly identified.

3   Conclusions

In this paper we have presented six examples where the evolution of minimal recurrent neural networks for embodied agents explores the dynamics of robot-environment interactions. We have seen how oscillations can be integrated by the body of a robot or even induced by the sensorimotor loop through the environment. Furthermore, in neural structures with equivalent dynamical properties transient effects resulting from robot-environment interactions are used for completely different tasks, such as the locomotion in walking machines and the filtering of auditory signals. Finally, through interactions with the environment, internal rhythms determining differing behavior patterns can emerge in individuals or even become synchronized within large robot groups. Only by thoroughly analyzing evolved RNNs in the context of robot-environment interactions it was possible to reveal the interrelation between internal and external mechanisms underlying the evolved robot behavior.

The dynamical systems approach to adaptive behavior is still at its beginning in the context of ER experiments (e.g., [20,2]). And only very few studies involve thorough analyses of the evolved neural mechanisms (e.g., [20]) which can help to better understand the dynamical mechanisms underlying complex behavior and to clarify which behavioral aspects can be accounted to internal dynamics or to properties emerging from the sensorimotor loop.

However, our approach does not only advance our understanding of these issues. It also enables us *to construct* highly efficient neural control systems by considering the sensorimotor loop to minimize the complexity required at the neurodynamics level. Our examples demonstrate that the evolution of minimal recurrent neural robot control enforces the development of simple networks (concerning their size, not their dynamics). This makes it possible to extract and set up basic neural structures together with their functions in a respective sensorimotor loop. Provided with such building blocks one then should be able to develop gradually more and more elaborated behavior control for autonomous robots with a richer sensomotoric equipment.

References

1. Pfeifer, R., Scheier, C.: Understanding Intelligence. MIT Press, Cambridge (2000)
2. Harvey, I., Di Paolo, E., Wood, R., Quinn, M., Tuci, E.: Evolutionary robotics: a new scientific tool for studying cognition. Artificial Life 11, 79–98 (2005)

3. Nolfi, S., Floreano, D.: Evolutionary Robotics: The Biology, Intelligence, and Technology of Self-Organizing Machines. MIT Press, Cambridge (2000)
4. Clark, A.: Being there: putting brain, body and world together again. MIT Press, Cambridge (1997)
5. Nolfi, S.: Using emergent modularity to develop control systems for mobile robots. Adaptive Behavior 5, 343–363 (1997)
6. Ziemke, T.: On 'parts' and 'wholes' of adaptive behavior: Functional modularity and dichronic structure in recurrent neural robot controllers. In: Meyer, J.A., et al. (eds.) Proc. of the 6th Int. Conf. on Simulation of Adaptive Behavior, pp. 115–124 (2000)
7. Suzuki, M., Floreano, D., Di Paolo, E.: The contribution of active body movement to visiual development in evolutionary robots. Neural Networks 18, 656–665 (2005)
8. Pfeifer, R., Gomez, G.: Interacting with the real world: design principles for intelligent systems. Artificial Life and Robotics 9, 1–6 (2005)
9. Pasemann, F.: Complex dynamics and the structure of small neural networks. Network: Computation in Neural Systems 13, 195–216 (2002)
10. Hülse, M., Wischmann, S., Pasemann, F.: Structure and function of evolved neurocontrollers for autonomous robots. Connection Science 16, 249–266 (2004)
11. Pasemann, F.: Dynamics of a single model neuron. International Journal of Bifurcation and Chaos 3, 271–278 (1993)
12. Morris, L., Hooper, S.: Muscle response to changing neuronal input in the lobster (panulirus interruptus) stomatogastric system: Slow muscle properties can transfrom rhythmic input into tonic output. The Journal of Neuroscience 18, 3433–3442 (1998)
13. Wischmann, S., Hülse, M., Pasemann, F.: (Co)Evolution of (de)centralized neural control for a gravitationally driven machine. In: Capcarrère, M.S., Freitas, A.A., Bentley, P.J., Johnson, C.G., Timmis, J. (eds.) ECAL 2005. LNCS (LNAI), vol. 3630, pp. 179–188. Springer, Heidelberg (2005)
14. von Twickel, A., Pasemann, F.: Reflex-Oscillations in Evolved Single Leg Neurocontrollers for Walking Machines. Natural Computing 6(3), 311–337 (2007)
15. Hülse, M., Pasemann, F.: Dynamical neural schmitt trigger for robot control. In: Dorronsoro, J.R. (ed.) ICANN 2002. LNCS, vol. 2415, pp. 783–788. Springer, Heidelberg (2002)
16. Manoonpong, P., Pasemann, F., Fischer, J., Roth, H.: Neural processing of auditory signals and modular neural control for sound tropism of walking machines. Int. Journal of Advanced Robotic Systems 2, 223–234 (2005)
17. Pasemann, F., Hild, M., Zahedi, K.: SO(2)-networks as neural oscillators. In: Mira, J.M., Álvarez, J.R. (eds.) IWANN 2003. LNCS, vol. 2686, pp. 144–151. Springer, Heidelberg (2003)
18. Wischmann, S., Hülse, M., Knabe, J., Pasemann, F.: Synchronization of internal neural rhythms in multi-robotic systems. Adaptive Behavior 14, 117–127 (2006)
19. Winfree, A.: The Geomerty of Biological Time. Springer, Heidelberg (1980)
20. Beer, R.: The dynamics of active categorical perception in an evolved model agent. Adaptive Behavior 11, 209–244 (2003)

Adaptive Behavior Control with Self-regulating Neurons

Keyan Zahedi[1] and Frank Pasemann[2]

[1] MPI for Mathematics in the Sciences, Inselstrasse 22, 04103 Leipzig, Germany
[2] Fraunhofer Institute IAIS, Schloss Birlinghoven, 53754 Sankt Augustin, Germany

Abstract. It is claimed that synaptic plasticity of neural controllers for autonomous robots can enhance the behavioral properties of these systems. Based on homeostatic properties of so called self-regulating neurons, the presented mechanism will vary the synaptic strength during the robot interaction with the environment, due to driving sensor inputs and motor outputs. This is exemplarily shown for an obstacle avoidance behavior in simulation.

1 Introduction

Despite the many impressing results AI has achieved over the last 50 years there are several shortcomings right from its foundation. This is mainly due to the fact that cognition was understood as a symbol processing mechanism; and that the digital computer and its operations became the leading metaphor for cognitive processes. Taking human capabilities of language and mathematical calculation as a model for intelligence, and building to a large extent on Shannon's theory of information and on formal logic, artificial intelligent systems were thought of as detached from the physical world, and were developed largely without reference to the abilities of intelligent biological organisms.

One of the often mentioned shortcomings of classical AI is lack of robustness of the developed systems, meaning here that this approach had difficulties to design systems which can react in a self-sustaining way to changing properties of an unstructured environment or to failures of its subsystems or sub-processes. Thus, desired systems should be adaptive and able to learn. We will take up this question for adaptivity and learning, following a dynamical systems approach [3] to embodied cognition [13], using recurrent neural networks [12] for the behavior control of autonomous robots.

Considering robots as acting in a sensori-motor loop, a learning process will have to generate solutions to problems which are posed by the environments in which these robots have to operate. In general the property of "being a solution" to these problems can not be predicted or explained at the level where the search is carried out. If one is using the dynamics of a recurrent neural network for behavior control of these mobile systems, its functionality is bounded by the richness of its attractor landscape, determined by a fixed set of network parameters. To attain new behaviors or functionalities a learning process must

M. Lungarella et al. (Eds.): 50 Years of AI, Festschrift, LNAI 4850, pp. 196–205, 2007.

change parameters like synaptic weights and bias terms of the networks, or even vary their underlying connectivity structure. And, in the given context, such a rearrangement of neural control has to be the result of the systems interaction with the physical environment.

Learning in this sense means the selectivity of reactions, and a process of selectivity may be generated, for instance, by strengthening excitatory neural pathways as well as by weakening inhibition in certain pathways. Using this *plasticity* of the neural control will allow to scan through the reservoir of potential attractor landscapes corresponding to a given connectivity structure, such that specific attractors vanish or reappear, or new types of attractors become a functionality which was not used before; i.e. *emergent behavior* is observed. It is this intricate balance of stability and instability which is an essential property of neural networks seen as complex adaptive systems.

As a first step towards such a learning mechanism we will concentrate on the synaptic plasticity of a neural controller without changing its connectivity structure. Defining a reasonable weight dynamics in the given context, where changing environments and unexpected situations have to be accounted for, is a difficult problem because there is still a lack of mathematical and theoretical insight into the dynamical properties of high-dimensional nonlinear systems; and because it is quite hard to define appropriate error functions, teacher signals or rewarding instances for a more or less unspecified behavior which enables survival in an unstructured environment.

Avoiding teachers and rewarding instances as well as fitness functions, we focus on the self-organizing properties of neural networks. Then, one of the possibilities is, that learning, i.e weight dynamics, is the result of local interactions of neurons; which means that their activity is driven by sensor inputs and motor outputs of the neural controller. We therefore assume that every neuron can control its inputs and outputs in such a way that it keeps itself in a desired activity state under external perturbations. This is of course the central property of a homeostat as it was conceptualized for instance by Cannon [4] and Ashby [1]. A neuron in this setup is able to regulate its so-called *transmitter strength* and *receptor strength* appropriately. Such a neuron, defined by its 3-dimensional homeostatic dynamics, will be called *self-regulating*. The choice of a desirable state is for the moment in some sense arbitrary, but as long as one is interested in the abundance of nonlinear effects there is a canonical choice to make.

The activity of a single neuron is communicated to other neurons of the network through the connections, with synaptic weights given by the product of pre-synaptic transmitter strength and post-synaptic receptor strength. Because recurrences may be composed of excitatory and inhibitory connections, one can assume that the asymptotic behavior of such a network is not only given by stationary states, but oscillations and chaotic dynamics can be available as well.

These networks, acting as controllers in a sensori-motor loop, will then be driven by sensor signals. But any activity injected into the neural system will be communicated to all its neurons and will therefore modulate the weights of the network. Thus, the whole system will stay plastic during the interaction of the robot with its environment.

The next section will introduce the 3-dimensional homeostatic dynamics of single neurons with standard strictly positive sigmoidal transfer function. Section 3 will demonstrate the resulting synaptic plasticity of controllers for an obstacle avoidance behavior using a simulated Khepera robot. A concluding section will discuss the results and give an outlook.

2   Synaptic Plasticity

In what follows a single neuron i is described by a parametrized 3-dimensional bounded dynamical system $f : \mathcal{R}^3 \to \mathcal{R}^3$ with state variables $(a_i, \xi_i, \eta_i) \in \mathcal{R}^3$, where a_i denotes its activity, and ξ_i and η_i its receptor strength and transmitter strength, respectively. The parameter θ_i is considered as a stationary (slow) external input. Given a network composed from these type of neurons let c denote the structure matrix of the network defined by $c_{ij} = \pm 1$ if there is an excitatory/inhibitory connection from neuron j to neuron i, otherwise $c_{ij} = 0$. The output of the neuron is given by a sigmoidal transfer function, and in the following the standard sigmoid $\sigma = (1 + e^{-x})^{-1}$ is chosen.

For this neuron a homeostatic property should be achieved; i.e. the control of receptor and transmitter strengths should result in convergence to a desirable activity state for the neuron. There are several "good" choices for such a desirable state. The most interesting dynamical effects arise for an activity a^* for which the non-linearity of the sigmoid σ is "maximal", i.e. $\sigma'''(a^*) = 0$. Because σ''' is a symmetric function there are two such values, $a^* = \pm 1.317$, and for these specific values we have

$$\sigma(a^*) = \frac{1}{2} \pm \sqrt{\frac{1}{12}}, \quad \sigma'(a^*) = \frac{1}{6}. \tag{1}$$

The homeostatic dynamics to be defined should be able to stabilize the desired states a^* at least for a certain range of input signals. The basic dynamical equations are then set up as follows:

$$a_i(t+1) = \theta_i + \xi_i(t) \sum_{j=1}^{n} c_{ij} \, \eta_j(t) \, \sigma(a_j(t)) \quad i = 1, \ldots, n. \tag{2}$$

This defines the standard activity dynamics of a neuron. The receptor strength ξ_i is always positive and should increase if the neurons activity $a_i(t)$ satisfies $|a_i(t)| < |a^*|$, otherwise it should decrease; this is realized by the following equation

$$\xi_i(t+1) = \xi_i(t) \left(1 + \beta \cdot g(a_i(t))\right), \quad 0 < \beta < 1, \tag{3}$$

with the function g given by

$$g(x) := \sigma'(x) - \sigma'(a^*). \tag{4}$$

Finally, the positive transmitter strength $\eta_i > 0$ is designed to impart the internal activation of the neuron to the neural network. The higher the activation of the neuron, the more transmitter are released. It has a constant decay rate $(1 - \gamma)$, and it increases with the activity of the neuron. Thus we have

$$\eta_i(t+1) = (1 - \gamma)\,\eta_i(t) + \delta \cdot \sigma(a_i(t))\,, \quad 0 < \gamma < \delta < 1 \tag{5}$$

The connection strength or weight w_{ij} from neuron j to neuron i is defined by

$$w_{ij}(t) := c_{ij}\,\xi_i(t)\,\eta_j(t)\,.$$

The weight change per time step is then given by $\Delta w_{ij}(t) := w_{ij}(t+1) - w_{ij}(t)$ or in terms of transmitter and receptor strengths

$$
\begin{aligned}
c_{ij}\,\Delta w_{ij}(t) &= c_{ij}(\xi_i(t+1)\,\eta_j(t+1) - \xi_i(t)\,\eta_j(t)) \\
&= \xi_i(t)\,\eta_j(t)[\,\beta\,g(a_i(t))(1 - \gamma) - \gamma] \\
&\quad + \xi_i(t)\,\delta\,[1 + \beta\,g(a_i(t))]\,\sigma(a_j(t))
\end{aligned}
\tag{6}
$$

with g denoting the function given by equation (4). The weight dynamics defined by equation (6) will be referred to as *SRN-plasticity* for short.

If all neurons have their preferred activity a^* then there should be no changes in their transmitter and receptor strengths, i.e. in the weights. This means there exists at least one stationary state $(a^*, \xi^*, \eta^*) \in \mathcal{R}^{3n}$ for the $3n$-dimensional network dynamics (which must not be stable!). For simplicity we will study this situation for a single neuron.

2.1   Behavior of a Single Neuron

The fixed points (a^*, η^*, ξ^*) of the SRN-dynamics for a single self-regulating neuron with self-weight w are then given by the non-trivial fixed point equations

$$a^* = \theta + c\,\xi^*\,\eta^*\,\sigma(a^*)\,, \quad c = \pm 1\,, \tag{7}$$

$$\eta^* = \frac{\delta}{\gamma}\,\sigma(a^*)\,. \tag{8}$$

From equations (7) and (8) it follows for the asymptotic receptor strength

$$\xi^* = \frac{\gamma\,c\,(a^* - \theta)}{\delta\,\sigma^2(a^*)}\,. \tag{9}$$

Hence, ξ^* depends explicitly on the bias term θ. Because it is assumed that ξ is always positive, from the last equation one derives the consistency condition

$$c\,(a^* - \theta) > 0\,, \tag{10}$$

and immediately realizes, that this condition can not be satisfied for all $\theta \in \mathcal{R}$; in general the consistency condition is satisfied for $\theta \in (-a^*, a^*)$ if $c = 1$, and $\theta \notin [-a^*, a^*]$ if $c = -1$. What can be deduced from these conditions is that the dynamics will have parameter domains where non-convergent behavior, e.g. oscillatory behavior, has to be expected. Furthermore, from equation (9) one immediately gets for the asymptotic self-weight

$$w^* = c\,\xi^*\,\eta^* = \frac{a^* - \theta}{\sigma(a^*)} \,, \tag{11}$$

which exists if the consistency condition (10) is satisfied.

2.2 Simulations

To get a first impression of the effects of the SRN-plasticity the behavior of a single neuron with self-connection is studied. To relate the results to experiments done for neurocontrollers, this neuron is connected via an inhibitory connection to an input neuron, i.e. a buffer neuron with $\xi = 1$. Then the bias term θ_1 of the input neuron is varied and the output neuron 2 can regulate its input (eq. (3)).

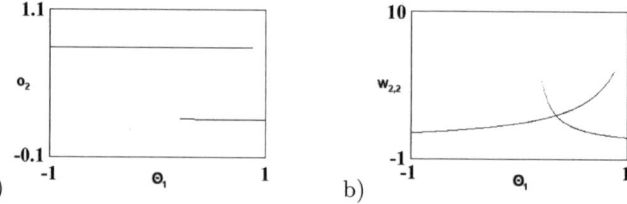

a)   b)

Fig. 1. Positive self-connection w, bias $\theta_2 = 0$, $\beta = 0.1$, $\gamma = 0.02$, $\delta = 0.01$. a) The output of the neuron shows an hysteresis effect, b) plasticity of the weight.

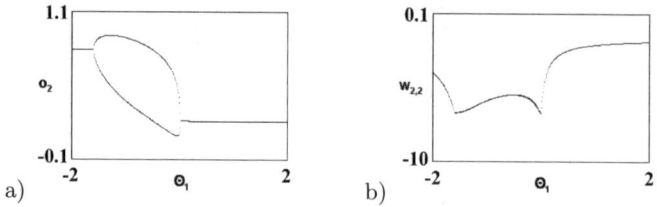

a)   b)

Fig. 2. Negative self-connection w: bias $\theta_2 = 0$, $\beta = 0.1$, $\gamma = 0.02$, $\delta = 0.01$. a) The neuron oscillates with period 2 over a certain interval; The ratio determines the width of the oscillator.

In figure 1a the bifurcation diagrams show the asymptotic states for varying input θ_1. A hysteresis effect is clearly observable over an interval contained in $[-1, 1]$. Hysteresis effects appear also for neurons with static excitatory self-connections $w > 4$ [11]. The asymptotic receptor strength ξ_2^* changes with the input θ_1 corresponding to equation (9), whereas the asymptotic transmitter strength η_2^* is constant over a large interval corresponding to equation (8). Over the hysteresis interval it jumps between two values, depending on which of the possible two asymptotic activity values is stabilized. The resulting asymptotic weight w_{22}^* of this process is shown in figure 1b.

Correspondingly, for a self-inhibitory ($c = -1$) the resulting behavior is shown in figure 2a. Here, we observe an oscillatory behavior (period 2) over a certain interval. Outside of this interval we again observe the stabilization of the desired activity values. The development of the corresponding asymptotic self-weight w_{22} can be seen in figure 2b.

The dynamic effects, not only for the single neurons, depend to a certain degree on the choice of the plasticity parameters β, γ, and δ. For instance, the width of the hysteresis and the oscillatory intervals depend on the quotient γ/δ, and fixed bias terms will shift these intervals. For the above experiments θ_2 was set to zero.

From these first simulations one can easily deduce that a network of self-regulating neurons, having a mixture of excitatory and inhibitory connections, will display non-trivial dynamics. But, with respect to neural control, up to now it is absolutely unclear on what structures behavior relevant dynamics can be implemented. A simple behavior for a simulated Khepera robot may give a first impression of the working SRN-plasticity.

3 Obstacle Avoidance Control

SRN-plasticity (6) will now be applied to the simple problem of obstacle avoidance for a Khepera robot using the standard 2D-simulator [9]. The activity and weight dynamics of the controlling network will be updated synchronously. SRN-plasticity is first applied to a simple recurrent two neuron network shown in figure 3a, where neurons 0 and 1 refer to the left and right sensor inputs, 2 and 3 to the left and right motor outputs, respectively. Input neurons are linear buffers with transmitter strength set to 1; they represent the mean values, denoted by $I_{0,1}$, of the three left and three right front distance sensors of the Khepera. This corresponds to the structure of a simple Braitenberg obstacle avoidance controller [2] with the additional feature, that motor neurons have an excitatory self-connection. The bias terms for the output neurons are both set to zero. The SRN-parameters for all neurons in this example are $\beta = 0.1$, $\gamma = 0.02$, $\delta = 0.01$.

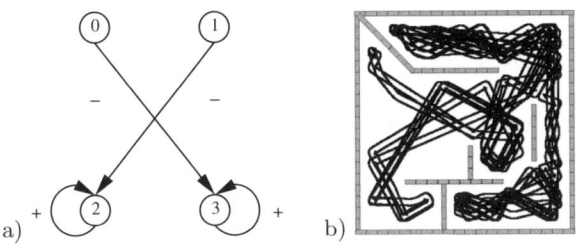

Fig. 3. SRN-plasticity generates obstacle avoidance behavior and exploration. a) the network structure, b) the path of the robot for 20.000 time steps.

Figure 3b demonstrates that the controller in fact leads very effectively to an obstacle avoidance behavior and it allows the robot to turn in sharp corners and dead ends. This behavior is comparable to the so called MRC (*minimal recurrent controller*) described in [8], which has static synapses, and *tanh* as transfer function. But the MRC needs an additional loop of inhibitory connections between the output neurons for comparable performance.

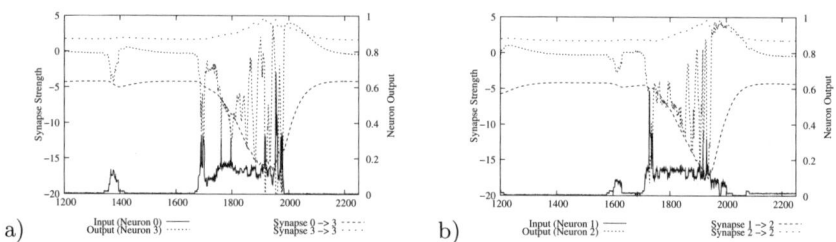

Fig. 4. SRN-plasticity during the interaction of the robot with the environment

The effects of the SRN-plasticity can be followed in figures 4a and 4b where the modulation of the synaptic strength during a run of the simulated robot is shown; the robot enters the left upper corner (from the top) of the environment (figure 3b). First an obstacle on the right side appears around time step 1380 in figure 4a, then on the left side of the robot around time step 1620 in figure 4b. During the time interval 1700 – 2000 the robot oscillates in the corner and the absolute strength of all synapses are increased. This leads to a turn of the robot which is large enough to leave the corner again. In free space the synaptic strength will approach the asymptotically stable values, as can be read from the figures after time step 2200. In the case of static synapses neither a pure Braitenberg controller nor additional undercritical self-connections of the motor neurons will enable the robot to leave sharp corners, unless high noise is added to the sensors or motors, or the self-connections are overcritical (see figure 6), resulting in a decrease of the performance of the exploration behavior. A fixed Braitenberg structure was parameter optimized by evolution and manually symmetrized. The resulting controller does not perform exploration as well, due to the large turning angle caused by the overcritical self-connections (see fig. 6b). Undercritical variations were also tested with noise on sensors and/or motors (see fig. 6c,e,f). These controllers are able to escape sharp corners (performance depending on the noise settings) but because of the high noise, are very likely to collide with walls (not shown). It can be followed that, compared to the discussed Braitenberg controllers, the SRN controller only produces noise, when required (sharp corners). This demonstrates that SRN-plasticity makes controllers more adaptive to different environmental situations.

The figures 5b and 6d shows the trajectory for the same controller but different parameter setting: $\beta = 0.1, \gamma = 0.1, \delta = 0.1$. The plots show, that the controller does not explore as well anymore, but the transients are comparable to those

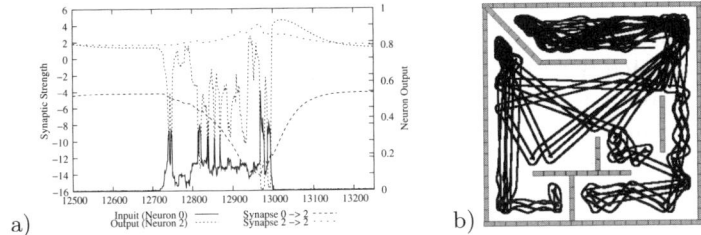

a)   b)

Fig. 5. SRN-plasticity during the interaction of the robot with the environment, c,d) SRN-plasticity with different parameter set $\beta = 0.1, \gamma = 0.1, \delta = 0.1$, trajectory and transient of one neuro-module.

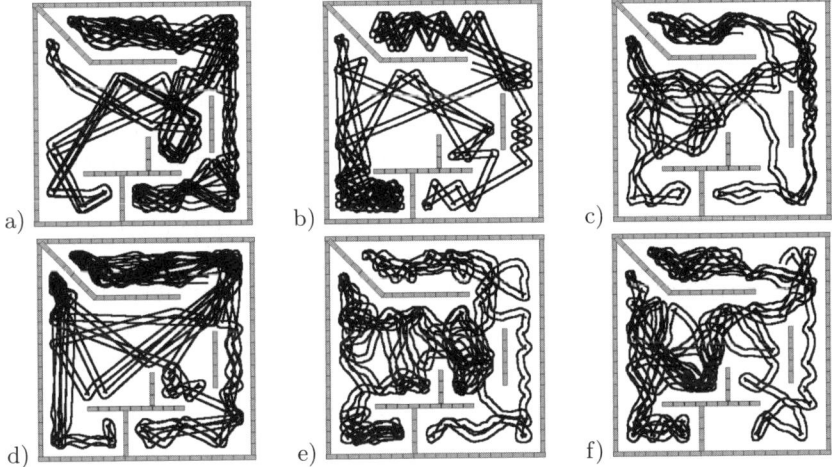

Fig. 6. Comparison of obstacle avoidance behavior for different configurations. a) SRN-Controller with $\beta = 0.1, \gamma = 0.02, \delta = 0.01$, d) with $\beta = 0.1, \gamma = 0.1, \delta = 0.1$, b) Static Braitenberg vehicle with $w_{ij} = -9, w_{ii} = 6$, c,e,f) Braitenberg vehicle with $w_{ij} = -9, w_{ii} = 2$ and c) 20% noise on sensors, e) 70% noise on motors, f) 15% noise on sensors, and 50% noise on motors (minimal required settings for obstacle avoidance).

of the original SRN-controller (fig. 4 and 5a). This implies, that the controller is robust to variations of the parameters over a certain interval, so that the behavior in this parameter interval depends on the structure only. A neural network generation method, such as artificial evolution could then be used to only generate the structure by filling the trinary values of the connection matrix, which is a reduction of the search space.

4   Discussion and Outlook

The term "homeostasis", first introduced in physiology [4], was recently brought to the realm of synaptic physiology again [14] and also to robot control [5], [15],

[6]. Here the SRN-plasticity rule is based on self-regulating neurons, which, as isolated units, can act as homeostatic elements on finite parameter domains. It was demonstrated that this type of neuron configures itself as a bi-stable element, allowing for hysteresis effects, if self-excitation is used. A self-inhibitory neuron acts as a period-2 oscillator on a certain input domain.

The described *Self-regulating Neuron* plasticity is able to generate an effective obstacle avoidance behavior for a simulated Khepera robot. Graphical monitoring of neuron activities and synaptic weights during a run of the robot revealed that synapses in fact vary during robot-environment interactions. The behavior will depend also on the learning parameters β, γ, and δ.

Although demonstrated here only for the trivial obstacle avoidance task, SRN-plasticity, when running on specific structures, leads to effective adaptive behavior. It gives already a first hint to what type of mechanisms may be at work in autonomous robots with adaptive behavior; or, perhaps, may underlie also brain-like processes like learning and memorizing.

At this stage of development there are at least two problems to be tackled. First, how can the desired activity value a^* for a neuron be determined? In an advanced formalism they might be generated by segmental signals produced in analogy to what hormones or other bio-chemical substances do with parts of the brain. It might as well be possible to derive these values from higher level neural control structures, using some specific input channels to the neurons.

Second, because the SRN-plasticity rule does not distinguish between the strength of inhibitory and excitatory transmitters and corresponding receptors, it is obvious that this type of dynamics can not be successfully implemented on arbitrary neural structures, but its functionality highly depends on the network topology. Different construction rules for the connection matrix might come up in a more advanced state of theoretical development. These could be that the entries of a row or a column of the connection matrix c must be of the same sign (Dale's rule) or sum up to zero. This is part of ongoing research. But as long as this is an open question, one may apply evolutionary techniques to get a large variety of such architectures, from which a convenient rule can be extracted. In fact, using an ALife approach to Evolutionary Robotics it is quite natural to apply evolutionary algorithms for structure generation; for instance the ESN^3-algorithm implemented in an evolution-simulation environment, called ISEE [8]. Current experiments focus on a light seeker, which adapts to dynamically changing environments for which it has no direct sensor (ambient light) and the control of sensor driven walking machines, which require coordinated control of a more complex morphology.

The final goal of a neurodynamical approach to embodied cognition, including also the weight dynamics of the underlying networks, like the presented SRN-plasticity, is to have machines with behavioral capacities resembling those of animals. Instantiated as autonomous robots they will be controlled by analogue computers which implement something like the (continuous-time version of the) neural dynamics presented here. On such machines there will run plastic neural networks with synaptic weights changing during the process of interaction with the environment. Because synaptic plasticity can be effective on different time scales, due to different SRN parameters, it may serve for short-time memory effects, for learning, for adaptation, or only for a smooth behavior which is more elegant.

Acknowledgments

This research was funded in part by DFG-grant PA480/2.

References

1. Ashby, W.R.: Design for a Brain: The Origin of Adaptive Behavior, 2nd edn. Chapman and Hall, London (1960)
2. Braitenberg, V.: Vehicles: Experiments in Synthetic Psychology. MIT Press, Cambridge, MA (1984)
3. Beer, R.: A dynamical systems perspective on agent environment interaction. Artificial Intelligence 72, 173–215 (1995)
4. Cannon, W.: The Wisdom of the Body. Norton, New York (1932)
5. Der, R., Liebscher, R.: True autonomy from self-organized adaptivity. In: Proceedings of the EPSRC/BBSRC International Workshop on Biologically-Inspired Robotics: The Legacy of Grey Walter, Bristol (2002)
6. Di Paolo, E.A.: Organismically-inspired robotics: Homeostatic adaptation and natural teleology beyond the closed sensorimotor loop. In: Murase, K., Asakura, T. (eds.) Dynamical Systems Approach to Embodiment and Sociality, Advanced Knowledge International, Adelaide, pp. 19–42 (2003)
7. Hülse, M., Pasemann, F.: Dynamical neural Schmitt trigger for robot control. In: Dorronsoro, J.R. (ed.) ICANN 2002. LNCS, vol. 2415, pp. 783–788. Springer, Heidelberg (2002)
8. Hülse, M., Wischmann, S., Pasemann, F.: Structure and function of evolved neurocontrollers for autonomous robots, Special Issue on Evolutionary Robotics. Connection Science 16, 249–266 (2004)
9. Michel, O.: Khepera Simulator Package version 2.0: Freeware Khepera simulator, downloadable at http://wwwi3s.unice.fr/~om/khep-sim.html
10. Nolfi, S., Floreano, D.: Evolutionary Robotics: The Biology, Intelligence, and Technology of Self-Organizing Machines. MIT Press, Cambridge, MA (2000)
11. Pasemann, F.: Dynamics of a single model neuron. International Journal of Bifurcation and Chaos 2, 271–278 (1993)
12. Pasemann, F.: Complex dynamics and the structure of small neural networks. Network: Computation in Neural Systems 13, 195–216 (2002)
13. Pfeifer, R., Bongard, J.: How the body shapes the way we think. MIT Press, Cambridge, MA (2007)
14. Turrigiano, G.G.: Homeostatic plasticity in neuronal networks: the more things change, the more they stay the same. Trends in Neuroscience 22, 221–227 (1999)
15. Williams, H.: Homeostatic plasticity in recurrent neuronal networks. In: Schaal, S., et al. (eds.) From Animals to Animats 8, MIT Press, Cambridge, MA (2004)

Brain Area V6A: A Cognitive Model for an Embodied Artificial Intelligence

Fattori Patrizia[1], Breveglieri Rossella[1], Marzocchi Nicoletta[1],
Maniadakis Michail[2], and Galletti Claudio[1]

[1] Dipartimento di Fisiologia Umana e Generale, Università di Bologna, I-40126 Bologna, Italy
[2] Institute of Computer Science, Foundation for Research & Technology - Hellas (FORTH),
Science and Technology Park of Crete, Vassilika Vouton, P.O. Box 1385,
GR 711 10 Heraklion, Crete, Greece
patrizia.fattori@unibo.it
Tel.: +39 051 2091749; Fax: +39 051 2091737

Abstract. We found that single neurons in the parietal area V6A of the macaque brain deal with all the components of reaching and grasping actions: locating in space the object target of action, directing the eyes toward it, sensing where the arm is in space, directing the arm toward the spatial location where the object is in order to reach and grasp it, adapting the grip to the object shape and size. The knowledge of how the brain codes simple visuomotor acts can be useful to build artificially-intelligent systems that have to interact with objects, localize them, direct their arm toward them, and grasp them with their gripper. Single cell recordings can also be useful in understanding how to perform more complex visuomotor tasks, like interacting with human beings, exchanging objects with them, and acting in an ever changing environment.

Keywords: neurophysiology, action, spatial perception, reaching and grasping.

An humanoid robot gazes at a dish that it should insert into a dishwasher. The robot reaches the dish with its arm and grasps it with its gripper, lifts the dish, and puts it into the appropriate place on the dishwasher plate.

The robot has produced a sequence of actions that every human being performs naturally and dexterously hundreds of times each day: locating a visual object in space, directing the eyes toward it, directing our arm toward the spatial location where the object is in order to reach and grasp it, adapting the grip to the object shape and size. In order to perform this task successfully, our brain, and also the cognitive architecture of a robot, should know where the eyes are directed, where is the hand in space, and where the goal of action is located in peripersonal space.

The prehension task is achieved by primates through a series of neural elaborations that are performed in the parietal and frontal lobes. Recording of spike trains from single neural cells and analysis of their modulations according to the different phases of the prehension task are the most used techniques through which we acquire knowledge of how the prehension task is achieved. This is the job of neurophysiologists (like us) who select a brain area (supposed to be involved in certain functions), and record the bioelectrical signals from single cells of that area with fine wire microelectrodes. The frequency of discharge of action potentials (spikes) changes

M. Lungarella et al. (Eds.): 50 Years of AI, Festschrift, LNAI 4850, pp. 206–220, 2007.

according to the signal that is processed by the neuron itself. The knowledge on how the brain codes the different phases of prehension task can be useful to build up artificially-intelligent systems, in particular to build embodied aspects of cognition. So we propose here a summary of our studies in order to solve some problems that scientists who are involved in artificial intelligence (AI) could encounter.

1 Neurophysiology of Prehension

We cannot record from the human brain (or can only occasionally, and for a very short time, during a neurosurgery), because of ethical reasons. If we want to know how the human brain works, we have to study the brain of an animal that is able to perform the same task we want to investigate in human. For studying the brain control of prehension, the most used animal is the macaque because its visuomotor functions are almost identical to the human being.

We have been studying for several years a region of the macaque brain known to be involved in visuomotor functions. In particular, we are currently studying the functional properties of neurons of a parietal area called V6A (V stands for visual, as it was originally identified for its visual properties [1]) which contains visual [2] as well as reaching [3,4] neurons. The visuomotor properties of this area have been intensively studied by our laboratory in the last decade [see 5 for a comprehensive review on this topic]. The following represents a summary of these studies.

2 The Parietal Area V6A

Area V6A is a brain cortical area located at the boundary between the occipital lobe, classically known to be devoted to the analysis of visual information, and the parietal lobe (Fig. 1). The anterior part of parietal lobe hosts the primary somatosensory cortex, which is the first cortical sector analysing sensory information from the body. The posterior parietal cortex contains several bimodal visual-somatosensory areas involved in the guidance of arm movements.

The role of V6A as a visuo-motor area was confirmed after neuro-anatomical studies performed in the macaque brain. This technique is based on the use of neuronal tracers, which are substances that, once injected in a brain region, are captured by the neurons and/or by the terminals of nerve cells, and are transported along the neuronal axon up to other brain areas. With this technique we can trace the information flow towards and from the injected region. In other words, we can discover the hard-wired connections between different modules performing a certain job.

As shown in Figure 2, we found a pathway connecting area V6A with both visual and motor cortices. The visual input to V6A derives from area V6, a higher order visual area of the dorsomedial visual stream directly connected with the primary visual area V1 [6]. Area V6A is also linked, directly, with the dorsal premotor cortex [7,8,see 9 for a review]. Therefore, there is a short route from vision to action (V1-V6-V6A-PM cortex) which is part of the so called dorsomedial visual stream, and which is thought to be useful for the on-line control of hand action [5].

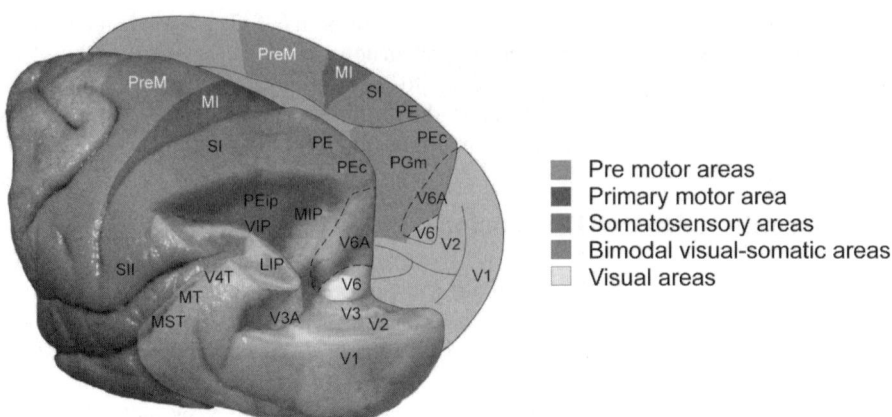

Fig. 1. Cortical areas of the macaque brain
Left: postero-lateral view of a partially dissected left hemisphere with a part of the inferior parietal lobule and of the occipital lobe cut away in order to show area V6A hidden in the parieto-occipital sulcus. *Right*: medial view of the right hemisphere.

Labels on different brain regions indicate cortical areas according to anatomical and functional criteria. Colors indicate sensory or motor properties of different regions of the brain. Note that area V6A is at the posterior end of the bimodal (visual/somatosensory) region, and borders the visual areas of the occipital pole.

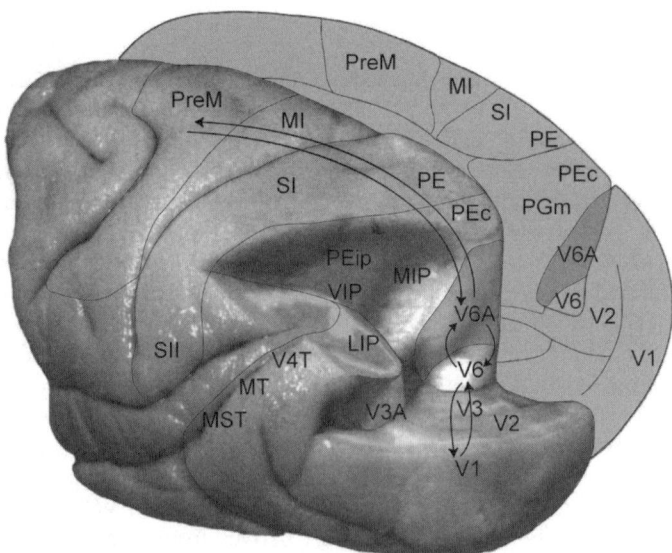

Fig. 2. The dorsomedial visuomotor pathway
Arrows indicate anatomical connections between different cortical areas. There is a dorsomedial visuomotor pathway connecting the primary visual area (V1) with the premotor areas of the frontal cortex.

Other details as in Figure 1.

All these cortical connections are strictly bidirectional, that is area V6A receives visual information from area V6, but also sends information to this same area; similarly, area V6A sends information to the premotor cortex, but also receives information from it. It is a cortical loop that includes visual, visuomotor, and motor areas. Area V6A is one of the areas nestled in this loop.

As can be inferred by its connections, area V6A has both visual and motor properties, features which can be useful for the visual guidance of prehension. These functional properties of V6A neurons have been investigated in a series of electrophysiological studies, summarized hereafter.

3 Visual Neurons Able to Localize Objects in Space

Physiologists use the term "receptive field" (RF) to indicate the region of visual field from which a visual neuron receives visual information: the RF of a visual neuron is its window on the world. Contrary to what is generally thought, this window is not able to localize an object in space, because the RF moves with the eyes (being a part of the retina) and therefore explores different spatial locations according to the direction of gaze.

In area V6A there are visual cells in which the visual response (the response to the visual stimulation of the RF) is modulated by the direction of gaze [10]. These neurons are able to code the location of objects in space because they discharge differently to the same object according to its spatial location.

Figure 3A shows an example of this gaze-dependency of visual responses. When the animal gazed at the center of the screen, we found a good visual response when the right part of the animal's visual field was stimulated. This meant that the "window" (the RF) of this neuron was located there. When the fixation point was displaced in another location, for instance in the top left corner of the screen, the same RF stimulation as before evoked a neural response much stronger than before. How could this happen? Evidently this cell was informed about the type of visual stimulus that activated its receptive field, but it was also informed about where the eyes were directed. Combining the information coming from the retina with that on the direction of gaze, different visual responses can be obtained according to the direction of gaze (gaze-dependent visual cell).

Figure 3B shows the analysis of gaze modulation in a V6A cell where a high number of gaze locations were tested. Curved lines link together spatial locations where the visual response were the same (iso-excitability lines). To obtain these data, we required the animal to fixate many different locations, thus displacing the neuron's receptive field on the screen many times. Each time we displaced the fixation point, we stimulated the RF with the same visual stimulus, obtaining a full field analysis of the neuronal visual responsiveness according to different angles of gaze. Different visual responses were obtained according to the direction of gaze. In other words, this type of cells transforms different spatial positions (those of the RF when the animal gazed at different locations) in different frequencies of discharge according to the direction of gaze. So they can inform us about the spatial location where the object is. This information can be used by the brain for many different purposes. Among them, that of directing the hand towards an object to be grasped.

A   B

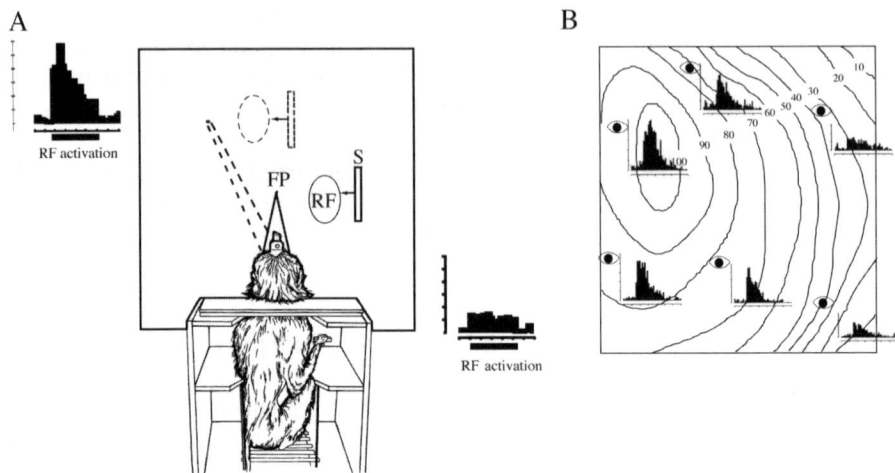

Fig. 3. Encoding of visual space by a gaze dependent visual neuron
A. Experimental set-up and visual responses of a gaze-dependent visual cell. The square represents the screen in front of the animal. FP is the fixation point of the monkey, RF the receptive field of the neuron, and S visual stimulus used to activate the cell (the stimulus was moved leftwards across the RF). The histogram to the right of the screen is the response of the cell when its RF was activated while the animal gazed at the centre of the screen. When the monkey directed its gaze to the top left part of the screen (dashed lines), the RF of the cell moved to the top part of the screen too. The same stimulus as before across the RF in this new screen position evoked a good response (displayed to the left of the screen). Thick lines under neural responses indicate the stimulation time.
B. Gaze modulation in a V6A gaze-dependent visual cell. The square represents the screen. Six different visual responses are shown, evoked by stimulating the RF of the cell with the same stimulus, while the gaze was directed towards 6 different directions (6 eye symbols). Curved lines are iso-excitability lines linking together spatial locations where the visual responses were the same. The cell encoded the visual space in frequency of discharge: when the stimulus activated the RF on the top left corner of the screen it evoked the maximum discharge frequency (100%) from the cell; when it activated the RF on the right top or bottom corners it evoked a very poor response (10%) from the cell.

We also found that in a minority of V6A neurons the RF remained stable in space despite changes in eye position. This finding contrasts dramatically with the behaviour of typical visual neurons, in which the RF is firmly anchored to the retina (being physically a part of it) and therefore move through space in tandem with the eyes (like for the neuron shown in Figure 3). The new type of visual cells receives (and encodes) visual information from different parts of the retina depending on the direction of gaze, but from a constant part of the visual space. We called them "real-position" cells [11]. Evidently, in the real-position cells, the gaze signal is used to gate the retinal locations from where visual information are picked up. This visual transformation, that has been then described also in other cortical areas of the parietal and frontal cortices [12,13], appears early in the dorsomedial visual stream in area V6A.

A: same retinotopic position

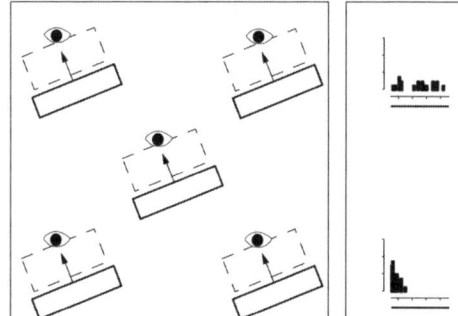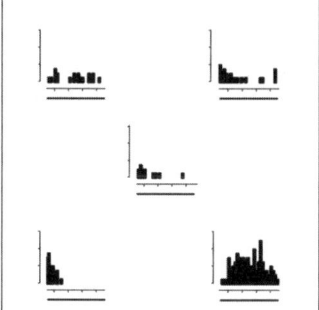

B: same spatial position

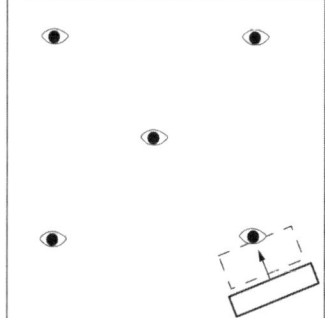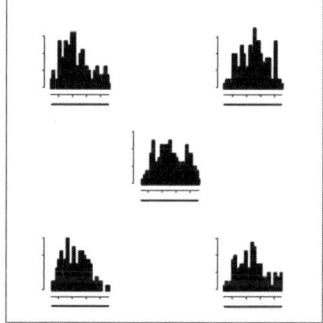

Fig. 4. Encoding of visual space by a real-position cell

Neural responses of a 'real-position' cell to the visual stimulation of the same retinotopic location (A) or the same spatial location (B), while the animal looked at different screen positions. Each large square represents the screen in front of the animal. *Left*: experimental paradigm. Fixation-point locations on the screen are indicated by eye symbols. Visual stimuli (full line rectangles) were moved across the receptive field (dashed line rectangles) in the direction indicated by the arrow. *Right*: Neural responses to visual stimulations reported at fixation point locations. Thick lines under neural responses indicate the stimulation time. Scales are 4 spikes per vertical division, and 300 ms per horizontal division. Note that there is a visual response only when the visual stimulus is in the right bottom corner of the screen, irrespective from the direction of gaze.

Other details as in Figure 3. Modified from Galletti and Fattori [14].

Figure 4 shows an example of such a type of cell. This cell had a visual RF just below the fovea. The visual stimulation of the RF evoked a good response from the cell when the animal looked towards the bottom, right part of the animal's field of view, whereas the stimulation of the same retinotopic position was not effective when the animal looked towards all the other spatial locations (Fig. 4A). As shown in the Fig. 4B, if the stimulus was on the bottom, right part of the animal's field of view, the cell was always strongly activated, no matter where the animal was looking at. In other words, the RF of this cell did not move in tandem with the eyes as in any other visual neuron. The cell encoded a specific part of the field of view regardless of the direction of gaze.

Neurons of this type encode *directly* the visual space. Each real-position cell encodes a different spatial location according to the spatial coordinates of its visually-responsive region. The cell in Figure 4 encodes the bottom right part of visual space. When this cell discharges, it means that the object activating the cell is in that particular region of the visual space, no matter where the animal was looking at.

A robot having to manipulate different objects placed in different positions around it or having to select the target of its actions according to its position in space could of course benefit from a mechanism like that of real-position cells. In addition to be used to direct movements towards visual targets, the output of real-position cells could be used to direct selective attention to relevant points in space for acquisition of stimuli in the immediate environment, either by gaze or manual reaching. To this regard, it is worthy noticing that when we reach toward a target that suddenly appears in the peripheral visual field, not only does the arm extend toward the object, but the eyes, head, and body also move in such a way that the image of the object falls on the fovea. Because the eyes start to foveate the object while the hand is still moving, the reaching target changes its retinal location from its appearance in the visual field (peripheral location) till the end of reaching execution (foveal location). Nevertheless, the hand goes straight towards the target, as whether the motor center controlling the arm movement 'knew' in advance the final position to be reached out in spatial coordinates. We suggested that area V6A, and in particular the real position cells of this area, could play an important role in all these visuomotor transformations [14].

4 Somatosensory Neurons Monitoring Arm Position in Space

The humanoid robot dealing with the dishwasher (and ourselves in similar countless actions that we perform throughout the day) must take into account the position of arm in space and with respect to the torso to correctly guide an arm movement. If for instance the robot has the arm near the torso, it has to extend it in order to grasp the dish on the table; but if it starts the movement from a position reached in a previous action, the movement could be different. For example, if it starts the reaching movement with the gripper inside the dishwasher, it could need to adduct and flex the arm in order to grasp the dish on the table.

Which are the brain's sources of information about the position of the arm? The most important is the so called "proprioception", that is information coming from proprioceptors located inside the arm, giving the internal feeling of the limb position. The same that we could feel when, with the eyes closed, we try to locate our arms or fingers in space. Proprioceptive information arises from receptors that signal the stretch of muscles or the angle of joints. This information is carried by sensory fibers that reach the primary somatosensory cortex (depicted in blue in figure 1), and from there several other cortical areas of the superior parietal lobule, including area V6A [15]. Fig. 5 shows the distribution of receptive fields of V6A proprioceptive neurons: note that they are located only in the upper limbs. V6A contains also tactile neurons, that is neurons informed about touches of the hair or the skin: note that they are located only in the upper limbs and in parts of the trunk adjacent to the limbs (see Fig. 5). In summary, area V6A is informed about position in space of the arms as well as their interaction (contact) with objects in extrapersonal space.

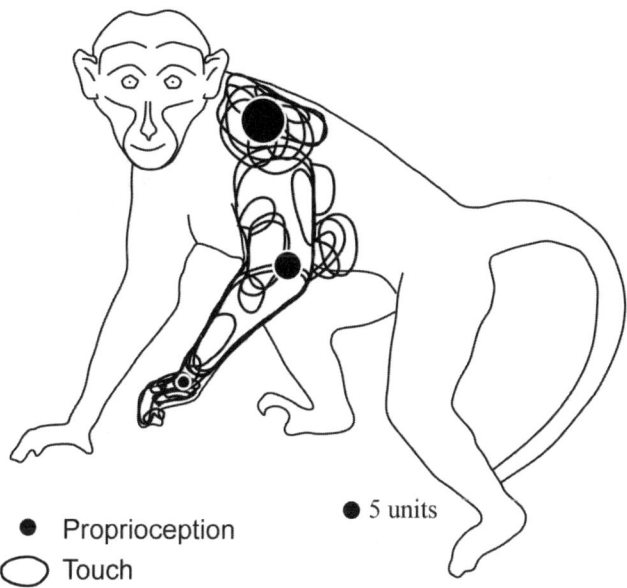

Fig. 5. Somatosensory representation by V6A neurons
Dots: proprioceptive V6A neurons modulated by the rotation of a joint. The modulating joints are indicated by the dot location. The size of the dot is proportional to the number of modulated units. Continuous lines: extent and location of tactile receptive fields of V6A neurons. In V6A, body representation is largely incomplete and the representation of the arm is emphasised. Modified from Breveglieri et al [15].

Note that the somatic representation in V6A is different with respect to that of the typical cortical somatosensory areas, as in V6A the somatic representation is restricted to the upper contralateral limb. The fact that only the arm is represented in area V6A suggests that this region is involved in the control of arm movements. Proprioceptive cells could provide useful information about the spatial position of arm and hand while performing different hand-object interactions. Tactile receptive fields located on the arm and hand could be useful in recognizing the physical interaction between the moving arm and the environment, or between the hand and the object that it is grasping. All these information confirm us the actual location and status of the arm, and in particular the ongoing interaction between the hand and the grasped object.

5   Neurons Encoding Planning and Execution of Reaching Movements

A direct involvement of area V6A in arm movement execution has been demonstrated by the use of a specifically designed reaching task. It is sketched in figure 6A.

In the task, the hand performs a reaching movement from a position near the body to a position in the peripersonal space in front of the body, trying to reproduce under controlled conditions the reaching movements performed in every day life when we reach out for objects.

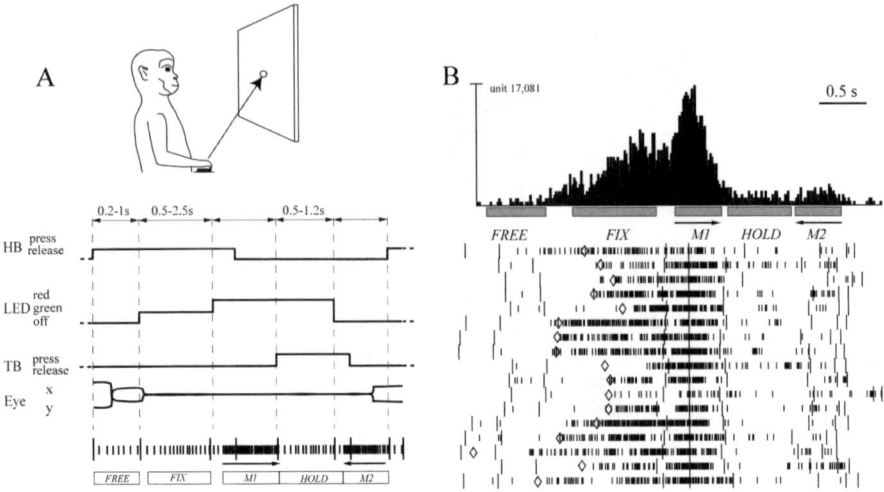

Fig. 6. Reaching activity in a V6A neuron

A. The reaching task. *Top*: Scheme of the experimental set-up. Reaching movements were performed in darkness, from a home-button (black rectangle) towards a target (open circle) located on a panel in front of the animal. *Bottom:* Time course of the task: the sequences of status of the home-button (HB), target button (TB), and of the colour the target button (LED) are shown. Lower and upper limits of time intervals are indicated above the scheme. Under the scheme, typical examples of eye-traces (X and Y components) and neural activity during a single trial are shown. Short vertical ticks are spikes. Long vertical ticks among spikes indicate the occurrence of behavioural events (markers). From left to right, the markers indicate: trial start (HB press), target appearance (LED light-on green), go-signal for outward movement (green to red change of LED light), start and end of outward movement (HB release and TB press, respectively), go-signal for inward movement (LED switching off), start and end of the inward movement (TB release and HB press, respectively), and end of data acquisition.

Rectangles under neural activity indicate the time epochs referred to behavioural events. FREE: reference activity at rest; FIX: delay preceding reaching movement where gaze direction and arm movement are constant; M1: outward reaching, indicated by the arrow pointing to the right; HOLD: time of hand holding on the reached target; M2: inward reaching, indicated by the arrow pointing to the left.

B. Example of a V6A neuron coding planning and execution of reaching movements. Neural activity is shown as cumulative time histogram and as raster activities. Diamonds in raster activities indicates the onset of fixation of the reaching target. Cell's activity is aligned with the onset of forward arm movement (M1).

Scales: vertical bar on histogram: 140 spikes/s; other details as in Figure 3. Modified from Fattori et al [4].

The first controlled condition in the task is that the eyes are fixed in a position (the fixation LED), which represents also the goal of the reaching movement. The second controlled condition is that the arm always starts the movement from a button near the chest and reaches a target placed on a panel in front of the monkey with a direct, ballistic movement that follows a precise time sequence (summarized in figure 6A) decided by the experimenter. The last controlled condition is that the task is executed in complete darkness, with the only exception of the fixation light, which was a LED the brightness of which was reduced so that it was barely visible during the task. This was chosen to avoid to evoke visual activation during the execution of the task other than the fixation point. It is evident that in these experimental conditions any neural modulation during the task must be ascribed to arm-related activity.

Using this task, we found that many neurons in V6A were modulated during the preparation and execution of reaching movements [4]. An example of these reach-related cells is shown in Figure 6 B. This neuron strongly discharged during the execution of the reaching movement toward the target (M1). The neuron shows also an increase in its firing rate in the delay preceding the movement (FIX period). In this epoch, the monkey is already fixating the target of reaching and no arm movement is occurring. Therefore, the neural discharge cannot be explained by arm-movements nor by oculomotor behaviour. We suggested that the neural discharge is a preparatory signal for the impending reaching movement [4].

The figure 7 shows two V6A reach-related cells studied with the same task. The first (Fig. 7A) is activated by movements of the arm directed toward the visual target (M1, outward reaching); the second, for arm movements directed away from the target, towards the body (M2, inward reaching).

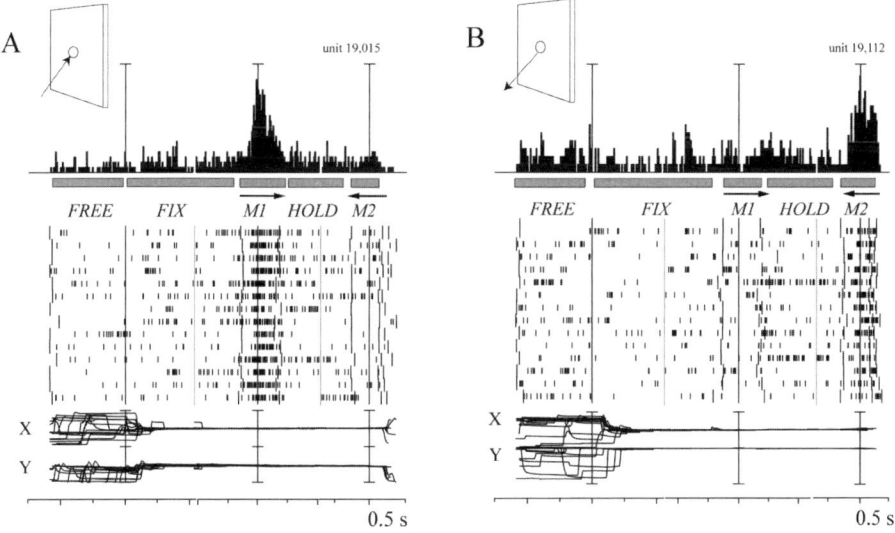

Fig. 7. Two V6A neurons modulated by reaching movements
A: Neuron modulated by outward reaches. From top to bottom: cumulative time histogram of neural activity, time epochs, raster displays of impulse activity, recordings of X and Y components of eye positions. Neural activity and eye traces are aligned three times for each neuron: with the LED appearance (1st), with the onset of outward (2nd) reaching movements, and with the onset of inward (3rd) reaching movements.

Peri-event time histograms: binwidth = 15 ms; scalebar = 100 sp/s. Eyetraces: scalebar = 60 degrees.
B: Neuron modulated by inward reaches.

Scalebar in peri-event time histograms: = 65 sp/s (**B**).
Other details as in Fig. 6. Modified from Fattori et al [16].

To reach the dish on the table, the humanoid robot has to know in which direction to move the hand. A variance of the task allowed us to test whether the direction of reaching movements influenced the discharge of V6A reaching neurons. Monkeys were required to reach visual targets placed in different spatial locations while gazing at them. In other words, the animal performed reaching movements toward different spatial locations while maintaining the target of reaching under foveal control. We found that the direction of reaching strongly modulates the activity of V6A cells [16]. An example of this behaviour is shown in Figure 8.

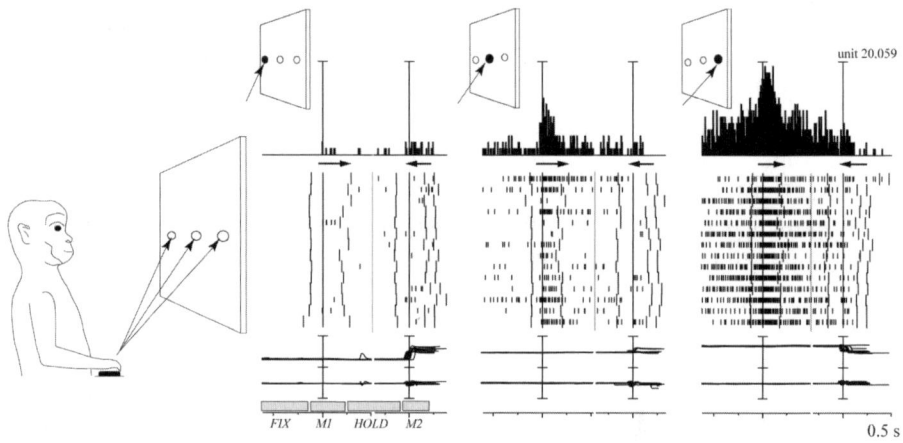

Fig. 8. V6A neuron coding directions of reaching
Neuron preferring rightward M1 movements and rightward gaze directions.
Sketches on the left and top illustrate the different directions of reaching tested in this neuron. Neural activity and eye-traces are aligned twice in each inset, with the onsets of outward (1st) and inward (2nd) reach movements, respectively. The mean duration of epochs FIX, M1, HOLD, M2 is indicated in the bottom left inset.
Scalebar in peri-event time histograms: = 70 sp/s. All other details as in Fig. 7.
Modified from Fattori et al [16].

The unit in Figure 8 discharged for reaches directed to the visual target only when it was straight ahead or in the right part of space. The cell was not activated at all by reaching movements directed to the same target placed in the left part of space. The neuron was also strongly affected by the direction of gaze, being the cell strongly activated when the animal looked straight ahead or to the right without performing any arm movement (see activity during FIX epoch). This cell signalled the occurrence of both rightward ocular and reaching movements. Other V6A cells signalled only the direction of gaze or the direction of reaching movement. In the whole, V6A cells were able to encode the entire set of directions we tested in the workspace [16].

6 Neurons Encoding the Grasping Phase of Prehension

The act of prehension includes the reaching movement, that is the transport of the hand towards the object to be grasped, and grasping movements, which involve more distal parts of the arm, as the wrist, hand and fingers. Recently, we have began to study whether V6A is involved in the control of grasping, by training monkeys to perform reach-to-grasp movements under controlled conditions [17]. In Fig. 9, the behaviours of a cell to reaching and to grasping movements are compared. The direction of reaching was the same (straight ahead) in reach-to-point (Fig. 9A) and reach-to-grasp (Fig. 9 B) tasks, but in the latter the monkey had to preshape the hand to grasp an handle and to flex its fingers to secure the grasp of the object. Therefore, any difference in neural activity in the reach-to-grasp with respect to reach-to-point task must be attributed to the grasping action, as the transport phase of reaching movement was the same in the two experimental situations. This cell was not activated during the execution of the reach-to-point movement, but was excited during reach-to-grasp action. In this action, the finger extended to embrace the handle and then flexed to acquire it. Many V6A cells behaved like that shown in Figure 9, and V6A seems to have a role in coding also distal, besides proximal, components of the act of prehension.

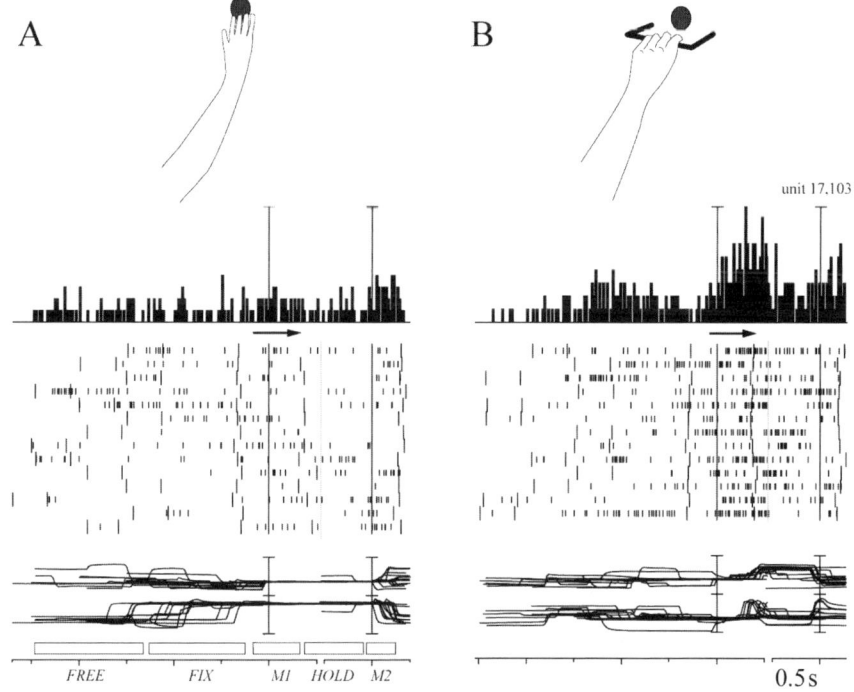

Fig. 9. Reach-to-point and reach-to-grasp activities in a V6A neuron
Top: sketch of the final hand position in the reach-to-point (A) and reach-to-grasp (B) tasks. *Bottom*: Activity has been aligned twice, with the onset of forward and backward arm movements, respectively. Peri-event time histograms: binwidth = 20 ms; vertical bar on histograms: 55 spikes/s; eye traces: 60 degrees/division. All other details as in Figure 6.
Note that the cell was clearly more activated in the reach-to-grasp than in the reach-to-point task, and the handle in (B) was in the same spatial location as the LED in (A). Therefore, the arm movements performed in (B) and (A) were similar in trajectory amplitude and direction, but only in (B) they did include the grasp.
Modified from Fattori et al [17].

In summary, the picture emerging from data reviewed here is that area V6A is a cortical region involved in the control of many components of the act of prehension under visual guidance: in localizing objects in space, in sensing where our arm is and what our hand touches, in transporting the hand towards the target object, in coding the direction of reaching and in adapting the hand to the object features. Area V6A could be useful in the online control of arm movement by elaborating sensory inputs and motor outputs to represent the internal body state for the purpose of sensorimotor integration.

7 Links to Robotics Research

The design of artificial systems having efficient reaching and grasping capabilities is currently a very hot topic in the field of robotics. New impetus can be given to this area by examining the working principles of cortical area V6A being directly involved in the solution of the underlying tasks. In the following we highlight some working principles of V6A [5] that worth consideration in robotics research, aiming at improving the efficacy of contemporary robotic systems.

At first, we note the specialization of V6A in processing sensory information coming mainly from the peripersonal space of the animal. This type of limited spatial perception, filters environmental sensory input, providing to the relevant motor system only the information that can be useful during action.

Furthermore, the simultaneous utilization of many different coordinate systems for encoding information, seems to be necessary for efficient reaching and grasping movements. In particular, V6A is directly involved in the transformation of retinotopic stimuli to the executional motion direction, encoding how the same information is referenced in all the intermediate coordinate systems. In other words, it is important to combine information referred to both an action-irrelevant coordinate system like the retinotopic, and also an action specific coordinate system, as it is the case with arm motion-direction cells and with cells like the real-position cells that code directly the visual space. This is necessary for making direct and effective modifications to the executed action when sensory changes are identified.

Additionally, the reciprocal connectivity of V6A, with both the sensory and the motor areas, seems to be very important. Specifically, the brain pathway responsible for reaching and grasping movements does not operate in a feed-forward way, but it rather follows recurrent connectivity in all stages of information processing. During every phase of the action, multi-modal information is integrated in order to confirm that the execution proceeds in the appropriate way.

Finally, the encoding of both hand and visual information in the same associative brain area implies that in producing artificial intelligent systems that perform prehension actions, these two information must operate in a coordinated manner rather than in isolation from one another.

In addition to the above general principles that could be proved beneficial for the design of novel robotic systems, neurophysiological studies can provide further input to robotics research. This can be done in a first approach by abstracting higher level cognitive information addressing the connectivity of brain areas and the type of information they convey to each other. Furthermore, in a more detailed and practical level, biological data can be utilized by well known computational methods designing artificial systems that approximate the structural and functional characteristics of biological ones. Initial attempts towards this brain approximation direction have been done in [18]. Robotics and neurophysiology could also be met in a new research field where bioelectrical signals obtained by single cell recordings are utilized for driving robotic devices, formulating a new alternative approach on Brain-Computer Interface studies [19].

8 Conclusions

Neural behaviours like those here described for V6A have been described in many areas of the brain, but what is unique for V6A so far is that this region contains *all* these neural behaviours. Cells encoding the visual space cohexist in V6A with cells controlling grasping movements and with somatosensory cells signalling what our arm is doing. We are currently working on how these different neurons interplay together.

Area V6A can be a good model and could be "copied" in building the cognitive architecture of artificially-intelligent systems that have to interact with objects, localize them, direct toward them their arm and grasp them with their gripper.

Combing the neurophysiological expertise with the engineering and computer science ones can be a way to implement the evolution of humanoid robots performing more efficiently the dishwasher task and even harder tasks like those of interacting with human beings, exchanging objects with them, and acting in an ever changing environment.

Aknowledgements

This work was supported by European Union Commission, FP6-IST-027574-MATHESIS and by MIUR and Fondazione del Monte di Bologna e Ravenna, Italy.

References

1. Galletti, C., Fattori, P., Battaglini, P.P., Shipp, S., Zeki, S.: Functional demarcation of a border between areas V6 and V6A in the superior parietal gyrus of the macaque monkey. Eur. J. Neurosci. 8(1), 30–52 (1996)
2. Galletti, C., Fattori, P., Kutz, D.F., Gamberini, M.: Brain location and visual topography of cortical area V6A in the macaque monkey. Eur. J. Neurosci. 11(2), 575–582 (1999)
3. Galletti, C., Fattori, P., Kutz, D.F., Battaglini, P.P.: Arm movement-related neurons in the visual area V6A of the macaque superior parietal lobule. Eur. J. Neurosci. 9(2), 410–413 (1997)
4. Fattori, P., Gamberini, M., Kutz, D.F., Galletti, C.: 'Arm-reaching' neurons in the parietal area V6A of the macaque monkey. Eur. J. Neurosci. 13(12), 2309–2313 (2001)
5. Galletti, C., Kutz, D.F., Gamberini, M., Breveglieri, R., Fattori, P.: Role of the medial parieto-occipital cortex in the control of reaching and grasping movements. Exp. Brain. Res. 153(2), 158–170 (2003)
6. Galletti, C., Gamberini, M., Kutz, D.F., Fattori, P., Luppino, G., Matelli, M.: The cortical connections of area V6: an occipito-parietal network processing visual information. Eur. J. Neurosci. 13(8), 1572–1588 (2001)

7. Matelli, M., Govoni, P., Galletti, C., Kutz, D.F., Luppino, G.: Superior area 6 afferents from the superior parietal lobule in the macaque monkey. J. Comp. Neurol. 402(3), 327–352 (1998)
8. Shipp, S., Blanton, M., Zeki, S.: A visuo-somatomotor pathway through superior parietal cortex in the macaque monkey: cortical connections of areas V6 and V6A. Eur. J. Neurosci. 10(10), 3171–3193 (1998)
9. Galletti, C., Fattori, P., Gamberini, M., Kutz, D.F.: The most direct visual pathway to the frontal cortex. Cortex 40(1), 216–217 (2004)
10. Galletti, C., Battaglini, P.P., Fattori, P.: Eye position influence on the parietooccipital area PO (V6) of the macaque monkey. Eur. J. Neurosci. 7(12), 2486–2501 (1995)
11. Galletti, C., Battaglini, P.P., Fattori, P.: Parietal neurons encoding spatial locations in craniotopic coordinates. Exp. Brain Res. 96(2), 221–229 (1993)
12. Fogassi, L., Gallese, V., Fadiga, L., Luppino, G., Matelli, M., Rizzolatti, G.: Coding of peripersonal space in inferior premotor cortex (area F4). J. Neurophysiol. 76, 141–157 (1996)
13. Duhamel, J.R., Bremmer, F., BenHamed, S., Graf, W.: Spatial invariance of visual receptive fields in parietal cortex neurons. Nature 389, 845–848 (1997)
14. Galletti, C., Fattori, P.: Posterior parietal networks encoding visual space. In: Karnath, H.-O., Milner, A.D., Vallar, G. (eds.) The Cognitive and Neural Bases of Spatial Neglect, pp. 59–69. Oxford University Press, Oxford, New York (2002)
15. Breveglieri, R., Kutz, D.F., Fattori, P., Gamberini, M., Galletti, C.: Somatosensory cells in the parieto-occipital area V6A of the macaque. Neuroreport 13(16), 2113–2116 (2002)
16. Fattori, P., Kutz, D.F., Breveglieri, R., Marzocchi, N., Galletti, C.: Spatial tuning of reaching activity in the medial parieto-occipital cortex (area V6A) of macaque monkey. Eur. J. Neurosci. 22(4), 956–972 (2005)
17. Fattori, P., Breveglieri, R., Amoroso, K., Galletti, C.: Evidence for both reaching and grasping activity in the medial parieto-occipital cortex of the macaque. Eur. J. Neurosci. 20(9), 2457–2466 (2004)
18. Maniadakis, M., Hourdakis, E., Trahanias, P.: Modeling Overlapping Execution/ Observation Brain Pathways. In: The International Joint Conference on Neural Networks (IJCNN) (accepted for publication, 2007)
19. Birbaumer, N., Weber, C., Neuper, C., Buch, E., Haagen, K., Cohen, L.: Physiological regulation of thinking: brain-computer interface (BCI) research. Progress in Brain Research 19, 369–391 (2006)

The Man-Machine Interaction: The Influence of Artificial Intelligence on Rehabilitation Robotics

Alejandro Hernández Arieta[1], Ryu Kato[1], Wenwei Yu[2],
and Hiroshi Yokoi[1]

[1] Developmental Cognitive Machines Laboratory, Precision Engineering Department,
University of Tokyo, Tokyo, Japan
[2] Lab. of Bioinstrumentation and Biomechatronics. University of Chiba, Tokyo, Japan
{alex,katoh,hyokoi}@robot.t.u-tokyo.ac.jp, yuwill@faculty.chiba-u.jp

Abstract. We are leaving in a world where the interaction with intelligent machines is an every day life event. The advances in artificial intelligence had allowed the development of adaptive machines that can modify its internal parameters to adjust their behavior according to the changing environment. One field that has profit from this is rehabilitation and prosthetics. In this respect, is our interest to evaluate the effects that this interaction has on the user. In this study, we use an f-MRI (functional Magnetic Resonance Imaging) device to measure the changes on the motor and sensory cortex of a right hand amputee's using an EMG controlled Adaptable prosthetic hand with tactile feedback. Our results show the improvement in the adaptation to the prosthetic device, also, our experiments point to a possible modification of the body schema, generating an illusion of belonging of the robot hand to the human body.

Keywords: AI,fMRI, EMG, Electrical Stimulation.

1 Introduction

Nowadays, we are living in a world where the interaction between human and machines is more common. This interaction has been changing our society, what once was considered or revered with awe, nowadays our children use as an everyday tool. Now is difficult to imagine our every day lives without the direct interaction with new technologies. In this study, we focus on the applications for rehabilitation engineering, where the interaction with "intelligent" machines had the most impact. In the field of prosthetics, this influence affects the way the person perceives its own body. With the new recognition technologies, a person can control a robot hand close to the way he used to control his original hand. This is important especially for amputees, whose brains suffered cortical reorganization[1] after the amputation, leading to the effect known as phantom pain [2]. However when an amputee uses an EMG (electromyography) controlled hand, the neural paths used to control the robot hand are the same to those used before to control his original hand. This allows the brain to revert the cortical reorganization that occurred due to the lack of sensorial information [3]. However,

M. Lungarella et al. (Eds.): 50 Years of AI, Festschrift, LNAI 4850, pp. 221–231, 2007.

the cortical reorganization is not complete since, the patients does not receive any sensation from the prosthetic device. The systems interacting with the human body, all face the same big challenge to overcome the individual changes between users, the signals from one person are not the same to those of the next one.In the case of the Electromyography (EMG) controlled prosthetic hands, the problem does not only involve the differences from one person to the other, but also, the complexity increases when we consider that the EMG signals do change with time. Therefore, the system does not only need to adapt to different users, but also, compensate the changes in the same user [4]. In this aspect, the field of artificial intelligence has provided with several tools to overcome this problem, where, the Artificial Neural Networks (ANN) show the more versatility. One of the main advantages of the ANN is their robustness in discrimination tasks. So now we question, it is possible to include a prosthetic hand as part of the patient's body schema?. In this respect we conducted a series of experiments using an fMRI device to measure the changes in the brain of a right hand amputee using a individually adaptable EMG controlled prosthetic system. We used electrical stimulation to transferee the tactile sensation from the robot hand to the amputee's body.

2   The Body and Its Surroundings

All this new changes in our actual society are due to the new trend of adaptive machines, which make use of artificial intelligence to exploit the physical characteristics, as well as, computation possibilities of the new technologies. Our bodies affect the way we think[5], our perception of the world is dominated by our physical characteristics. The sensorimotor relationship generated by the continuous interaction between the muscular and somatosensory systems develops what is called the body schema/image, which allow us to control our bodies. However, when the control loop is disrupted, we lose control over our bodies, then what we used to do without effort, becomes an imposing task. Is it possible to include external objects as part of our bodies?. We can mention the work from Holmes [6] where he describes an integrated neural representation of the body (body schema) and of the space around the body (peripersonal space). Also the work from Iriki et al. [7][8], who introduce interesting insights on the neural mechanisms behind tool use. On his experiments with monkeys he explore the dynamics in the brain when using a "T" shaped tool to handle objects (in this case, food). The experiment consisted on moving the visual stimuli (food) either toward or away from the monkey's hand in a centripetal or centrifugal fashion. After that, the monkey was trained to use a 'T' shaped bar to retrieve the food. The experiments results shows a displacement on putative visual receptive field. The neurons in the monkey's brain are activated when the visual receptive field includes the point of the tool after 5 minutes of use. We could conclude that the monkey's brain identify the branch (as longs as the task is performed) as part of the monkey's body. Graziano [9][10]later demonstrated the effects of certain objects entering the peripersonal space. These results point out the brain capability to include external objects inside its body schema. With these studies in mind, we face new possibilities for rehabilitation technologies. Galfano et al. [11] shows an interesting study concerning modifications to the body schema. In his study, the hands of the participant are hidden from view, two rubber hands are placed in the same position as the hands of the test subject. A light emitting diode (LED) is placed next to each hand for visual feedback. An electrical vibrator is placed on the fingertips of the test subject. The experiment consists of a set of test with synchronous and asynchronous vibrations along with the lighting of the LED's synchronously and asynchronously with the vibrations. The position of the rubber hands varies in possible and impossible proprioceptive positions. Galfano results shows that the visual, tactile and proprioceptive information are necessary for the correlation in the brain to generate the illusion of the rubber hands as part of the participant's body. As O'Regan et al [12] has shown, the human sensations are not fixed mechanisms, but actually, are continually updated to the changes in the body. In Pavani's experiments[11], we can see the importance of the role of visual feedback to develop an illusion of "ownership" of external objects. But, visual feedback is not enough to "fool" the body into believe that what is seeing is part of it. Here the importance of tactile and proprioceptive information. For example, if the "extra" hand is put in an impossible proprioceptive position, the body will "reject" such object as part of it. In prosthetic applications, even though the advances in their control have improve since their introduction in 1960 [13], we still face the disembodiment problem, that is, the prosthetic device can be controlled, but it is not recognized as part of the body, requiring extra effort from the patient to use the "extra" limb. If we provide with feedback to the prosthetic hand's user[14], we can close the sensorimotor loop, promoting the adaptation to the robot hand, developing a modification on the person's body schema, which makes us think that the person accepts the robot hand as part of his/her body[15].

3   Body Schema Modification

In order to get a more profound understanding of these mechanisms behind the modification of the body schema, we performed some experiments using a functional magnetic resonance imaging device to measure the neural activity in the brain from using an EMG controlled prosthetic hand, with electrical stimulation applied by a pair of electrodes to the upper left arm to provide tactile feedback translated from the information acquired by the pressure sensors positioned at the finger tips and on the palm of the robot hand.

3.1   FMRI Settings

Cerebral activity was measured with a functional Magnetic Resonance Imaging device using blood oxygen level-dependent contrast [16]. After automatic shimming, a time course series of 59 volumes was obtained using single-shot gradient-refocused echo-planar imaging (TR = 4000 msec, TE = 60 msec, flip angle = 90 degree, inter-scan interval 8 sec, in-plane resolution 3.44 x 3.44 mm,

FOV = 22 cm, contiguous 4-mm slices to cover the entire brain) with a 1.5T MAGNETOM Vision plus MR scanner (Siemens, Erlangen, Germany) using the standard head coil. Head motion was minimized by placing tight but comfortable foam padding around the subject's head. The first five volumes of each f-MRI scan were discarded because of non-steady magnetization, with the remaining 54 volumes used for the analysis.

The f-MRI protocol was a block design with one epoch of the task conditions and the rest condition. Each epoch lasted 24 seconds equivalent to 3 whole-brain fMRI volume acquisitions. Data were analyzed with Statistical Parametric Mapping software 2 [17]. The functional magnetic resonance test was set to 8 seconds, with a scan time of 3 seconds, and a rest time of 5 seconds between scan. 54 scans were acquired for each test. The scans were realigned and transformed to the standard stereotactic space of Talairach using an EPI template [18]. Data were then smoothed in a spatial domain (full width at half-maxim = 8 x 8 x 8 mm) to improve the signal to noise ratio. After specifying the appropriate design matrix, delayed box-car function as a reference waveform, the condition, slow hemodynamic fluctuation unrelated to the task and subject effects were estimated according to a general linear model taking temporal smoothness into account. Global normalization was performed with proportional scaling. To test hypotheses about regionally specific condition effects, the estimates were compared by means of linear contrasts of each rest and task period. The resulting set of voxel values for each contrast constituted a statistical parametric map of the t statistic SPMt. For analysis of the each session, voxels and clusters of significant voxels were given a threshold of P < 0.005, not corrected for multiple comparisons.

3.2   Facing Individual Changes

We discussed the importance of transferring the intention from the patient to the actual movement of a prosthetic device in order to revert the cortical reorganization[3]. Still we face a big challenge with the changes of individual characteristics, and since we are using EMG signals, this too face changes in time. In order to overcome this challenges, we use the system developed by Kato et al.[4], for the adaptation to individual characteristics and changes over time of the EMG signals. The system use a competitive learning method to adjust to gradual and drastic changes in the patient's individual characteristics. Using a three layers neural network for the discrimination of movement intended by the patient(figure 2), this method implement three functions for learning data: automatic elimination (AE), automatic addition (AA), and selective addition (SA) to solve the problems of large data sets, which makes the discrimination process slow, and the decrease in the discrimination rate due to signal changes over time that challenge conventional approaches. AE and AA judge the discrimination state by monitoring the discrimination results all the time, and adjust to the gradual change in the characteristics by eliminationg/adding learning data according with the continuity of the motion. SA helps to adjust to drastic changes by adding new learning data sets instructed from the user(Figure 3).

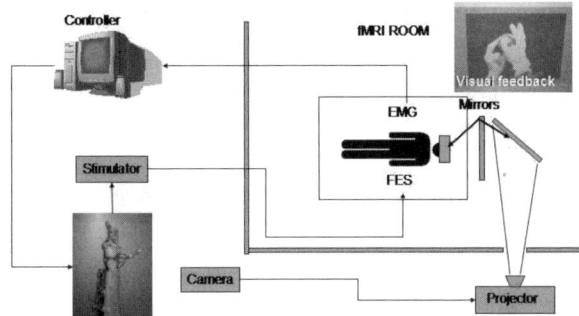

Fig. 1. f-MRI room distribution. Due to strong magnetic fields inside the room, no equipment can be placed inside the scanning room, therefore we have to take the signals through shielded cables, and present the robot hand to the experiment participants using a projector and some mirrors.

3.3 Evaluation Procedure

Three persons participated in our experiments, 2 healthy persons, and 1 right forearm amputee, with 5 years from the amputation. We used a projector and a set of mirrors as shown in figure 1 to show the robot hand as visual feedback inside the f-MRI room. The experiments were divided in two phases; first we applied electrical stimulation with 2 different intensities on the left upper arm suppressing any other stimulation as much as possible (no visual or auditive feedback was applied). To reduce the effects of the noise generated during the fMRI scan, the participants used a set of headphones to reduce the noise, and enable them to receive instructions from the control room. The participants were required to remain as still as possible for the duration of the experiment. The passive stimulation was done by sending direct commands to the electrical stimulator. The stimulation was done using a frequency of 4kHz for a biphasic balanced square signal. The voltage amplitude was of 9V. We adjusted the duty rate of the square signal to control the intensity of the stimulation. For each participant we set two levels (weak and strong) of stimulation. The second phase, we ask the participants to close the robot hand when they see a ball coming near the robot hand on the screen. The participants train the control of the robot hand before entering the room. During the experiments the participants are required to remain still, and only move the hand with the EMG sensors. The participants received an explanation of the experiment before starting. To produce the tactile feedback, one pair of electrodes were placed on the left upper arm. We placed 6 FSR pressure sensors on the fingertips and the palm. When any of the sensors is activated higher than the preset threshold, the stimulator send the stimulation signal to the arm using the parameters described above with strong stimulation. In order to evaluate the symbiosis between both systems (intelligent robot hand and human brain), we measured again the brain activation using the same

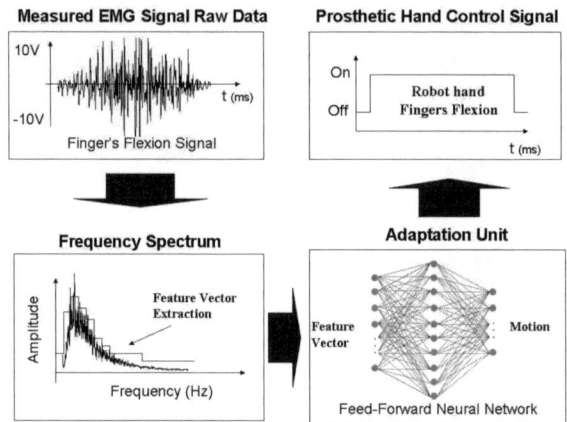

Fig. 2. EMG signal discrimination process. The EMG raw signal is extracted using dry-type sensors, the signal is processed by FFT producing later the feature vector, which is inserted as input of the classification unit (feed-forward neural network), obtaining the appropriate command for the prosthetic hand.

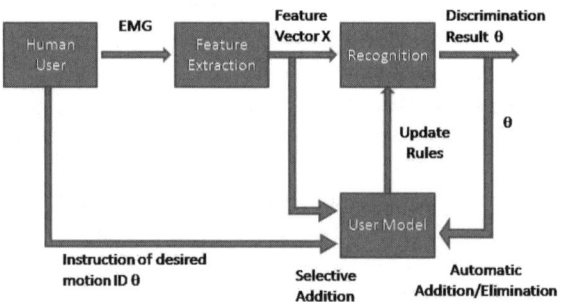

Fig. 3. Overview of the classification system. The EMG signal is processed and included in the learning data set. Once the Neural network has been trained, the system monitors the output, correcting the learning data set to adjust to the individual changes over time.

conditions than the previous experiment. In the time between experiments we asked the amputee to use the robot hand at home.

3.4 Results

We found that when applying surface electrical stimulation only, the results from the test subject presented activation on the frontal lobe, denoting the processing of a new sensation, but the activation on the somatosensory and parietal area was localized outside of the somatosensory area S1. However, when the person was asked to close the robot hand over the ball, we see that the brain activation is located on the primary motor are (M1) and primary somatosensory area S1 of the brain. Is important to consider that this effect occurs on both, healthy and amputee participants, regardless of the fact that the amputee does not have the right hand (Figure 4). Using the Talairach daemon client developed by Lancaster et al. [19] we confirmed the activation of brodmann areas 4 (M1) and 3 (S1) for Talairach coordinates x=-36 y=-17 z=56 with a cube range of 7mm. In the case of the amputee Brodmann areas 4(M1) and 3(S1) were activated for Talairach coordinates x=-38 y=-32 z=62 with a cube range of 7mm.

The f-MRI scan showed some interesting results after 3 months of continuos use of the prosthetic system. We found that the brain activation was reduced, leaving only certain "nodes" of activity. Figure 5 shows the reorganization that the brain suffered after 3 months of continuous use of the robot hand with tactile feedback.

4 Discussion

Can the artificial intelligence help us in our lives? Yes, we can say, not only with better "machines" like the dishwashers or the cleaning machines, but in a more direct interaction into our bodies. We are coming into a new age where intelligent machines work in a new type of "symbiosis", leading into a more intuitive and easy to use devices to help us in the interaction with our environment. This symbiosis can help us enhance our perception of the world, or can restore some lost functionality. Now in our specific case, when the electrical stimulation is applied in concordance with the action of grabbing an object, we have three channels working altogether, the intention from the subject, the visual feedback, and the stimulus provided by the electrical stimulation. This action is done several times during the scanning process (8 min), which allows the brain to correlate this information as a simultaneous and repetitive event. The fMRI resulting shows how the brain changes the perception of the electrical stimulation applied on the left arm. Now, the primary somatosensory area (S1) related to the hand presents an activation high enough to be detected in the fMRI image ($p¡0.005$, $T=2.69$). After an amputation, the body suffers some modifications on its body schema, what is called "cortical reorganization", because the neurons does not longer receive any signal from the corresponding sensor neuron, it start making new connections with neighboring neurons. This effect causes the spread cerebral activation observed at the beginning of our experiments (Figure 4). But the continual use of EMG based robot hands has shown that this process can be reversed. This is an example of two intelligences working together in a new "symbiosis" between man and machine. For this symbiosis to work is necessary that the machine can interpret the "intention" from its user, thus evolving into collaboration between them. The uses of artificial intelligence have shown great applications in the development of more intuitive machines that can react to the environment and predict accordingly behaviors to deal with the changing conditions. The brain still recognizes that the stimulation is done on the left

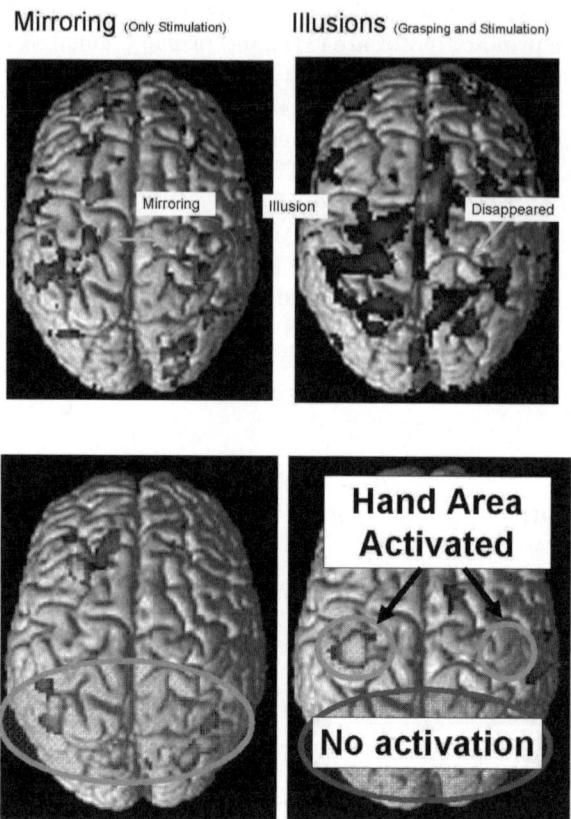

Fig. 4. Sensation Illusion. Upper f-MRI scans are from the amputee participant. Left image: Cerebral activity due to the application of electrical stimulation alone. We can see some mirroring effect between both hemispheres. Right image: Cerebral activity when the person grabs a ball using the robot hand. The mirroring effect between hemispheres disappears. The illusion of "ownership" occurs, that is, the person "feels" as if the robot hand is part of his/hers body. Lower f-MRI scans are from a healthy participant. Left Image: Cerebral activity due to the application of electrical stimulation alone, there is no activation on the arm or hand somatosensory area. Right image: Cerebral activity when the person grabs a ball using the robot hand. The mirroring effect between hemispheres disappears. The illusion of "ownership" occurs even on a healthy person.

arm. Although, when we compared the results to those of the stimulation alone, we found the activation related to the motion of the hand on the motor cortex, but also, we found the activation of the sensory area related to the right hand. This makes us think that the brain is correlating the multi-sensorial input as a single event, localizing it in to the right hand (in this case, the prosthetic hand). It is important to notice that the subjects do not interact directly with the object in question, but through the robot hand. They receive only visual

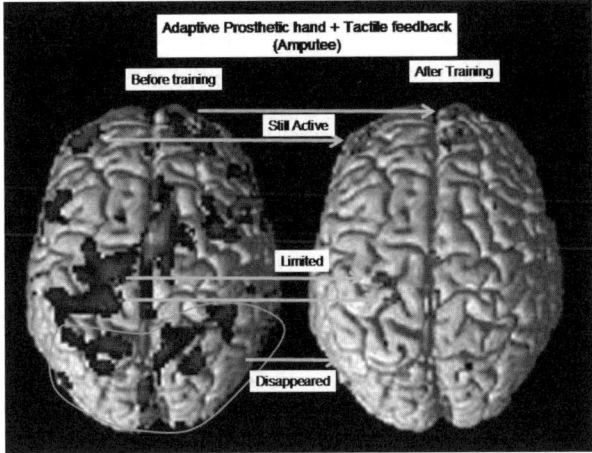

Fig. 5. Training effects. This image shows the change on the brain cortical activation levels of the amputee after using the robot hand for 3 months. The brain shows reduction on the cortical activation due to the adaptation to the prosthetic hand.

feedback through the video display and the stimulation on the left arm. These results show the possibility to use the brain plasticity into the generation of new communication channels with the robotic system. The f-MRI proved a useful tool to measure objectively the changes in the cortical activation due to the use of the prosthetic system, and allowing a more detailed feedback on the workings of the amputee brain. This allow for a more detailed medical evaluation for the rehabilitation process of an amputee. The use of the myoelectric prosthetic hands helps to stop the cortical reorganization that takes place when a limb is amputated from the body. The simultaneous application of electrical stimulation along with a system that follows the user's intention allow for the development of an "illusion" into the brain, that allows to the amputee to reconstruct its missing limb body image. All these results present a promising possibility for the development of new man-machine interfaces that allow the subconscious control of an external device, in this case a prosthetic device.

5   Conclusions

Artificial intelligence has shown during its 50 years several faces, from the deeper understanding of how we think we think (not yet there), to more mundane applications as in the dish washers. But also, artificial intelligence opened a door to more important applications, such as the medical applications, where it helps restore some function to those who for some reason lost it. When dealing with the human body, we face a highly complex system that present several challenges in the form individual differences in biological signals that change with time. The Artificial Intelligence permitted the development of adaptive tools to deal with these and other complex problems. Now, here we can also get into the subject of two intelligences working together, as we mention in the chapter above, the symbiosis between man and machine. This "symbiosis" open the door for several new applications into the life sciences, and bring hope for all those who have lost partial function of their bodies. The possibilities in this evolving field are still to be seen.

References

1. Karl, A., Birbaumer, N., Lutzenberger, W., Cohen, L.G., Flor, H.: Reorganization of motor and somatosensory cortex in upper extremity amputees with phantom limb pain. J. Neurosci. 21(10), 3609–3618 (2001)
2. Flor, H., Elbert, T., Mühlnickel, W., Pantev, C., Wienbruch, C., Taub, E.: Cortical reorganization and phantom phenomena in congenital and traumatic upper-extremity amputees. Exp. Brain Res. 119(2), 205–212 (1998)
3. Lotze, M., Flor, H., Grodd, W., Larbig, W., Birbaumer, N.: Phantom movements and pain. an fmri study in upper limb amputees. Brain 124(Pt 11), 2268–2277 (2001)
4. Kato, R., Yokoi, H., Arai, T.: Competitive learning method for robust emg-to-motion classifier. In: Arai, T. (ed.) IAS-9. Proceedings of the 9th International Conference on Intelligent Autonomous Systems (2006)
5. Pfeifer, R., Bongard, J.C.: How the body shapes the way we think: a new view of intelligence. MIT Press, Cambridge, MA (2007)
6. Holmes, N.P., Spence, C.: The body schema and the multisensory representation(s) of peripersonal space. Cogn. Process 5(2), 94–105 (2004)
7. Iriki, A., Tanaka, M., Iwamura, Y.: Coding of modified body schema during tool use by macaque postcentral neurones. Neuroreport 7(14), 2325–2330 (1996)
8. Iriki, A., Tanaka, M., Obayashi, S., Iwamura, Y.: Self-images in the video monitor coded by monkey intraparietal neurons. Neurosci. Res. 40(2), 163–173 (2001)
9. Graziano, M.S.: Where is my arm? the relative role of vision and proprioception in the neuronal representation of limb position. Proc. Natl. Acad. Sci. USA 96(18), 10418–10421 (1999)
10. Graziano, M.S.A., Gross, C.G.: The representation of extrapersonal space: a possible role for bimodal, visual-tactile neurons. The cognitive neurosciences, pp. 1021–1034 (1994)
11. Galfano, G., Pavani, F.: Long-lasting capture of tactile attention by body shadows. Exp. Brain Res. 166(3-4), 518–527 (2005)
12. O'Regan, J.K., Myin, E., No, A.: Skill, corporality and alerting capacity in an account of sensory consciousness. Prog. Brain Res. 150, 55–68 (2005)
13. Kobrinski: Problems of bioelectric control. In: Coles, J.F. (ed.) Procceedings. 1st ifac international congress. Automatic and Remote Control, vol. 2, p. 619 (1960)
14. Hernández Arieta, A., Yokoi, H., Arai, T., Wenwei, Y.: Fes as biofeedback for an emg controlled prosthetic hand. In: TENCON 2005, 2005 IEEE Region 10, pp. 1–6 (November 2005)
15. Hernández Arieta, A., Kato, R., Yokoi, H., Arai, T.: A fmri study of the cross-modal interaction in the brain with an adaptable emg prosthetic hand with biofeedback. In: Engineering in Medicine and Biology Society, 2006. EMBS 2006. 28th Annual International Conference of the IEEE, pp. 1280–1284 (August 2006)

16. Logothetis, N.K., Wandell, B.A.: Interpreting the bold signal. Annu. Rev. Physiol. 66, 735–769 (2004)
17. Missimer, J., Knorr, U., Maguire, R.P., Herzog, H., Seitz, R.J., Tellman, L., Leenders, K.L.: On two methods of statistical image analysis. Hum. Brain Mapp. 8(4), 245–258 (1999)
18. Talairach, J., Tournoux, P.: Co-Planar Stereotaxic Atlas of he Human Brain: 3-dimensional Proportional System: an Approach to Cerebral Imaging. Thieme Medical Publishers, Inc. (1988), ISBN 0865772932
19. Lancaster, J.L., Woldorff, M.G., Parsons, L.M., Liotti, M., Freitas, C.S., Rainey, L., Kochunov, P.V., Nickerson, D., Mikiten, S.A., Fox, P.T.: Automated talairach atlas labels for functional brain mapping. Hum. Brain Mapp. 10(3), 120–131 (2000)

Tests of Machine Intelligence

Shane Legg[1] and Marcus Hutter[2]

[1] IDSIA
Galleria 2, Manno-Lugano CH-6928, Switzerland
shane@idsia.ch
www.idsia.ch/~shane
[2] RSISE/ANU/NICTA
Canberra, ACT, 0200, Australia
marcus@hutter1.net
www.hutter1.net

Abstract. Although the definition and measurement of intelligence is clearly of fundamental importance to the field of artificial intelligence, no general survey of definitions and tests of machine intelligence exists. Indeed few researchers are even aware of alternatives to the Turing test and its many derivatives. In this paper we fill this gap by providing a short survey of the many tests of machine intelligence that have been proposed.

1   Introduction

Despite solid progress on many fronts over the last 50 years, artificial intelligence is still a very young field with many of its greatest achievements, and some of its most fundamental problems, yet to be tackled. From a theoretical perspective, one of the most fundamental problems in the field is that the very concept of intelligence remains rather murky. This is somewhat true in the context of humans, but it is especially true when we consider machines which may have completely different sensors, bodies, cognitive capacities and live in different environments to ourselves. What does "intelligence" mean for a machine? Perhaps the first attempt to answer this question, and certainly the only attempt that most researchers are aware of, is Alan Turing's famous imitation game [33]. Turing recognised how difficult it would be to explicitly define intelligence and thus attempted to sidestep the issue completely. Although this was a clever move, it leaves us with a test of machine intelligence that tells us almost nothing about what intelligence actually is, and thus is of little use as a foundation, either theoretical or practical, for our research.

Since then, a few bold researchers have tried to tackle this difficult problem in a more satisfactory way by proposing various definitions and tests of machine intelligence. By and large, these proposals have been ignored by the community. Indeed to the best of our knowledge, no general survey of tests and definitions of intelligence for machines has ever been published.

We feel that to ignore a question as fundamental as the definition of machine intelligence is a serious mistake. In any science, issues surrounding fundamental

M. Lungarella et al. (Eds.): 50 Years of AI, Festschrift, LNAI 4850, pp. 232–242, 2007.

definitions and methods of measurement play a central role and form the foundation on which theoretical advances are constructed and practical advances are measured. If we are to truly advance as a field over the next 50 years, we will need to return to this most central of problems in order to secure what artificial intelligence is and what it aims for. As a first step in this direction, it is necessary that researchers are at least aware of the many alternatives to Turing's tests that have been proposed. In this paper we hope to partly meet this need by providing the first general survey of tests and definitions of machine intelligence.

2   Turing Test and Derivatives

The classic approach to determining whether a machine is intelligent is the so called Turing test [33] which has been extensively debated over the last 50 years [26]. Turing realised how difficult it would be to directly definite intelligence and thus attempted to side step the issue by setting up his now famous imitation game: If human judges cannot effectively discriminate between a computer and a human through teletyped conversation, then we must conclude that the computer is intelligent.

Though simple and clever, the test has attracted much criticism. Block and Searle argue that passing the test is not *sufficient* to establish intelligence [3,28,7]. Essentially they both argue that a machine could appear to be intelligent without having any "real intelligence", perhaps by using a very large table of answers to questions. While such a machine might be impossible in practice due to the vast size of the table required, it is not logically impossible. In which case an unintelligent machine could, at least in theory, consistently pass the Turing test. Some consider this to bring the validity of the test into question. In response to these challenges, even more demanding versions of the Turing test have been proposed such as the Total Turing test [11], the Truly Total Turing test [27] and the inverted Turing test [35]. Dowe argues that the Turing test should be extended by ensuring that the agent has a compressed representation of the domain area, thus ruling out look-up table counter arguments [6]. Of course these attacks on the Turing test can be applied to any test of intelligence that considers only a system's external behaviour, that is, most intelligence tests.

A more common criticism is that passing the Turing test is not *necessary* to establish intelligence. Usually this argument is based on the fact that the test requires a machine to have a highly detailed model of human knowledge and patterns of thought, making it a test of humanness rather than intelligence [9,8]. Indeed even small things like pretending to be *unable* to perform complex arithmetic quickly and faking human typing errors become important, something which clearly goes against the purpose of the test.

The Turing test has other problems as well. Current AI systems are a long way from being able to pass an unrestricted Turing test. From a practical point of view this means that the full Turing test is unable to offer much guidance to our work. Indeed, even though the Turing test is the most famous test of machine intelligence, almost no current research in artificial intelligence is specifically directed toward being able to pass it. Unfortunately, simply restricting the domain of conversation in the Turing test to make the test easier, as is done in the Loebner competition [22], is not sufficient. With restricted conversation possibilities the most successful Loebner entrants are even more focused on faking human fallibility, rather than anything resembling intelligence [15]. Perhaps a better alternative then is to test whether a machine can imitate a child (see for example the tests described in Sections 4 and 5). Finally, the Turing test returns different results depending on who the human judges are. Its unreliability has in some cases lead to clearly unintelligent machines being classified as human, and at least one instance of a human actually failing a Turing test. When queried about the latter, one of the judges explained that "no human being would have that amount of knowledge about Shakespeare"[29].

3   Compression Tests

Mahoney has proposed a particularly simple solution to the binary pass or fail problem with the Turing test: Replace the Turing test with a text compression test [23]. In essence this is somewhat similar to a "Cloze test" where an individual's comprehension and knowledge in a domain is estimated by having them guess missing words from a passage of text.

While simple text compression can be performed with symbol frequencies, the resulting compression is relatively poor. By using more complex models that capture higher level features such as aspects of grammar, the best compressors are able to compress text to about 1.5 bits per character for English. However humans, which can also make use of general world knowledge, the logical structure of the argument etc., are able to reduce this down to about 1 bit per character. Thus the compression statistic provides an easily computed measure of how complete a machine's model of language, reasoning and domain knowledge are, relative to a human.

To see the connection to the Turing test, consider a compression test based on a very large corpus of dialogue. If a compressor could perform extremely well on such a test, this is mathematically equivalent to being able to determine which sentences are probable at a given point in a dialogue, and which are not (for the equivalence of compression and prediction see [2]). Thus, as failing a Turing test occurs when a machine (or person!) generates a sentence which would be improbable for a human, extremely good performance on dialogue compression implies the ability to pass a Turing test.

A recent development in this area is the Hutter Prize [17]. In this test the corpus is a 100 MB extract from Wikipedia. The idea is that this should represent a reasonable sample of world knowledge and thus any compressor that can perform very well on this test must have a good model of not just English, but also world knowledge in general.

One criticism of compression tests is that it is not clear whether a powerful compressor would easily translate into a general purpose artificial intelligence.

4 Linguistic Complexity

A more linguistic approach is taken by the HAL project at the company Artificial Intelligence NV [32]. They propose to measure a system's level of conversational ability by using techniques developed to measure the linguistic ability of children. These methods examine things such as vocabulary size, length of utterances, response types, syntactic complexity and so on. This would allow systems to be "... assigned an age or a maturity level beside their binary Turing test assessment of 'intelligent' or 'not intelligent' "[31]. As they consider communication to be the basis of intelligence, and the Turing test to be a valid test of machine intelligence, in their view the best way to develop intelligence is to retrace the way in which human linguistic development occurs. Although they do not explicitly refer to their linguistic measure as a test of intelligence, because it measures progress towards what they consider to be a valid intelligence test, it acts as one.

5 Multiple Cognitive Abilities

A broader developmental approach is being taken by IBM's Joshua Blue project [1]. In this project they measure the performance of their system by considering a broad range of linguistic, social, association and learning tests. Their goal is to first pass what they call a "toddler Turing test", that is, to develop an AI system that can pass as a young child in a similar setup to the Turing test. As yet, this test is not fully specified.

Another company pursuing a similar developmental approach based on measuring system performance through a broad range of cognitive tests is the a2i2 project at Adaptive AI [34]. Rather than toddler level intelligence, their current goal to is work toward a level of cognitive performance similar to that of a small mammal. The idea being that even a small mammal has many of the key cognitive abilities required for human level intelligence working together in an integrated way. While this might be useful to guide the development of moderate intelligence, it is unknown whether it will scale to higher levels of intelligence. The specific tests being used have not been published.

6 Competitive Games

The Turing Ratio method of Masum et al. has more emphasis on tasks and games rather than cognitive tests. They propose that "... doing well at a broad range of tasks is an empirical definition of 'intelligence'."[24] To quantify this they seek to identify tasks that measure important abilities, admit a series of strategies that are qualitatively different, and are reproducible and relevant over an extended period of time. They suggest a system of measuring performance through pairwise comparisons between AI systems that is similar to that used to rate players in the international chess rating system. The key difficulty however, which the authors acknowledge is an open challenge, is to work out what these tasks should be, and to quantify just how broad, important and relevant each is. In our view these are some of the most central problems that must be solved when attempting to construct an intelligence test and thus this approach is incomplete in its current state.

7   Collection of Psychometric Tests

An approach called Psychometric AI tries to address the problem of what to test for in a pragmatic way. In the view of Bringsjord and Schimanski, "Some agent is intelligent if and only if it excels at all established, validated tests of [human] intelligence."[4] They later broaden this to also include "tests of artistic and literary creativity, mechanical ability, and so on." With this as their goal, their research is focused on building robots that can perform well on standard psychometric tests designed for humans, such as the Wechsler Adult Intelligent Scale and Raven Progressive Matrices.

As effective as these tests are for humans, they seem inadequate for measuring machine intelligence as they are highly anthropocentric and embody basic assumptions about the test subject that are likely to be violated by computers. For example, consider the fundamental assumption that the test subject is not simply a collection of specialised algorithms designed only for answering common IQ test questions. While this is obviously true of a human, or even an ape, it may not be true of a computer. The computer could be nothing more than a collection of specific algorithms designed to identify patterns in shapes, predict number sequences, write poems on a given subject or solve verbal analogy problems — all things that AI researchers have worked on. Such a machine might be able to obtain a respectable IQ score [25], even though outside of these specific test problems it would be next to useless. If we try to correct for these limitations by expanding beyond standard tests, as Bringsjord and Schimanski seem to suggest, this once again opens up the difficulty of exactly what, and what not, to test for. Psychometric AI, at least as it is currently formulated, only partially addresses this central question.

8   Smith's Test

The basic structure of Smith's test is that an agent faces a series of problems that are generated by an algorithm [30]. In each iteration the agent must try to produce the correct response to the problem that it has been given. The problem generator then responds with a score of how good the agent's answer was. If the agent so desires it can submit another answer to the same problem. At some point the agent requests to the problem generator to move onto the next problem and the score that the agent received for its last answer to the current problem is then added to its cumulative score. Each interaction cycle counts as one time step and the agent's intelligence is then its total cumulative score considered as a function of time. In order to keep things feasible, the problems must all be in P, i.e. the solution must be verifiable in polynomial time.

We have two main criticisms of Smith's definition. Firstly, while for practical reasons it might make sense to restrict problems to be in P, we do not see why this practical restriction should be a part of the very definition of intelligence as Smith suggests. If some breakthrough meant that agents could solve difficult problems in not just P but sometimes in NP as well, then surely these new agents would be more intelligent?

Secondly, while the definition is somewhat formally defined, it still leaves open the important question of what exactly the tests should be. Smith suggests that researchers should dream up tests and then contribute them to some common pool of tests. As such, this is not a fully specified test.

9   C-Test

One perspective among psychologists who support the g-factor view of intelligence, is that intelligence is "the ability to deal with complexity"[10]. Thus in a test of intelligence the most difficult questions are the ones that are the most complex because these will, by definition, require the most intelligence to solve. It follows then that if we could formally define and measure the complexity of test problems we could construct a formal test of intelligence. The possibility of doing this was perhaps first suggested by the complexity theorist Chaitin [5]. While this path requires numerous difficulties to be dealt with, we believe that it is the most natural and offers many advantages: It is formally motivated, precisely defined and potentially could be used to measure the performance of both computers and biological systems on the same scale without the problem of bias towards any particular species or culture.

One intelligence test that is based on formal complexity theory is the C-Test from Hernández [13,14]. This test consists of a number of sequence prediction and abduction problems similar to those that appear in many standard IQ tests. Similar to standard IQ tests, the C-Test always ensures that each question has an unambiguous answer in the sense that there is always one hypothesis that is consistent with the observed pattern that has significantly lower complexity than the alternatives. The key difference to sequence problems that appear in standard intelligence tests is that the questions are based on a formally expressed measure of complexity, namely Levin's computable Kt complexity [20] (rather than Kolmogorov's incomputable complexity [21]) to get a practical test. In order to retain the invariance property of Kolmogorov complexity, Levin complexity requires the additional assumption that the universal Turing machines are able to simulate each other in linear time.

The test has been successfully applied to humans with intuitively reasonable results [14,12]. As far as we know, this is the only formal definition of intelligence that has so far produced a usable test of intelligence.

One criticism of the C-Test and Smith's tests is that the way intelligence is measured is essentially static, that is, the environments are passive. We believe that dynamic testing in active environments is a better measure of a system's intelligence. To put this argument another way: Succeeding in the real world requires you to be more than an insightful spectator! One must carefully choose actions knowing that these may affect the future.

10   Universal Intelligence

Another complexity based test is the *universal intelligence* test [19]. Unlike the C-Test and Smith's test, universal intelligence tests the performance of an agent in a fully interactive environment. This is done by using the reinforcement learning framework in which the agent sends its *actions* to the environment and receives *observations* and *rewards* back. The agent tries to maximise the amount of reward it receives by learning about the structure of the environment and the goals it needs to accomplish in order to receive rewards.

Formally, the process of interaction produces an increasing history $o_1 r_1 a_1 o_2 r_2 a_2 o_3 r_3 a_3 o_4 \ldots$ of observations o, rewards $r \geq 0$, and actions a. The agent is simply a function, denoted by π, which is a probability measure over actions conditioned on the current history, for example, $\pi(a_3|o_1 r_1 a_1 o_2 r_2)$. The environment, denoted μ, is similarly defined: $\mu(o_k r_k|o_1 r_1 a_1 o_2 r_2 a_2 \ldots o_{k-1} r_{k-1} a_{k-1})$. The performance of agent π in environment μ can be measured by its total expected reward $V_\mu^\pi := \mathbf{E}[\sum_{i=1}^\infty r_i | \mu, \pi]$, called value. The largest interesting class of environments is the class E of all computable probability distributions μ. For technical reasons, the values are assumed to be bounded by some constant c.

To get a single performance measure V_μ^π is averaged over all $\mu \in E$. As there are an infinite number of environments, with no bound on their complexity, it is impossible to take the expected value with respect to a uniform distribution — some environments must be weighted more heavily than others. Considering the agent's perspective on the problem, it is the same as asking: Given several different hypotheses which are consistent with the observations, which hypothesis should be considered the most likely? This is a fundamental problem in inductive inference for which the standard solution is to invoke Occam's razor: *Given multiple hypotheses which are consistent with the data, simpler ones should be preferred.* As this is generally considered the most intelligent thing to do, one should test agents in such a way that they are, at least on average, rewarded for correctly applying Occam's razor. This means that the a priori distribution over environments should be weighted towards simpler environments.

As each environment μ is described by a computable measure, their complexity can be measured with Kolmogorov complexity $K(\mu)$, which is simply the length of the shortest program that computes μ [21]. The right a priori weight for μ is $2^{-K(\mu)}$. We can now define the *universal intelligence* of an agent π to simply be its expected performance,

$$\Upsilon(\pi) := \sum_{\mu \in E} 2^{-K(\mu)} V_\mu^\pi.$$

By construction, universal intelligence measures the general ability of an agent to perform well in a very wide range of environments, similar to the essence of many informal definitions of intelligence [18]. The definition places no restrictions on the internal workings of the agent; it only requires that the agent is capable of generating output and receiving input which includes a reward signal. If we wish to bias the test to reflect world knowledge then we can condition the complexity measure. For example, use $K(\mu|D)$ where D is some set of background knowledge such as Wikipedia.

By considering V_μ^π for a number of basic environments, such as small MDPs, and agents with simple but very general optimisation strategies, it is clear that Υ correctly orders the relative intelligence of these agents in a natural way. A very high value of Υ would imply that an agent is able to perform well in many environments. The maximal agent with respect to Υ is the theoretical AIXI agent which has been shown to have many strong optimality properties [16]. These results confirm that agents with high universal intelligence are indeed very powerful and adaptable. Universal intelligence spans simple adaptive agents right up to super intelligent agents like AIXI. The test is completely formally specified in terms of fundamental concepts such as universal Turing computation and complexity and thus is not anthropocentric.

A test based on Υ would evaluate the performance of an agent on a large sample of simulated environments, and then combine the agent's performance in each environment into an overall intelligence value. The key challenge that needs to be dealt with is to find a suitable replacement for the incomputable Kolmogorov complexity function, possibly Levin's Kt complexity [20], as is done by the C-Test.

11   Summary

We end this survey with a comparison of the various tests considered. Table 1 rates each test according to the properties described below. Although we have attempted to be as fair as possible, some of the scores we give on this table will naturally be debatable. Nevertheless, we hope that it provides a rough overview of the relative strengths and weaknesses of the proposals.

Valid: A test of intelligence should capture intelligence and not some related quantity. *Informative*: The result should be a scalar value, or perhaps a vector. *Wide range*: A test should cover low levels of intelligence up to super intelligence. *General*: Ideally we would like to have a very general test that could be applied to everything from a fly to a machine learning algorithm. *Dynamic*: A test should directly take into account the ability to learn and adapt over time. *Unbiased*: A test should not be biased towards any particular culture, species, etc. *Fundamental*: We do not want a test that needs to be changed from time to time due to changing technology and knowledge. *Formal*: The test should be precisely defined, ideally using mathematics. *Objective*: The test should not appeal to subjective assessments such as the opinions of human judges. *Fully Defined*: Has the test been fully defined, or are parts still unspecified? *Universal*: Is the test universal, or is it anthropocentric? *Practical*: A test should be able to be performed quickly and automatically. *Test vs. Def*: Finally we note whether the proposal is more of a test, more of a definition, or something in between.

Table 1. In the table ● means "yes", • means "debatable", · means "no", and ? means unknown. When something is rated as unknown that is usually because the test in question is not sufficiently specified.

Intelligence Test	Valid	Informative	Wide Range	General	Dynamic	Unbiased	Fundamental	Formal	Objective	Fully Defined	Universal	Practical	Test vs. Def.
Turing Test	●	·	·	·	●	·	·	·	·	●	·	●	T
Total Turing Test	●	·	·	·	●	·	·	·	·	●	·	·	T
Inverted Turing Test	●	•	·	·	●	·	·	·	·	●	·	●	T
Toddler Turing Test	●	·	·	·	●	·	·	·	·	·	·	●	T
Linguistic Complexity	●	●	•	·	·	·	·	●	●	·	●	●	T
Text Compression Test	●	●	●	•	·	•	•	●	●	●	•	●	T
Turing Ratio	●	●	●	●	?	?	?	?	?	·	?	?	T/D
Psychometric AI	●	●	•	●	?	•	·	●	•	•	·	·	T/D
Smith's Test	●	●	●	•	·	?	●	●	●	·	?	•	T/D
C-Test	●	●	●	•	·	●	●	●	●	●	●	●	T/D
Universal Intelligence	●	●	●	●	●	●	●	●	●	●	●	·	D

Acknowledgements. This work was supported by SNF grant 200020-107616.

References

1. Alvarado, N., Adams, S., Burbeck, S., Latta, C.: Beyond the Turing test: Performance metrics for evaluating a computer simulation of the human mind. In: Performance Metrics for Intelligent Systems Workshop, Gaithersburg, MD, USA, North-Holland, Amsterdam (2002)
2. Bell, T.C., Cleary, J.G., Witten, I.H.: Text compression. Prentice-Hall, Englewood Cliffs (1990)
3. Block, N.: Psychologism and behaviorism. Philosophical Review 90, 5–43 (1981)
4. Bringsjord, S., Schimanski, B.: What is artificial intelligence? Psychometric AI as an answer. In: Eighteenth International Joint Conference on Artificial Intelligence, vol. 18, pp. 887–893 (2003)
5. Chaitin, G.J.: Gödel's theorem and information. International Journal of Theoretical Physics 22, 941–954 (1982)
6. Dowe, D.L., Hajek, A.R.: A non-behavioural, computational extension to the Turing test. In: ICCIMA 1998. International Conference on Computational Intelligence & Multimedia Applications, pp. 101–106. Gippsland, Australia (1998)
7. Eisner, J.: Cognitive science and the search for intelligence. Invited paper presented to the Socratic Society, University of Cape Town (1991)
8. Ford, K.M., Hayes, P.J.: On computational wings: Rethinking the goals of artificial intelligence. Scientific American, Special edn. (4) (1998)
9. French, R.M.: Subcognition and the limits of the Turing test. Mind 99, 53–65 (1990)
10. Gottfredson, L.S.: Why g matters: The complexity of everyday life. Intelligence 24(1), 79–132 (1997)
11. Harnad, S.: Minds, machines and Searle. Journal of Theoretical and Experimental Artificial Intelligence 1, 5–25 (1989)

12. Hernández-Orallo, J.: Beyond the Turing test. Journal of Logic, Language and Information 9(4), 447–466 (2000)
13. Hernández-Orallo, J.: On the computational measurement of intelligence factors. In: Performance Metrics for Intelligent Systems Workshop, Gaithersburg, MD, USA, pp. 1–8 (2000)
14. Hernández-Orallo, J., Minaya-Collado, N.: A formal definition of intelligence based on an intensional variant of Kolmogorov complexity. In: EIS 1998. Proceedings of the International Symposium of Engineering of Intelligent Systems, pp. 146–163. ICSC Press (1998)
15. Hutchens, J.L.: How to pass the Turing test by cheating (1996), www.cs.umbc.edu/471/current/papers/hutchens.pdf
16. Hutter, M.: Universal Artificial Intelligence: Sequential Decisions based on Algorithmic Probability, p. 300 Springer, Berlin (2005), http://www.hutter1.net/ai/uaibook.htm
17. Hutter, M.: The Human knowledge compression prize (2006), http://prize.hutter1.net
18. Legg, S., Hutter, M.: A collection of definitions of intelligence. In: Goertzel, B. (ed.) Proc. 1st Annual artificial general intelligence workshop (to appear), Online version www.idsia.ch/ shane/intelligence.html
19. Legg, S., Hutter, M.: A formal measure of machine intelligence. In: Benelearn 2006. Annual Machine Learning Conference of Belgium and The Netherlands, Ghent (2006)
20. Levin, L.A.: Universal sequential search problems. Problems of Information Transmission 9, 265–266 (1973)
21. Li, M., Vitányi, P.M.B.: An introduction to Kolmogorov complexity and its applications, 2nd edn. Springer, Heidelberg (1997)
22. Loebner, H.G.: The Loebner prize — The first Turing test (1990), http://www.loebner.net/Prizef/loebner-prize.html
23. Mahoney, M.V.: Text compression as a test for artificial intelligence. In: AAAI/IAAI (1999)
24. Masum, H., Christensen, S., Oppacher, F.: The Turing ratio: Metrics for open-ended tasks. In: GECCO 2002. Proceedings of the Genetic and Evolutionary Computation Conference, pp. 973–980. Morgan Kaufmann Publishers, New York (2002)
25. Sanghi, P., Dowe, D.L.: A computer program capable of passing I.Q. tests. In: Sloot, P.M.A., Abramson, D., Bogdanov, A.V., Gorbachev, Y.E., Dongarra, J.J., Zomaya, A.Y. (eds.) ICCS 2003. LNCS, vol. 2657, pp. 570–575. Springer, Heidelberg (2003)
26. Saygin, A., Cicekli, I., Akman, V.: Turing test: 50 years later. Minds and Machines 10 (2000)
27. Schweizer, P.: The truly total Turing test. Minds and Machines 8, 263–272 (1998)
28. Searle, J.: Minds, brains, and programs. Behavioral & Brain Sciences 3, 417–458 (1980)
29. Shieber, S.: Lessons from a restricted Turing test. CACM: Communications of the ACM 37 (1994)
30. Smith, W.D.: Mathematical definition of "intelligence" (and consequences) (2006), http://math.temple.edu/~wds/homepage/works.html
31. Treister-Goren, A., Dunietz, J., Hutchens, J.L.: The developmental approach to evaluating artificial intelligence – a proposal. In: Performance Metrics for Intelligence Systems (2000)
32. Treister-Goren, A., Hutchens, J.L.: Creating AI: A unique interplay between the development of learning algorithms and their education. In: Proceeding of the First International Workshop on Epigenetic Robotics (2001)

33. Turing, A.M.: Computing machinery and intelligence. Mind (October 1950)
34. Voss, P.: Essentials of general intelligence: The direct path to AGI. In: Goertzel, B., Pennachin, C. (eds.) Artificial General Intelligence, Springer, Heidelberg (2005)
35. Watt, S.: Naive psychology and the inverted Turing test. Psycoloquy 7(14) (1996)

A Hierarchical Concept Oriented Representation for Spatial Cognition in Mobile Robots

Shrihari Vasudevan, Stefan Gächter, Ahad Harati, and Roland Siegwart

Autonomous Systems Laboratory,
Swiss Federal Institute of Technology Zürich,
8092 Zürich, Switzerland
{vasudevs,stefanga,haratia,rsiegwart}@ethz.ch

Abstract. Robots are rapidly evolving from factory work-horses to robot-companions. The future of robots, as our companions, is highly dependent on their abilities to understand, interpret and represent the environment in an efficient and consistent fashion, in a way that is compatible to humans. The work presented here is oriented in this direction. It suggests a hierarchical, concept oriented, probabilistic representation of space for mobile robots. A salient aspect of the proposed approach is that it is holistic - it attempts to create a consistent link from the sensory information the robot acquires to the human-compatible spatial concepts that the robot subsequently forms, while taking into account both uncertainty and incompleteness of perceived information. The approach is aimed at increasing spatial awareness in robots.

1 Introduction

Robotics today, is visibly and very rapidly moving beyond the realm of factory floors. Robots are working their way into our homes in an attempt to fulfill our needs for household servants, pets and other cognitive robot companions. If this "robotic-revolution" is to succeed, it is going to warrant a very powerful repertoire of skills on the part of the robot. Apart from navigation and manipulation, the robot will have to understand, interpret and represent the environment in an efficient and consistent fashion. It will also have to interact and communicate in human-compatible ways. Each of these is a very hard problem. These problems are made difficult by a multitude of reasons including the extensive amount of information, the huge number of types of data (multi-modality), the presence of entities in the environment which change with time, to name a few. Adding to all of these problems are two simple facts - everything is uncertain and at any time, only partial knowledge of the environment is available.

The underlying representation of the robot is probably the single most critical component in that it constitutes the very foundation for all things we might expect the robot to do, these include the many complex tasks mentioned above. Thus, the extent to which robots will evolve from factory work-horses to robot-companions will in some ways, albeit indirectly, be decided by the way they represent their surroundings. This chapter is thus dedicated towards finding an appropriate representation that will make today's dream, tomorrow's reality.

M. Lungarella et al. (Eds.): 50 Years of AI, Festschrift, LNAI 4850, pp. 243–256, 2007.

2   A Brief History in AI

Knowledge representation has been a very pivotal component of AI research. This has yielded a significant number of representation methodologies, ontologies and programming languages oriented towards representing knowledge. In the context of robotics, there have been broadly two schools of thought. The first school of thought believed in the more conventional AI based approach of using a representation as the basis of all forms of artificial intelligence. Works centered on this philosophy relied on a formal perception-representation-action loop with a reliable interface between the modules within the system. This approach exhibited two weaknesses - slow progress with the state-of-the-art being dominated by symbolic (not grounded) results and slow speed due to the use of a centralized controller mechanism. These issues consequently heralded the formation of a new paradigm for intelligence - one that would do away with the use of a formal representation as previously understood by the AI community. A very representative work of this new basis for intelligence was [1]. Brooks argued against the use of a formal representation as he believed that the appropriate formulation of one was an almost intractable problem. His work prescribed the real world as being its own model and suggested that various action producing modules directly interface with the real world rather than between themselves. This behavior based /subsumption / reactive methodology produced situated results in robotic platforms within a very short span of time. While this approach has had tangible success in the context of lower level robot sensory-motor skills, we believe that higher level cognitive capabilities such as natural language interaction, manipulation, spatial cognition etc. would require a more powerful basis (an appropriate representation) to realize. Similar reflections can be obtained from more recent works [2]. The need for a suitable representational basis forms the central motivation for the work presented here and relates it to past AI research. However, taking inspiration from prior research, an attempt has been made to address the concerns of representation based approaches. Our methodology prescribes a ground-up formation of the representation. The requirement of a consistent link from sensory data to the more abstract concepts is very strictly enforced in our approach. Thus, the approaches proposed here yield situated embodied intelligent agents. We also take into account two fundamental ubiquities that such agents have to deal with - incompleteness and uncertainty. Further, rather than addressing the 'knowledge representation for intelligence' problem in general, our approach focuses on a representation that is suited for mobile robots in the context of spatial cognition and navigation. The proposed representation is aimed at making robots more spatially aware of their surroundings. Thus, issues such as speed, scalability and performance would be more easily dealt with.

3   State of the Art

Robot mapping is a well researched problem, however, with many very interesting challenges yet to be solved. An excellent and fairly comprehensive survey of robot mapping has been presented in [3]. Robot mapping has traditionally been classified into two broad categories - metric and topological. Metric mapping ([4] & [5]) tries to map the environment using geometric features present in it. A related concept in this context is that of the relative map [6] - a map state with quantities invariant to rotation and translation of the robot. Topological mapping ([7] & [8]) usually involves encoding place related data and information on how to get from one place to another. The more recent scheme of hybrid mapping ([9] & [10]) typically uses both a metric map for precision navigation in a local space and a global topological map for moving between places.

The one similarity between all these representations is that all of them are navigation-oriented, i.e. all of them are built around the single application of robot-navigation. These maps fail to encode the semantics of the environment. This leaves them with little scope for use in more complex and interactive tasks. This is also the reason that the level of spatial awareness in current robot systems is quite modest. The focus of this work is to address this deficiency. A single unified representation that is multi-resolution, multi-modal, probabilistic and consistent is still a vision of the future and is the aspiration of this work.

Typically, humans seem to perceive space in terms of objects, states and descriptions, relationships etc. This seems both intuitive and is also subsequently validated through user studies that were conducted as a part of this work [11]. Thus, a *cognitive* or human compatible spatial representation could be expected to encode similar information. The major issues that need to be addressed towards having a mobile robot do this include high level feature[1] extraction (HLFE), representation (assimilation and modeling of the information) and cognition (reasoning and understanding through the acquired representation). Each of these issues are addressed in the approach suggested here.

The representation presented here takes inspiration from the way we believe humans represent space and the notion of a hierarchical representation of space. Ref. [12] suggests one such hierarchy for environment modeling. In [13], Kuipers put forward a *Spatial Semantic Hierarchy* which models space in layers comprising respectively of sensorimotor, view-based, place-related and metric information. Since the introduction of the term *Cognitive Map* in Tolman's seminal work [14], many research efforts have attempted to understand and conceptualize a cognitive map. The most relevant works include those of Kuipers [15] and Yeap [16]. The former viewed the cognitive map as having five different kinds of information (topological, metric, routes, fixed features and observations) each with its own representation. Yeap et al. in [16], review prior research on *early cognitive mapping* and classify representations as being space based or object based. The proposed approach attempts to take the best of both worlds.

Object classification, an instance of HLFE, is a hard problem because of the challenges that accrue from the objects in question (appearance change across views of object and objects within class), the environment (occlusion and clutter), and the sensor in use (various forms of noise). Representations for

---

[1] Objects, doors, walls etc. are considered high-level features contrasting with lines, corners etc. which are considered low-level ones.

classification span prototypical models (class based or generic models) to exemplar-based models (template or appearance-based models). Historically, approaches to object classification / recognition moved from generic to exemplar based approaches [17]. However, current efforts are being redirected towards generic ones. In particular, one important representation, that is also the basis for the approach presented here, is the functionality of the object. One of the more influential concepts in psychology about an object's function was introduced by Gibson [18]. It put forward the notion of affordance, which can be defined as the functionality an object offers to an agent. Thus, the function an object can afford, not only depends on the physical structure of the object but also on the action of the agent on the object. For example, a chair's function depends on whether an agent intends to use it to sit at a table or to to climb on it and use it as a ladder. However, in computer vision literature, functionality as used in the contexts of representation, classification and recognition has typically referred to semantic annotations of the object's structure. A good survey of techniques developed in this context has been presented in [19].

Several previous works ([20], [21] and [22]) inspire our approach towards functional object classification. The general approach undertaken in these most representative works comprised of the following elements. A functionality was generally defined as a combination of functional parts, which in turn were understood as a set of object-parts with associated attributes. This is in accordance with a school of thought that proposed to associate a correspondence between functionality and object structure. Different forms of segmentation including planes and surface patches were used. Learning and representation included the use of histograms, multi-variate Gaussians and also more simplistic models. The classification process itself used diverse methods including verification trees, Bayes-nets, probabilistic grammars, voting methods and graph based search algorithms. Despite these noteworthy contributions, two aspects warrant further research - the first being a consistent probabilistic framework for functional object classification that works in real world environments. The other is the adaptation of these techniques to suit the complexities that plague mobile robotics - uncertainty and incompleteness. The approach presented in this work shows some of the steps being taken towards this objective.

Another aspect of the sought representation is the extraction of structural elements including doors, walls, ceilings and so on. Several works have attempted to model and detect doors. The explored techniques range from modeling the door opening [23] to those that model / estimate door parameters [24] and to those like [25], based on algorithms such as boosting. While there are numerous works in mobile robotics that detect the presence of structural elements through simplified methods, towards their larger objectives, few works exist that appropriately model structural aspects of the environment in order to enable a robot to make semantically meaningful inferences on the structure of its surroundings - this is the motivation for the proposed approach. Recent inspiring contributions to our approach include [26], which generated a structural model of indoor environments by segmenting and matching planar patches generated using a

3D laser scanner against a coarse semantic description which captures aspects such as parallelism, orthogonality etc. between structural elements; [27] which proposed a similar model (for outdoor environments) but with a more detailed semantic description and [28], which generated a structural model by classifying each data point as being part of a floor, object or ceiling - the salient aspect being that segmentation and labeling were performed simultaneously.

Increasingly intelligent robots are tending to be more-and-more socially interactive [29]. In the future, intelligence and the ability to meaningfully communicate will be critically important factors determining the compatibility and acceptability of robots in our homes. Most works in mobile robotics have until now restricted themselves to navigation related problems. Thus, few works evaluate their concepts in human centered experiments. A recent work which attempted to understand the acceptability of robots among people through a user study is done in [30]. This work was done on the sidelines of [31], which was a recent large scale demonstration of the remarkable growth of personal and service robotics. The representation proposed in this work promises to enable robots to not only perform navigation related tasks but also to be more spatially aware and human-compatible machines that could inhabit our homes alongside us. With the rapid increase in the importance of human robot interaction, the need for evaluating the work through human centered experiments was felt necessary. Further, it was felt that such experiments could contribute positively to the enhancement of the work itself. With this view, an elaborate user study was conducted to understand human perception and representation of spaces. This has been detailed later in section 4.4.

4   Approach

The proposed approach is shown in fig. 1. The principle idea is that by adding concepts (for instance, based on functionality) in the representation, semantics can be embedded in a purely navigation oriented spatial representation. The resulting representation can be understood as a hierarchical, functional representation of space. The following sub-sections elicit three mutually independent but complimentary directions of work which are integrable under the framework of the general approach and are aimed at addressing the issues raised earlier.

4.1   Towards an Object Based Representation of Space

The representation put forward here is a hierarchical one that is composed of places which are connected to each other through doors (structural elements) and are themselves represented by local probabilistic object graphs (probabilistic graph encoding the objects and relationships between them). This work attempts to research the kinds of information that could be incorporated in the representation and the manner in which this information may be used towards adding more semantic information, in the form of increasingly abstract concepts, in the representation. The extraction of the high level features would be supported through parallel ongoing efforts detailed in sections 4.2 and 4.3.

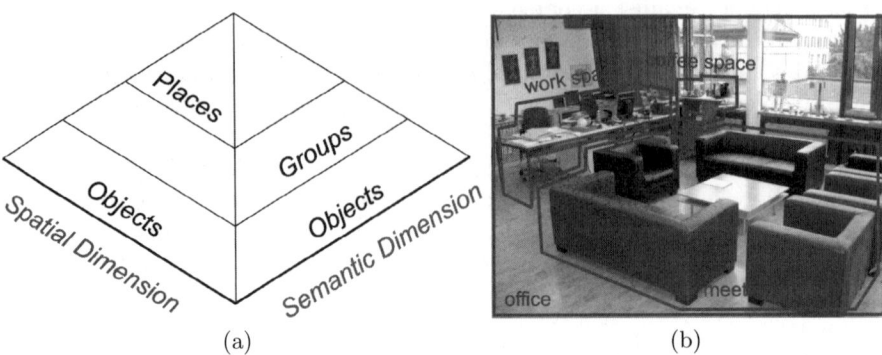

(a)   (b)

Fig. 1. **(a)** General approach - A robot uses the sensory information it perceives to identify high level features such as objects, doors etc. These objects are grouped into abstractions along two dimensions - spatial and semantic. Along the semantic dimension, objects are clustered into groups so as to capture the spatial semantics. Along the spatial dimension, places are formed as a collection of groups of objects. Spatial abstractions are primarily perceptual formations (occurrence of walls, doors etc.) whereas semantic or functional abstractions are primarily conceptual formations (similarity of purpose / functionality ; spatial arrangement). The representation is a single hierarchy composed of sensory information being mapped to increasingly abstract concepts. **(b)** An example scenario - The figure depicts a typical office setting. The approach proposed in this work would would enable a robot to recognize various objects, cluster the respective objects into meaningful semantic entities such as a meeting space and a work space, infer the presence of a being in a room which has a cuboidal shape and even understand that the place is an office because of the presence of a place to work and one to conduct meetings.

The detailed approach is elicited in [32]. The perception system included methods for object recognition and door detection. For this work, a SIFT based object recognition system was developed along the lines of [33]. A stereo camera was used to recognize the object and to obtain its coordinates in 3D space. Doors were used in this work in the context of place formation. A method of door detection based on line extraction and the application of certain heuristics, was used. The sensor of choice was the laser range finder. Knowing the robots pose (using odometry) relative to a local reference, these objects and doors are identified in the local frame of reference. Using this information, a probabilistic graphical representation encoding the objects and the relative spatial information between them is formed as a local representation for the place. The local representations of different places were connected through the doors that link them. In this way, the formed representation could be understood either as an extended relative metric representation (from the design perspective) or as a hierarchical metric-topological-semantic representation of space where the topological information is given by the places and the semantic content is encoded by using objects and their properties. Figure 2(a) depicts a 2D representation of the resultant object based representation - a section of which is shown in fig. 2(b).

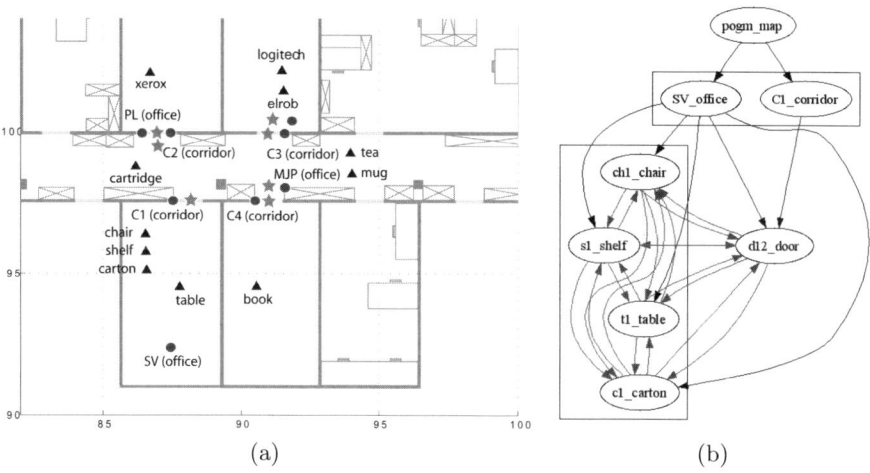

(a)   (b)

Fig. 2. (a) Object map produced as a result of exploring the test environment. Red circles are the place references, blue triangles are the objects and the green stars are the doors. **(b)** Probabilistic object graph representation of a single room and its connection to an adjacent one through the door. Places such as SV_office correspond respectively to the SV(office) shown in figure (a). Each place has a set of "children" objects, these correspond directly to the objects mapped in the respective place. Black lines link the place to the objects within it. Red lines (lighter; between objects and door(s) within a place) represent inter-object relationships. Blue lines (darker; between place nodes and doors) show the topological connection between the places through the doors.

In this work, spatial cognition was demonstrated using place classification and place recognition. While the latter has been addressed in the mobile robotics community ([34] and [8]), the former requires the robot to actually build a conceptual model of a place and is more general and harder, a problem. With the aim of improving on an initial solution proposed in [32] and towards the incorporation of more semantics in the representation, a Bayesian approach towards conceptualization of space has been proposed. Some preliminary results are shown in fig. 3. The process involves a conceptual clustering approach that uses a distance metric and a maximum-a-posteriori (MAP) estimate of the concept indicated by the incoming object, together with a naive Bayesian classifier based conceptualizer that actually infers the presence or absence of different concepts. Until now, models based on object occurrences (multiple occurrences) have been studied for the conceptualization process. Inter object relationships are currently being incorporated in the framework to further enhance it.

4.2 Towards Object Classification

In a recent effort towards functional object classification, a range camera (SwissRanger) has been used for probabilistic incremental object part detection. The detailed report of methods used may be found in [35]. Very briefly though, the objective of the work was to sequentially register range-data as obtained from

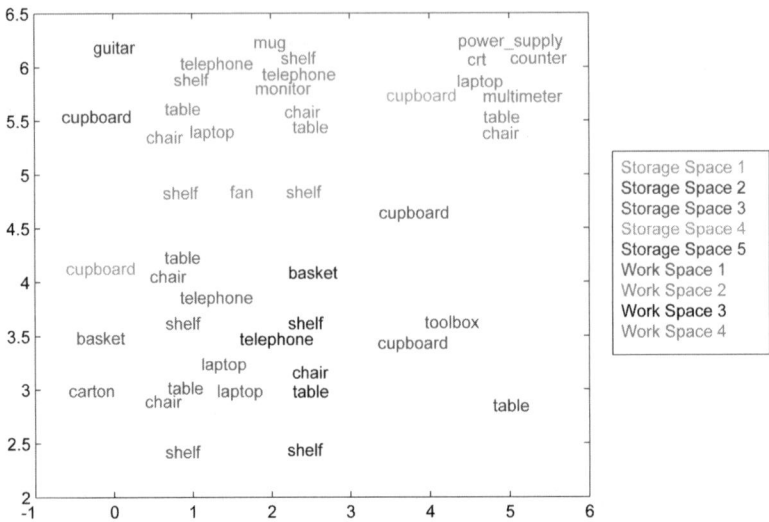

Fig. 3. Bayesian Conceptualization of an office - the objects are clustered and each cluster is shown with a different color. Each cluster is subsequently conceptualized into a functional grouping such as work space, storage space and meeting space. These concepts are in turn used to infer that the place is an office. This is aimed towards robust place classification and also representing space along the lines of fig. 1(a).

the range camera and to segment the resultant 3D model into object parts which would subsequently be used towards classifying objects. The system was probabilistic in that it took into consideration sensor and segmentation errors. The segmentation of the parts was done using morphological operators. The object parts were detected using a particle filter to track the state of each segmented part as being a known part or noise. The key idea was to accumulate evidence incrementally over several frames, for a particular object part, while taking into account the errors generated due to sensor and segmentation faults. Figure 4 shows some of the results obtained. The key significance of this work is in the development of a system that uses novel sensory information and that also takes into account the fact that these algorithms would have to function on a mobile robot platform - the existence of uncertain and incomplete information radically changes the application of most previously performed static approaches.

4.3   Towards a Structure Based Representation of Space

Given the framework shown in fig. 1(a), a key question that remains to be answered is - how can structural information be extracted and meaningfully understood in a consistent and probabilistic fashion towards a structural representation of space ? The approach adopted here uses a nodding SICK laser scanner to obtain a 3D point cloud of the indoor environment. The range image is first segmented into smooth areas by a fast edge based approach using

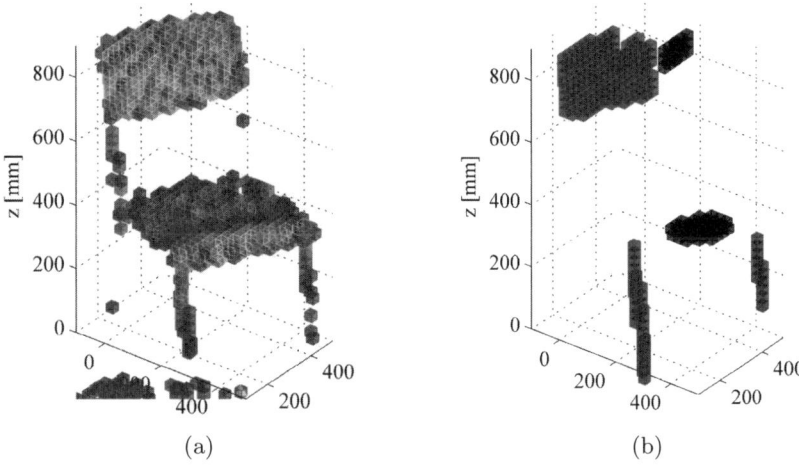

(a)   (b)

Fig. 4. (a) A voxel set of a chair generated from ten aligned and quantized point clouds acquired with a range camera. Voxels with lower and higher point density are depicted in blue and red, respectively. **(b)** Detected object parts of a chair. The color indicates the part category: red for leg, green for back, and blue for seat. The shading of the color indicates the probability of being a noisy part.

directional bearing angles [36]. This approach to segmentation also delivers boundary information and a map of depth discontinuities (laser beam jumps) which is later used to infer some information about the presence of holes, connection of the rooms and corridors, etc. Principal Component Analysis (PCA) is applied on the segmentation output to select planar patches, with boundaries being coded as 2D polygons . These polygons are later simplified using the information gathered, such as the adjacency of the planes (fig. 5(a)).

In addition to planar patches, the map also contains 3D corners which are formed by considering major orthogonal planes. These are relatively big planes with a large number of supporting points and are perpendicular to one axis of the building coordinate system. In each step, such planes are used to re-adjust the robot orientation. Then, 3D corners in the current observation are formed and matched with the corners in the map to find the translation of the robot between successive steps. Simple heuristics are used to recognize some parts of indoor structure within the mapped data, like ceiling, floor, walls, doors and windows (fig. 5(b)). This helps in creating sub-maps compatible with building parts like corridors and rooms, which eventually leads to a more compact representation of gathered data in terms of structural hierarchies and semantically annotated maps. Figure 6 shows the preliminary results obtained when this structural information is applied towards solving the simultaneous localization and mapping (SLAM) problem.

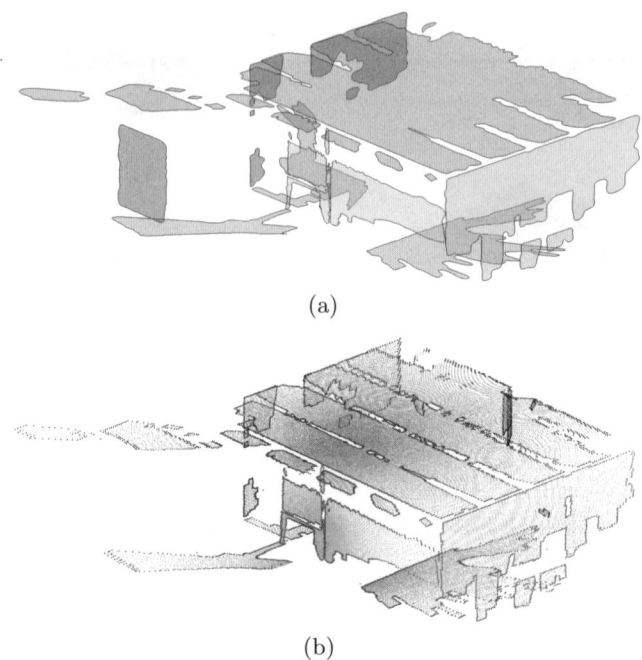

(a)

(b)

Fig. 5. (a) The major planar patches extracted from one scene with simplified boundaries. (b) The structure recognized includes the walls, the ceiling and the floor. The
structural elements are currently extracted from the planar patches shown in (a) by
applying various heuristics.

4.4   Perspectives from a User Study

The broad aim of the study was to validate the proposed representation in a
cognitive sense. The aim was to verify our approach and to find out what other
details (kinds of features / data) the proposed representation could encode. The
survey comprised of a questionnaire posed to fifty-two people who were taken
through a course within our premises, wherein they were exposed day-to-day
things and places. While the detailed survey including the methods adopted and
the results / analysis are presented in [11], some of the salient aspects that could
be concluded from the work are mentioned here. They support various aspects
of all three approaches presented above and the overall framework within which
these works are integrable.

The study concluded that an object based representation was indeed useful
for robots to develop a human compatible representation of space. Objects were
clustered into groups or concepts - these are the semantic / functional abstractions in space. They were mostly formed by similarities in purpose, functionality
and also by the relative spatial arrangements of objects. Places could be understood as spatial abstractions which were typically formed by bounding elements
such as walls and doors whereas semantic abstractions were most often formed
as a result of relative spatial arrangements between objects and/or similarities

Fig. 6. The outcome of the point cloud registration process performed using the structural elements extracted as shown in fig. 5. The experiment was carried out with six observations obtained using a nodding SICK laser scanner in an office.

in purpose or functionality. The survey also brought out to a significant extent, the various properties, functionalities that may be relevant towards enhancing the representation being sought. Although a more comprehensive proof is requisite, there was a clear indication that spatial abstractions contain the semantic ones. In the realm of objects, structural information (of objects) was found to be critical towards their representation or description.

5   Future Work

Building on the promising results obtained, a lot more work is ongoing or planned for the near future. Current work is focused on conceptualizing space. While preliminary results seem assuring, both clustering and conceptualization need further research. The concepts so formed, would then be used towards making the representation richer in semantic information and yet, more scalable. This would have to be supported by suitable advances on the object classification front and from the structural dimension as well. While preliminary results towards functional object classification look encouraging, a consistent probabilistic framework towards functional object classification is still the subject of ongoing research. The envisioned representation would also provide a firm basis to research and represent objects by classifying them through functionality as interpreted in terms of action-recognition [37] and by augmenting the context information in the classification process [38]. Along the structural dimension, ongoing work is oriented towards improving the structure based representation of space. This includes developing robust probabilistic algorithms towards identifying structural elements and analyzing their use towards solving the SLAM problem.

6   Conclusion

This chapter described an endeavor to create a hierarchical probabilistic multi-modal representation of space. A lot of relevant work has been carried out in the past years by the AI and Robotics communities. This chapter revisited some of these contributions; it described the current steps being undertaken and the recent advances made. This work may be understood as a conscientious and situated attempt to bridge the gap between AI and robotics. It elicited a three pronged effort being adopted towards aggressively dealing with the open challenges in this domain. The approach prescribed the use of sensory data to extract high level features such as objects, doors etc. These features were grouped along two dimensions - spatial to include the structural definition of space and semantic to include a conceptual / semantic description of it. The representation thus formed and the current results on conceptualization were found to be human compatible; they were adequately supported with results from an elaborate user study. As a result of these efforts, a clear increase in the degree of spatial awareness of robots was observed. The methods adopted exhibited a clear link from the sensory information acquired by a robot to the human compatible spatial concepts that the robot infers thereof - in this sense, the approach is holistic. Notwithstanding all of this, several issues still remain to be addressed. It is hoped that these efforts will inspire and bear tangible contributions that would eventually help realize the next generation of spatially cognizant robots.

Acknowledgments

This work has been supported by the EU Integrated Project COGNIRON (The Cognitive Robot Companion), funded by contract FP6-IST-002020 and the Swiss National Science Foundation (Grant No. 200021-101886).

References

1. Brooks, R.A.: Intelligence without representation. Artificial Intelligence 47, 139–159 (1991)
2. Steels, L.: Intelligence with representation. Philosophical Transactions of the Royal Society A 361(1811), 2381–2395 (2003)
3. Thrun, S.: Robotic Mapping: A Survey. In: Exploring Artificial Intelligence in the New Millenium, Morgan Kaufmann, San Francisco (2002)
4. Chatila, R., Laumond, J.P.: Position referencing and consistent world modeling for mobile robots. In: IEEE Int. Conf. on Robotics and Automation (ICRA), IEEE Computer Society Press, Los Alamitos (1985)
5. Arras, K.O.: Feature-Based Robot Navigation in Known and Unknown Environments. PhD thesis, Swiss Federal Institute of Technology Lausanne (EPFL), Thesis number 2765 (2003)
6. Martinelli, A., Svensson, A., Tomatis, N., Siegwart, R.: SLAM based on quantities invariant of the robot's configuration. In: IFAC Symposium on Intelligent Autonomous Vehicles (IAV) (2004)

7. Choset, H., Nagatani, K.: Topological Simultaneous Localization and Mapping (SLAM): Toward Exact Localization Without Explicit Localization. IEEE Transactions on Robotics and Automation 17, 125–137 (2001)
8. Tapus, A.: Topological SLAM - Simultaneous Localization And Mapping with fingerprints of places. PhD thesis, Swiss Federal Institute of Technology Lausanne (EPFL), Thesis Number 3357 (2005)
9. Thrun, S.: Learning Metric-Topological Maps for Indoor Mobile Robot Navigation. Artificial Intelligence 99 (1), 21–71 (1998)
10. Tomatis, N., Nourbakhsh, I., Siegwart, R.: Hybrid Simultaneous Localization And Map building: A natural integration of Topological and Metric. Robotics and Autonomous Systems 44, 3–14 (2003)
11. Vasudevan, S., Gächter, S., Siegwart, R.: Cognitive Spatial Representations for Mobile Robots - Perspectives from a user study. In: IEEE Int. Conf. on Robotics and Automation (ICRA) Workshop on Semantic Information in Robotics, Rome, Italy (April 2007)
12. Martinelli, A., Tapus, A., Arras, K.O., Siegwart, R.: Multi-resolution SLAM for Real World Navigation. In: International Symposium of Robotics Research (ISRR), Siena, Italy (2003)
13. Kuipers, B.: The Spatial Semantic Hierarchy. Artificial Intelligence 119, 191–233 (2000)
14. Tolman, E.C.: Cognitive Maps in Rats and Men. Psychological Review 55, 189–208 (1948)
15. Kuipers, B.J.: The cognitive map: Could it have been any other way? In: Spatial Orientation: Theory, Research, and Application, pp. 345–359. Plenum Press, New York (1983)
16. Yeap, W.K., Jefferies, M.E.: On early cognitive mapping. Spatial Cognition and Computation 2(2), 85–116 (2001)
17. Keselman, Y., Dickinson, S.J.: Generic Model Abstraction from Examples. In: Sensor Based Intelligent Robots, pp. 1–24 (2000)
18. Gibson, J.J.: The Ecological Approach To Visual Perception, 1st edn. Lawrence Erlbaum, Mahwah (1986)
19. Bicici, E., Amant, R.S.: Reasoning about the functionality of tools and physical artifacts. Technical Report TR-2003-22, Department of Computer Science, North Carolina State University (NCSU) (2003)
20. Stark, L., Bowyer, K.W.: Generic Object Recognition using Form and Function. World Scientific Publishing, Singapore (1996)
21. Pechuk, M., Soldea, O., Rivlin, E.: Function-Based Classification from 3D Data via Generic and Symbolic Models. In: Twentieth National Conference on Artificial Intelligence (AAAI), Pittsburgh, Pennsylvania, USA (July 2005)
22. Aycinena, M.A.: Probabilistic Geometric Grammars for Object Recognition. Master's thesis, Department of Electrical Engineering and Computer Science (EECS), Massachusetts Institute of Technology (MIT) (September 2005)
23. Kortenkamp, D., Baker, L.D., Weymouth, T.: Using Gateways to Build a Route Map. In: IEEE/RSJ International Conference on Intelligent Robots and Systems (IROS) (1992)
24. Anguelov, D., Koller, D., Parker, E., Thrun, S.: Detecting and modeling doors with mobile robots. In: IEEE International Conference on Robotics and Automation (ICRA), New Orleans, Louisiana, USA (April 2004)

25. Mozos, Ó.M., Stachniss, C., Burgard, W.: Supervised Learning of Places from Range Data using AdaBoost. In: IEEE International Conference on Robotics & Automation (ICRA), Barcelona, Spain, pp. 1742–1747. IEEE Computer Society Press, Los Alamitos (2005)
26. Nüchter, A., Surmann, H., Lingemann, K., Hertzberg, J.: Semantic Scene Analysis of Scanned 3D Indoor Environments. In: Vision, Modeling and Visualization (VMV) (2003)
27. Grau, O.: A Scene Analysis System for the Generation of 3-D Models. In: International Conference on Recent Advances in 3-D Digital Imaging and Modeling, Ottawa, Canada, pp. 221–228 (1997)
28. Nüchter, A., Wulf, O., Lingemann, K., Hertzberg, J., Wagner, B., Surmann, H.: 3D Mapping with Semantic Knowledge. In: RoboCup International Symposium, Osaka, Japan (July 2005)
29. Fong, T., Nourbakhsh, I., Dautenhahn, K.: A survey of socially interactive robots. Robotics and Autonomous Systems 42, 143–166 (2003)
30. Arras, K.O., Cerqui, D.: Do we want to share our lives and bodies with robots? A 2000-people survey. Technical Report 0605-001, Swiss Federal Institute of Technology Lausanne (EPFL) (June 2005)
31. Siegwart, R., Arras, K.O., Bouabdallah, S., Burnier, D., Froidevaux, G., Greppin, X., Jensen, B., Lorotte, A., Mayor, L., Meisser, M., Philippsen, R., Piguet, R., Ramel, G., Terrien, G., Tomatis, N.: Robox at Expo.02: A large-scale installation of personal robots. Robotics and Autonomous Systems 42, 203–222 (2003)
32. Vasudevan, S., Gächter, S., Nguyen, V.T., Siegwart, R.: Cognitive Maps for Mobile Robots - An object based approach. Robotics and Autonomous Systems 55(5), 359–371 (2007)
33. Lowe, D.G.: Distinctive image features from scale-invariant key-points. International Journal of Computer Vision 60(2), 91–110 (2004)
34. Ulrich, I., Nourbakhsh, I.: Appearance-Based Place Recognition for Topological Localization. In: IEEE International Conference on Robotics and Automation (ICRA), San Francisco, CA, USA, pp. 1023–1029. IEEE Computer Society Press, Los Alamitos (2000)
35. Gächter, S., Siegwart, R.: Incremental Object Part Detection with a Range Camera. Technical Report ETHZ-ASL-2006-12- Version 2.0.1, Swiss Federal Institute of Technology Zürich (ETHZ) (2006)
36. Harati, A., Gächter, S., Siegwart, R.: Fast Range Image Segmentation for Indoor 3D-SLAM. In: 6th IFAC Symposium on Intelligent Autonomous Vehicles, Toulouse, France (September 2006)
37. Veloso, M., von Hundelshausen, F., Rybski, P.E.: Learning visual object definitions by observing human activities. In: The proceedings of IEEE-RAS International Conference on Humanoid Robots (HUMANOIDS), Japan (December 2005)
38. Ramel, G.: Context Analysis with Probabilistic Methods for Human Robot Interaction (In French). PhD thesis, Swiss Federal Institute of Technology Lausanne (EPFL), Thesis number 3477 (2006)

Anticipation and Future-Oriented Capabilities in Natural and Artificial Cognition

Giovanni Pezzulo

Istituto di Scienze e Tecnologie della Cognizione - CNR
Via S. Martino della Battaglia, 44 - 00185 Roma, Italy
giovanni.pezzulo@istc.cnr.it

Abstract. Empirical evidence indicates that *anticipatory representations* grounded in the sensorimotor neural apparatus are crucially involved in several low and high level cognitive functions, including attention, motor control, planning, and goal-oriented behavior. A unitary theoretical framework is emerging that emphasizes how *simulative* capabilities enable social abilities, too, including joint attention, imitation, perspective taking and communication. We argue that anticipation will be a key element for bootstrapping high level cognitive functions in cognitive robotics, too. We thus propose the challenge of understanding how anticipatory representations, that serve for *coordinating with the future* and not only with the present, develop in situated agents[1].

1   Introduction

Anticipation has the potential to become a key issue in designing and developing the artificial cognitive systems of the future. In this paper we review evidence of the roles of anticipation in enabling several cognitive functions, bootstrapping high level cognitive functions, and developing a truly autonomous mental life. We will then argue that understanding anticipation and the development of increasingly sophisticated anticipatory capabilities in natural cognition permits to design artificial anticipatory cognitive embodied systems capable of coordinating their current actions with future outcomes, planning in view of their future needs, and finally formulating and achieving abstract goals.

The situated approach now dominant in the 'new AI' [8,12,56] focuses on reactive mechanisms and agent-environment engagement. It has produced many results, most notably a clarification of the roots of cognition in sensorimotor interactions, and the relevance of embodied, situated and emerging aspects of behavior. At the same time, the emphasis on reactive behavior has drastically reduced the efforts in understanding future-directed behaviors which are widespread in natural cognition. Now that the situated, embodied approach is widely accepted in robotics, it is time to study how to deal in this theoretical framework

---

[1] Work funded by the EU project **MindRACES** (FP6-511931).

M. Lungarella et al. (Eds.): 50 Years of AI, Festschrift, LNAI 4850, pp. 257–270, 2007.

with the variety of anticipatory behaviors in natural systems, and how they can emerge from more primitive forms of coordination and interaction. We think that a theoretical, empirical and computational investigation of anticipation, and in particular of *simulative* theories [3,33,35], will permit an 'evolutionary leap' in cognitive robotics: from reactive to anticipatory cognitive embodied systems.

A popular definition of anticipatory system is provided by Rosen [64]: *A system containing a predictive model of itself and/or its environment, which allows it to change state at an instant in accord with the model's predictions pertaining to a latter instant.* Behavior which is not simply reactive, or driven by stimuli which are here-and-now, but includes an (implicit or explicit) evaluation of future states of affairs is surprisingly widespread in natural cognition, ranging from sensorimotor interaction to higher-level cognitive abilities only available to humans and possibly to other mammals, such as reasoning, imitation, and social learning. Even behaviors which seem to be simple acts of coordination, in fact, often require an estimation of future states of affairs, as reported for example in the motor preparation of the prey-catching behavior of the jumping spider [68], for compensating the dinamicity of the environment. The influence of expected stimuli for orienting attention has been reported not only in humans (see [2]), but also in pigeons [63] and monkeys [15]. On these basis, models of the visual apparatus including (hierarchical) predictions have been proposed such as *predictive coding* [58]. Increasingly sophisticated capabilities are enabled by the capability to process in advance expected stimuli; for example, Hesslow [35] describes how rats are able to 'plan in simulation' and compare alternative paths in a T-maze *before acting in practice.* Cognitive agents are in fact not only able to exploit *affordances* [31] which are immediately perceptible in the environment. They also act in the world to force it to show its hidden affordances, or even to produce affordances for future use. For example, monkeys can fulfill complex tasks requiring to discard the most immediate affordances of the environment (e.g., going directly toward food), and invent creative ways to use objects such as sticks, breaking *functional fixity* (i.e., the incapacity to exploit an object other than for its default function). Overall, a significant part of natural agents behavior is not present-oriented and stimuli-driven, but future-oriented in nature, and it is motivated by *goals* (i.e., anticipatory representations of future, desired state of affairs that have the potential to prescribe and regulate our actions). This surprising 'inversion' of the direction of causality, from future to present, from goals to actions, is acknowledged in psychology by James's ideomotor principle [39], in cybernetics [65], and in control theory by Adams [1], who argues that goals serve as reference signals "from the future".

Anticipation is an important component of human cognition, too. Damasio [17] describes how during decision making humans engage in 'what-if' simulated loops of interaction with the environment in order to evaluate in advance, via *somatic markers*, possible future negative consequences of their actions. Increasingly complex uses of anticipation exist which lead us to *disengage* more and more from current sensorimotor cycle. Many species can build up and manipulate representations of future courses of events, for example in simulated planning; but typically they do that in order to satisfy their *present* drives or goals. As far as we know, only humans can endogenously formulate novel goals and planning in view of *future* needs; this includes abstract and distal ones such as having fun or becoming famous. Many of our individual and social practices ultimately serve to deal with future needs, and we accept immediately negative outcomes in view of distal positive ones (e.g., studying today for having a job tomorrow) –and the possibility to anticipate oneself could have lead to the capability to coordinate one's own actions in the present and in the future, and to have a sense of 'persisting self'. We can formulate expectations at an increasingly high level of abstraction, and to use them for regulating our actions. For example, we can decide whether or not to apply for a job depending on our expectations about the satisfaction it will provide us, the salary, the free time, etc. Not only we formulate such abstract expectations, but we also can 'match' them with imaginary futures and select among them. Another capability that is typical of human beings is *substitution* [57]. Probably several animal species are able to work on their *internal models* of the phenomena before (or instead of) acting in the real world, but we humans use that ability systematically. A mechanic can assemble and dismantle a motor in his mind before doing it in practice, and an architect can propose us different plans for restructuring our house. Thanks to anticipation it is possible to deal with entities also when they are not indeed present as stimuli: an ability that is crucial for defining an agent's *autonomy* [10]. One striking novelty of human cognition is our tendency to heavily modify and adapt the environment to us, and not only vice-versa. While other species adjust their representations for fitting the actual world, we often *act in the world in order to make it fit our representations of what we want, our goals.* Several animal species have the capability to adapt their environments, such as to build up nests, but typically they do that in a stereotyped way. We humans do not have not this limitation, and have heavily modified our environment to fit our present and future goals. Another feature of human cognition, that is its extremely sophisticated social life, depends on abilities that could be based on anticipation, too, such as perspective taking, imitation, and language [30,38,61].

In [55] we have argued that these capabilities are related. We have an unprecedented capability to endogenously produce internal representations of the (possible) future, and to flexibly manipulate them: selecting which ones to achieve (forming goal states), and deciding how to achieve them by only working on internal representations of our possible actions and their outcomes. Anticipation is the key mechanism for bootstrapping increasingly complex cognitive functions, and for this reason it has to be investigated in a unitary, developmental perspective. In this paper we put forward this perspective. We introduce anticipation from the theoretical point of view, we review how it is addressed both in the empirical literature, and we conclude by proposing the study of anticipation as a crucial challenge for cognitive robotics.

2   Anticipation: Coordinating with the Future

We propose to conceive *the mind as an essentially anticipatory device* [11,55] which serves for future-oriented behavior. It is nowadays widely assumed in the situated cognition literature that adaptive behavior both in natural and artificial systems depends crucially on the dynamics of interactions between brain, body and environment [12,51,77]. However, as above discussed, cognitive agents can break the boundaries of sensorimotor engagement. While adaptivity serves to coordinate with the present, *anticipation serves to coordinate with the future*.

Anticipatory behavior is not all-or-nothing ability, but comes in grades: the range of anticipatory capabilities is ample and new capabilities can be evolved on the basis of old ones. As a demarcation criterion we propose that the true mental life of a cognitive agent begins when it is able to *endogenously generate representations* which are not totally determined by actual sensed stimuli but derive from internal models, and to *use them in order to regulate its present behavior* (and in some species even future behavior). As pointed out by many researchers [14,55,73] even if connected causally to the environment, internal processes for dealing with representations do not share its dynamics. This permits representations to *detach* from the current sensorimotor cycle and to be used *instead of* the environment itself, for example when the environment is too noisy, or too rich of stimuli, or if not all the relevant information is (already) there, as in the case of future-oriented actions, which we stress here. In its more complex forms, detached representations are conceptual and not only perceptual, permitting more complex capabilities such as representing the non existent and reasoning.

This idea of cognition entails a notion of representation that is intimately anticipatory. As discussed by Roy [66], representations are related to the environment with a double-sided relationship: causation (from environment to agent) and anticipation (from agent to environment). For example, concepts for 'reachable' or 'graspable' objects are grounded by schemas which regulate agent behavior and include predictions of the consequences of expected interactions. Once in place for regulating present-directed actions, those anticipation-based representations offer an unique evolutionary advantage to cognitive agents: *to work on them before, or instead of, working on external reality*, leading to future-oriented capabilities such as formulating, pursuing and reasoning about distal goals. The ability that defines a true mind, as opposed to a merely adaptive systems, is in fact that of building up representations of the non-existent, of what is not currently (yet) true or perceivable, for the sake of acting on them.

Implicit, or behavioral, anticipation. This is not to say that all anticipation depends on *explicit* representations of future states of affairs. Some anticipatory capabilities, which we refer as *behavioral* or *implicit* anticipation, are selected by evolution and encoded into the sensorimotor apparatus. Consider as an example a *grasshopper* apparently reacting to a noise and escaping. In this case the grasshopper's behavior has been selected by evolution not to react to the noise itself, but to predators. It reacts now to a danger in the future: this behavior, even if realized by a reactive mechanism, is functionally anticipatory.

Gibson [31] firstly proposed that vision is an active process in which anticipation is implicitly produced by learned patterns of sensorimotor transformations; there is no need of representing anything, neither present not future states, since the environment is used as *the best representation of itself.* Brooks [8] points out that, strictly speaking, continuously coupled reactive agents are not memoryless, since their memory is in the environment; and we would say that also their expectations are in the environment and in the dynamics of agent-environment interaction. Recently O'Regan and Noe [52] also propose to conceive all perception as coordination of an agent's perceptual apparatus with the dynamical structure of sensory stimuli. In their sensorimotor view the organism shows an anticipatory behavior, that is to attend to the next relevant stimuli, by learning the patterns of transformation of sensory stimuli depending on its motor operations, without an explicit representation of the next incoming stimulus.

Explicit, or representational, anticipation. In nature there is thus a range of behaviors which maintains a reactive appearance but is functionally anticipatory; but how many anticipatory capabilities can be explained without resorting to anticipatory representations? In the next Section we will review empirical findings indicating that *explicit* anticipatory representations are actually involved in many anticipatory behaviors. Also many theories, such as Clark's *minimal representationalism*, point in the same direction without denying the embodied and dynamical nature of cognition: *minds may be essentially embodied and embedded and still depend crucially on brains which compute and represent* [13].

Some capabilities seem to be out of the scope of behavioral anticipation: for example, acting both for the present and for distal goals (to coordinate both with the present and the future), or considering both own and other's perspective, or taking into account multiple possibilities for action. If a unique, non representation-mediated mechanism is in play, these activities should interfere, but we know that conceiving *now* the future does not hinder the possibility to act in the present, and conceiving the other's perspective does not imply losing one's own. All these phenomena seem then to be based on representations: for example, internal, emulating models can be in play in several cognitive operations, running on-line and off-line and providing a credible substratum for representational activity [14] (but see [43] for an account of how non-representational systems can deal with distal behavior). Moreover, the possibility to *internalize* external structures to work on them, including maps but also language and cultural practices, seem to be a distinctive trait of high level cognition (see [55]). In a sense, this is an old story coming back to attention. Even in the past, in fact, many studies in traditional AI were focused on resolving problems by working on 'internal' or 'small-scale' models before acting in the world. For example, Craik [16] discusses how internal representations permit to generate imaginary experiences and 'mental simulations' of external reality, and Tolman [78] discusses the role of imagination for learning 'as if' experience had really happened. However, the problem is that often representations as used in AI are *disembodied* and *not grounded* [34,56]. How to develop a concept of representation that it is integrated in a naturalistic framework and in continuity with situated action?

Recently representations begin to be seen in a different way in cognitive science, which we could call a *motor-based* perspective. As suggested in particular by the discovery of *mirror neurons* [62], they are mainly *action-oriented*, originate in the sensorimotor apparatus and remain intimately related with it [13,14,55]. As an example of the coupling of representation and action, Gallese [29] argues that *the goal is represented as a goal-state, namely, as a successfully terminated action pattern.* The *ideomotor principle* [39], which recently received a number of empirical confirmations, [37,44], suggests in a similar way that *action planning takes place in terms of anticipated features of the intended goal.* It is thus not surprising that in this action-based view of cognition anticipation has assumed a crucial role, since it is a bridge between representation as traditionally conceived, and situated action. Anticipation permits in fact to look at representations not as abstract and disembodied symbols, as it was the case in traditional AI, but as structures enabling agent-environment coordination that arise for the sake of guiding behavior and remain intimately coupled with the sensorimomtor apparatus. Interactivism [4], for example, describes representations as *ways for setting up indications of further interactive potentialities*: representations serve thus for future interactions. This approach is reminiscent of the Kantian *productive* perspective of cognition, according to which we do not passively process environmental stimuli but actively produce representations by means our categorical apparatus; the novelty is the emphasis on the situated and action-based origin and nature of representations.

3   Anticipation in Natural and Artificial Cognition

How does the brain formulate expectations? Which brain structures and which mechanisms are involved? There are currently multiple directions of research, which emphasize different aspects. The ideomotor principle and related models [39,37,44] stress the formation of associative, *action-effect* rules, mediated by a common neural coding. Similarly, stimulus-stimulus associative links (e.g., lightning - thunder) might be involved the prediction of regularities of the environment. The reenactment of sensorimotor structures used for the control of action is instead stressed in the literature on the mirror system [29,62], that codes for both observed and executed actions. Reenactment and generative capabilities are central in the literature on internal forward models [42,81], that actively produce expectations and do not only explain statistical regularities in the stimuli, but include hidden states. Different anticipatory mechanisms and brain structures could then co-exist and have complementary powers and limitations [22]. For example, important distinctions are among prediction of events that we can or can not produce ourselves [69] and among prediction of action effects or behavioral goals [48]. See also [25] for a recent, comprehensive review of the neural correlates of anticipation in the mammalian brain.

In artificial systems research, one very influential model is that of Wolpert, Kawato and colleagues [42,81], that has the advantage to be well grounded in standard control theory. They propose that the brain uses *internal models*, which mimic the behavior of external processes, for motor control of action. In particular, *forward models* permit to generate expectations about the next sensed stimuli, given the actual state and motor command. *Inverse models* instead take as input actual stimuli and the goal state and provide as output the motor commands necessary to reach the desired state. Taken together, inverse and forward models permit not only to perform motor plans but also to control it and in general to regulate behavior in noisy and dynamic environments. In neurosciences forward models have been claimed to be involved in compensating for delays in sensory feedback, cancel the self-produced part of the input from sensory stimuli, etc. [81,82] and empirical evidence exists for their involvement in visuomotor tasks [46]. Similar structures have also been claimed to be involved in visual attention [2] and imagery [40].

Thanks to these findings *motor* and *simulative* theories of cognition are now widespread: for example, according to the *emulation theory* of Grush [33], representation is the ability to emulate internally part of external reality by means of internal models such as Kalman filters, that can also be nested to obtain abstraction. Similarly, the *simulation hypothesis* of Hesslow [35] argues that representing is engaging in simulated interaction with the environment by means of internal predictive models which can be chained and form 'loops'. Barsalou proposes the *perceptual symbol system* theory; arguing against amodal and disembodied notions of representations, proposes a situated view in which they retain part of their original sensorimotor structure. On the basis of perceptual systems, Barsalou proposes that concepts emerge as productive, simulative structures that can be used by the agent in order to *simulate* actual or expected sensorimotor engagements on the basis of past situated action, producing understanding of perceptual and abstract concepts. Similarly Gallese [27] claims that *looking at objects means to unconsciously 'simulate' a potential action*.

Internal models for simulating actual sensorimotor engagement can also be exploited for increasingly complex future-oriented activities. For example, this approach explains quite naturally one of the distinctive points of human cognition, the possibility to test potential actions 'in simulation' and for example avoid dangers [17]. Recent evidence suggests in fact that imagined and performed actions share a common timing and neural substratum [18]. Moreover, a comprehensive model of how impairments in anticipating the consequences of one's own actions produces diseases such as the 'anarchic hand' and schizophrenia has been proposed [26] that unifies a number of empirical findings under a common, anticipatory framework. Another capability that is often associated to internal, generative models is formulating and comparing in simulation multiple alternative courses of actions. It has been proposed in [50] that the possible neural substratum is a 'loop' between the *cerebellum*, which is supposed to be able to produce sensory predictions (see e.g., [6]), and the *basal ganglia*, that are involved in action selection and movement initiation (see e.g., [60]). Recently the chemical basis of such neural predictions have been investigated, too. For example a role of dopamine has been advocated in reward prediction [70] and signaling unpredictability of actions [59].

The use of anticipatory and simulative capabilities also extends to the social sphere. The neural substrates involved in performing, observing, simulating and imitating actions in fact largely overlaps, and evidence exists for a role of mirror neurons and simulative processes for imitation [38], distinguishing self from others [19], mind reading [28], language production and understanding [61]. Several researchers have proposed that the same generative mechanisms for controlling action can be reenacted endogenously for perceiving, understanding and imitating actions performed by other agents, for understanding behavior, and for inferring intentions from actions [5,38,40,80]. According to Rizzolatti and Arbib, *Individuals recognize actions made by others because the neural pattern elicited in their premotor areas during action observation is similar to that internally generated to produce that action* [61, p. 190]. Altogether, the beauty of the motor-based approach is in its power to unify spheres of cognition that are traditionally kept separated: action and perception, individual and social.

Anticipation in Artificial Systems. Recently anticipation has received attention in situated approaches to artificial systems, and principled design approaches have been proposed. For example, on the basis of the psychological literature, and in particular Hoffmann's theory of *anticipatory behavioral control* [37], a taxonomy of four kinds of anticipations for artificial systems is proposed in [9]: *implicit, payoff, sensorial* and *state* anticipation. At the same time, many mechanisms for predicting have been proposed in cognitive robotics, such as recurrent neural networks (RNN). For example, Jordan's type RNN [41] use expectations produced by forward models for 'vicarious' trial-and-error learning. Kalman filters have also been widely used; they incorporate many functionalities such as estimation and filtering, and for this reason Grush [33] considers them prototypical emulators. Bayesian predictors have also been used in the literature of motor control [82], and rule based systems such as the *schema mechanism* [24] and *anticipatory classifier systems* [9] have been shown to autonomously learn action-effect rules and chain them for planning and action control. The roles of reward prediction and surprise in action learning and in metalearning strategies such as curiosity are being studied [72], with convergent ideas between reinforcement learning and neuroscience [22,23]. Moreover, *predictive state representation* [45] has been proposed for substituting the classical concept of state.

Many cognitive functions related to anticipation, having different levels of complexity and sometimes nested in one another, are being studied in cognitive robotics. Anticipation plays in fact a crucial role in attention, conceived as 'selection of information relevant for action' [2,54]. The role of anticipation for the control of attentive strategies has also been demonstrated with hierarchical architectures combining the top-down contribute of expectation and the bottom-up one of incoming stimuli [53,58]. Several functions related to the *control of action* have been claimed to include anticipatory components, too, such as stabilizing perception [79], canceling the predictable part of the feedback [49], erasing stimuli produced e.g. by the body of the agent [21], producing a reference signal for the control of voluntary acts [1]. Many of them have been modeled in artificial systems, too. For example, the robot Murphy can [47] exploit efference copies of motor commands for generating simulated perceptual inputs and thus maneuver its arm robustly even in partial absence of sensory stimuli. Similar anticipatory strategies are widely used in the *Robocup* competition (e.g. [32]) for coordinating with the ball in dynamic environments: prediction is required for compensating the delays of the sensors. Combinations of *forward* and *inverse* internal models have also been widely used for action execution and control both in distributed, dynamic systems approaches [75] and in localist ones [20,54,76,82]. They permit to generate multiple competing motor plans, and select the one most appropriate to the context depending on its predictive accuracy. One example is choosing the most appropriate behavior to deal with 'full' or 'empty' glasses, the weight being the context [82]. Action understanding and imitation has also been demonstrated in artificial systems by running 'in simulation' the same generative mechanisms used for motor control [20].

Internal predictive models serving for the control of action have been used for other functionalities, increasingly disengaging from current sensorimotor cycle. In fact, if expectations produced by forward models are chained, and expected input is supplied in spite of actual input, it is possible to use the same machinery involved in online visual and motor planning for generating off-line, 'simulative' planning. In cognitive robotics this capability has been exploited for generating the sensory consequences of multiple possible plans and selecting the 'best' one [74,83], like in the *simulation hypothesis* [35]. It has also been used for generating long-term predictions related to the current course of actions in order to receive an evaluation from the future [71], like in the *somatic markers* hypothesis [17]. Another use of internal predictive models is understanding the boundaries of the personal sphere. Piaget [57] discusses how distinguishing self-produced motion from sensory stimuli which are caused by interaction with objects in the environment leads to develop a *body schema*; some of these ideas have been also used in robotics [7]. It has also been shown by schema-based architectures [24,54,66] that anticipation, as argued in constructivist theories [4,57], can bootstrap the acquisition of new concepts by interacting with the environment, as in the case of Drescher's *schema mechanism* [24] which learns *synthetic items*. Another related use is grounding concepts such as 'far', 'heavy', 'obstacle' or 'predator' as simulated interaction potentialities [36,66].

Computational studies have demonstrated that anticipatory mechanisms for the control of action can also be used for enabling social capabilities such as action understanding, imitation, joint attention, perspective taking (e.g., [20,67]). This fact parallels the huge empirical evidence that a common neural substrate, enabling anticipatory and simulative capabilities, is in place both for individual and social cognition [38,62]. The similarities between many of the above mentioned studies indicate that anticipatory capabilities are highly related both at the functional and at the mechanism level. This is an important reason for conceiving *anticipation as a unitary phenomenon*, which is fundamental in natural cognition and should inspire artificial systems design, too.

4   Conclusions

While the 'new AI' is nowadays mainly focused on reactive behavior, we argue that a crucial theoretical and computational challenge is putting in a naturalistic framework our ability to deal not only with the present but with anticipated, or desired futures, and make them happen for our sake. In artificial cognitive systems, as in natural ones, *implicit* and *explicit* anticipation permits an evolutionary leap from present-oriented to future-oriented capabilities, bootstrapping high level cognitive and social capabilities [55]. To imagine, to reason about the possible and the non existent, to evaluate in advance the results of one's own actions, to change the world according to one's own goals and, at the same time, to build up deceptive and illusory worlds, to dream and hallucinate: those are all features of a truly cognitive mind. We argue that a presupposition for autonomous mental life are anticipatory capabilities permitting to *disengage* from sensorimotor loops and to break the boundaries of adaptivity: (1) to *pursue autonomously generated goals*; (2) to regulate behavior according to *future and not only present potentialities for action*; (3) to *learn regularities in the environment* depending on the agent's (actual or possible) actions, and independent from them; (4) to *deal with entities even 'in their absentia'*, when they are not among the currently attended stimuli; (5) to *form conceptual representations* that are however grounded in (potential) interaction; (6) to *adapt the world to fit the agent's own goals*; (7) and *to boostrap high level cognition*.

Much theoretical, empirical and simulative work remains to be done in order to fully understand the phenomenon of anticipation in natural cognition, and how to endow artificial systems with future-oriented capabilities. We would conclude by anticipating some of the challenges that we envisage if we want to build anticipatory artificial systems. Perhaps the most basilar one is to understand the passage from reactive, to simple, and then increasingly complex anticipatory mechanisms, with a caveat: arguably, these mechanisms do not replace each other in full-fledged cognitive agents, but coexist and coordinate. Another crucial challenge is understanding which cognitive functions depend on anticipatory capabilities, and in particularly which ones are exaptations of the capability to predict. For example, the capability to conceive discrete objects and events could depend on the fact that we necessarily have to predict them at a high granularity, since fine-grained prediction fails when the time span is too long. Another relevant challenge is understanding if motor-based and simulative view of cognition (exemplified by the motto *the mind is an anticipatory device* [11,55]) will be really able to provide us with a unified perspective on cognition. Assuming that the motor apparatus of a robot can be used for predicting and understanding objects, events, and actions, the robot could internalize these predictions and use them independently of the current state of the world, for example reenacting them for planning, conceiving novel goals, comparing possible outcomes of its actions, or imagining the reaction of another robot to its actions. Several cognitive abilities could depend on the same anticipatory mechanisms.

Overall, we think that anticipation is a necessary condition in natural cognition for developing several individual and social abilities, and the motor-based, simulative view of cognition should inspire the way robots are designed, too [14,55]. Robots of the future should not simply adapt to their environment, but predict it, and their generative mechanisms will be the key for bootstrapping increasingly complex cognitive capabilities.

References

1. Adams, J.A.: A closed-loop theory of motor learning. Journal of Motor Behavior 3, 111–149 (1971)
2. Balkenius, C., Hulth, N.: Attention as selection-for-action: a scheme for active perception. In: Proceedinsg of EuRobot-1999, Zurich (1999)
3. Barsalou, L.W.: Perceptual symbol systems. Behavioral and Brain Sciences 22, 577–600 (1999)
4. Bickhard, M.H., Terveen, L.: Foundational Issues in Artificial Intelligence and Cognitive Science: Impasse and Solution. Elsevier Scientific, Amsterdam (1995)
5. Blakemore, S.-J., Decety, J.: From the perception of action to the understanding of intention. Nature Reviews Neuroscience 2, 561–567 (2001)
6. Blakemore, S.-J., Frith, C.D., Wolpert, D.M.: The cerebellum is involved in predicting the sensory consequences of action. Brain Imaging 12, 1879–1884 (2001)
7. Bongard, J., Zykov, V., Lipson, H.: Resilient machines through continuous self-modeling. Science 314(5802), 1118–1121 (2006)
8. Brooks, R.A.: Intelligence without representation. Artificial Intelligence 47(47), 139–159 (1991)
9. Butz, M.V., Sigaud, O., Gérard, P.: Internal models and anticipations in adaptive learning systems. In: Butz, M.V., Sigaud, O., Gérard, P. (eds.) Anticipatory Behavior in Adaptive Learning Systems: Foundations, Theories, and Systems, pp. 86–109. Springer, Heidelberg (2003)
10. Castelfranchi, C.: Guarantees for autonomy in cognitive agent architecture. In: Wooldridge, M.J., Jennings, N.R. (eds.) Intelligent Agents. LNCS, vol. 890, pp. 56–70. Springer, Heidelberg (1995)
11. Castelfranchi, C.: Mind as an anticipatory device: For a theory of expectations. In: De Gregorio, M., Di Maio, V., Frucci, M., Musio, C. (eds.) BVAI 2005. LNCS, vol. 3704, pp. 258–276. Springer, Heidelberg (2005)
12. Chiel, H.J., Beer, R.D.: The brain has a body: Adaptive behavior emerges from interactions of nervous system, body and environment. Trends in Neurosciences 20, 553–557 (1997)
13. Clark, A.: Being There. Putting Brain, Body, and World Together. MIT Press, Cambridge (1997)
14. Clark, A., Grush, R.: Towards a cognitive robotics. Adaptive Behavior 7(1), 5–16 (1999)
15. Colombo, M., Graziano, M.: Effects of auditory and visual interference on auditory-visual delayed matching to sample in monkeys (maca fascicularis). Behav. Neurosci. 108, 636–639 (1994)
16. Craik, K.: The Nature of Explanation. Cambridge University Press, Cambridge (1943)
17. Damasio, A.R.: Descartes' Error: Emotion, Reason and the Human Brain. Grosset/Putnam, New York (1994)
18. Decety, J.: Do imagined and executed actions share the same neural substrate? Brain Res. Cogn. 3, 87–93 (1996)

19. Decety, J., Chaminade, T.: When the self represents the other: A new cognitive neuroscience view on psychological identification. Consciousness and Cognition 12(20), 577–596 (2003)
20. Demiris, Y., Khadhouri, B.: Hierarchical attentive multiple models for execution and recognition (hammer). Robotics and Autonomous Systems Journal 54, 361–369 (2005)
21. Desmurget, M., Grafton, S.: Forward modeling allows feedback control for fast reaching movements. Trends Cogn. Sci. 4, 423–431 (2000)
22. Doya, K.: Complementary roles of basal ganglia and cerebellum in learning and motor control. Curr. Opin. Neurobiol. 10(6), 732–739 (2000)
23. Doya, K.: Metalearning and neuromodulation. Neural Netw. 15(4-6), 495–506 (2002)
24. Drescher, G.L.: Made-Up Minds: A Constructivist Approach to Artificial Intelligence. MIT Press, Cambridge, MA (1991)
25. Fleischer, J.G.: Neural correlates of anticipation in cerebellum, basal ganglia, and hippocampus. In: Butz, M., Sigaud, O., Pezzulo, G., Baldassarre, G. (eds.) Anticipatory Behavior in Adaptive Learning Systems: Advances in Anticipatory Processing. LNCS (LNAI), vol. 4520, Springer, Heidelberg (2007)
26. Frith, C.D., Blakemore, S.J., Wolpert, D.M.: Abnormalities in the awareness and control of action. Philos. Trans. R Soc. Lond. B Biol. Sci. 355(1404), 1771–1788 (2000)
27. Gallese, V.: The inner sense of action: agency and motor representations. Journal of Consciousness Studies 7, 23–40 (2000)
28. Gallese, V., Goldman, A.: Mirror neurons and the simulation theory of mind-reading. Trends in Cognitive Sciences 2(12), 493–501 (1998)
29. Gallese, V., Metzinger, T.: Motor ontology: The representational reality of goals, actions, and selves. Philosophical Psychology 13(3), 365–388 (2003)
30. Gardenfors, P., Orvath, M.: The evolution of anticipatory cognition as a precursor to symbolic communication. In: Brook, S. (ed.) Proceedings of the Morris Symposium on the Evolution of Language, NY, USA (2005)
31. Gibson, J.: The ecological approach to visual perception. Lawrence Erlbaum Associates, Inc., Mahwah, NJ (1979)
32. Gloye, A., Goektekin, C., Egorova, A., Tenchio, O., Rojas, R.: Fu-fighters small size 2004. In: Nardi, D., Riedmiller, M., Sammut, C., Santos-Victor, J. (eds.) RoboCup 2004. LNCS (LNAI), vol. 3276, Springer, Heidelberg (2005)
33. Grush, R.: The emulation theory of representation: motor control, imagery, and perception. Behavioral and Brain Sciences 27(3), 377–396 (2004)
34. Harnad, S.: The symbol grounding problem. Physica D: Nonlinear Phenomena 42, 335–346 (1990)
35. Hesslow, G.: Conscious thought as simulation of behaviour and perception. Trends in Cognitive Sciences 6, 242–247 (2002)
36. Hoffmann, H., Moller, R.: Action selection and mental transformation based on a chain of forward models. In: Schaal, S., Ijspeert, A., Billard, A., Vijayakumar, S., Hallam, J., Meyer, J.-A. (eds.) From Animals to Animats 8, Proceedings of the Eighth International Conference on the Simulation of Adaptive Behavior, Los Angeles, CA, pp. 213–222. MIT Press, Cambridge (2004)
37. Hoffmann, J.: Anticipatory behavioral control. In: Butz, M.V., Sigaud, O., Gérard, P. (eds.) Anticipatory Behavior in Adaptive Learning Systems. LNCS (LNAI), vol. 2684, pp. 44–65. Springer, Heidelberg (2003)

38. Iacoboni, M.: Understanding others: Imitation, language, empathy. In: Hurley, S., Chater, N. (eds.) Perspectives on Imitation: From Cognitive Neuroscience to Social Science, MIT Press, Cambridge, MA (2003)
39. James, W.: The Principles of Psychology. Dover Publications, New York (1890)
40. Jeannerod, M.: Neural simulation of action: A unifying mechanism for motor cognition. NeuroImage 14, S103–S109 (2001)
41. Jordan, M.I., Rumelhart: Forward models: Supervised learning with a distal teacher. Cognitive Science 16, 307–354 (1992)
42. Kawato, M.: Internal models for motor control and trajectory planning. Current Opinion in Neurobiology 9, 718–727 (1999)
43. Keijzer, F.: Representation and behavior. MIT Press, Cambridge, MA (2001)
44. Kunde, W., Koch, I., Hoffmann, J.: Anticipated action effects affect the selection, initiation and execution of actions. The Quarterly Journal of Experimental Psychology. Section A: Human Experimental Psychology 57(1), 87–106 (2004)
45. Littman, M., Sutton, R., Singh, S.: Predictive representations of state. In: Proc. NIPS 2002, Vancouver (2001)
46. Mehta, B., Schaal, S.: Forward models in visuomotor control. Journal of Neurophysiology 88, 942–953 (2002)
47. Mel, B.W.: Vision-based robot motion planning. Neural networks for control, 229–253 (1990)
48. Miall, R.C.: Connecting mirror neurons and forward models. Neuroreport 14(17), 2135–2137 (2003)
49. Miall, R.C., Wolpert, D.M.: Forward models for physiological motor control. Neural Networks 9(8), 1265–1279 (1996)
50. Middleton, F.A., Strick, P.L.: Basal ganglia output and cognition: evidence from anatomical, behavioral, and clinical studies. Brain Cogn. 42(2), 183–200 (2000)
51. Nolfi, S.: Behaviour as a complex adaptive system: On the role of self-organization in the development of individual and collective behaviour. ComplexUs 2, 195–203 (2005)
52. O'Regan, J., Noe, A.: A sensorimotor account of vision and visual consciousness. Behavioral and Brain Sciences 24(5), 883–917 (2001)
53. Pezzulo, G., Baldassarre, G., Butz, M.V., Castelfranchi, C., Hoffmann, J.: An analysis of the ideomotor principle and tote. In: Butz, M., Sigaud, O., Pezzulo, G., Baldassarre, G. (eds.) Proceedings of the Third Workshop on Anticipatory Behavior in Adaptive Learning Systems (ABiALS 2006). LNCS, vol. 4520, pp. 73–93. Springer, Heidelberg (2006)
54. Pezzulo, G., Calvi, G.: A schema based model of the praying mantis. In: Nolfi, S., Baldassarre, G., Calabretta, R., Hallam, J.C.T., Marocco, D., Meyer, J.-A., Miglino, O., Parisi, D. (eds.) SAB 2006. LNCS (LNAI), vol. 4095, pp. 211–223. Springer, Heidelberg (2006)
55. Pezzulo, G., Castelfranchi, C.: The symbol detachment problem. Cognitive Processing 8(2), 115–131 (2007)
56. Pfeifer, R., Scheier, C.: Understanding Intelligence. MIT Press, Cambridge, MA (1999)
57. Piaget, J.: The Construction of Reality in the Child. Ballentine (1954)
58. Rao, R.P., Ballard, D.H.: Predictive coding in the visual cortex: a functional interpretation of some extra-classical receptive-field effects. Nat. Neurosci. 2(1), 79–87 (1999)
59. Redgrave, P., Gurney, K.: The short-latency dopamine signal: a role in discovering novel actions? Nature Reviews Neuroscience 7, 967–975 (2006)
60. Redgrave, P., Prescott, T.J., Gurney, K.: The basal ganglia: a vertebrate solution to the selection problem? Neuroscience 89, 1009–1023 (1999)

61. Rizzolatti, G., Arbib, M.A.: Language within our grasp. Trends in Neurosciences 21(5), 188–194 (1998)
62. Rizzolatti, G., Fadiga, L., Gallese, V., Fogassi, L.: Premotor cortex and the recognition of motor actions. Cognitive Brain Research 3 (1996)
63. Roitblat, H.: Codes and coding processes in pigeon short-term memory. Anim. Learn Behav. 8, 341–351 (1980)
64. Rosen, R.: Anticipatory Systems. Pergamon Press, Oxford (1985)
65. Rosenblueth, A., Wiener, N., Bigelow, J.: Behavior, purpose and teleology. Philosophy of Science 10(1), 18–24 (1943)
66. Roy, D.: Semiotic schemas: a framework for grounding language in action and perception. Artificial Intelligence 167(1-2), 170–205 (2005)
67. Roy, D., yuh Hsiao, K., Mavridis, N., Gorniak, P.: Ripley, hand me the cup: Sensorimotor representations for grounding word meaning. In: Int. Conf. of Automatic Speech Recognition and Understanding (2006)
68. Schomaker, L.: Anticipation in cybernetic systems: A case against mindless antirepresentationalism. In: IEEE International Conference on Systems, Man and Cybernetics (2004)
69. Schubotz, R.I.: Prediction of external events with our motor system: towards a new framework. Trends in Cognitive Sciences 11(5), 211–218 (2007)
70. Schultz, W., Dayan, P., Montague, P.: A neural substrate of prediction and reward. Science 275, 1593–1599 (1997)
71. Shanahan, M.: Cognition, action selection, and inner rehearsal. In: Proceedings IJCAI 2005 Workshop on Modelling Natural Action Selection, pp. 92–99 (2005)
72. Singh, S., Barto, A.G., Chentanez, N.: Intrinsically motivated reinforcement learning. In: Saul, L.K., Weiss, Y., Bottou, L. (eds.) Advances in Neural Information Processing Systems 17, pp. 1281–1288. MIT Press, Cambridge, MA (2005)
73. Steels, L.: Intelligence - dynamics and representations. In: Steels, L. (ed.) The Biology and Technology of Intelligent Autonomous Agents, Springer, Berlin (1995)
74. Stephan, V., Gross, H.-M.: Visuomotor anticipation - a powerful approach to behavior-driven perception. Künstliche Intelligenz 17(2), 12–17 (2003)
75. Tani, J.: Learning to generate articulated behavior through the bottom-up and the top-down interaction processes. Neural Netw. 16(1), 11–23 (2003)
76. Tani, J., Nolfi, S.: Learning to perceive the world as articulated: an approach for hierarchical learning in sensory-motor systems. Neural Netw. 12(7-8), 1131–1141 (1999)
77. Thelen, E., Smith, L.B.: A Dynamic Systems Approach to the Development of Perception and Action. MIT Press, Cambridge (1994)
78. Tolman, E.: Purposive Behavior in Animals and Men. Appleton-Century-Crofts, New York (1932)
79. von Holst, E., Mittelstaedt, H.: Das reafferenzprinzip. Naturwissenschaften 37, 464–476 (1950)
80. Wolpert, D.M., Doya, K., Kawato, M.: A unifying computational framework for motor control and social interaction. Philos Trans. R Soc. Lond B Biol. Sci. 358(1431), 593–602 (2003)
81. Wolpert, D.M., Gharamani, Z., Jordan, M.: An internal model for sensorimotor integration. Science 269, 1179–1182 (1995)
82. Wolpert, D.M., Kawato, M.: Multiple paired forward and inverse models for motor control. Neural Networks 11(7-8), 1317–1329 (1998)
83. Ziemke, T., Jirenhed, D.-A., Hesslow, G.: Internal simulation of perception: a minimal neuro-robotic model. Neurocomputing 68, 85–104 (2005)

Computer-Supported Human-Human Multilingual Communication

Alex Waibel[1,2], Keni Bernardin[1], and Matthias Wölfel[1]

InterACT
International Center for Advanced Communication Technology
[1] Universität Karlsruhe (TH), Karlsruhe, Germany
[2] Carnegie Mellon University, Pittsburgh, PA, USA
waibel@cs.cmu.edu

Abstract. Computers have become an essential part of modern life, providing services in a multiplicity of ways. Access to these services, however, comes at a price: human attention is bound and directed toward a technical artifact in a human-machine interaction setting at the expense of time and attention for other humans. This paper explores a new class of computer services that support human-*human* interaction and communication *implicitly* and *transparently*. Computers in the Human Interaction Loop (CHIL), require consideration of all communication modalities, multimodal integration and more robust performance. We review the technologies and several CHIL services providing human-human support. Among them, we specifically highlight advanced computer services for *cross-lingual* communication.

1 Introduction

It is a common experience in our modern world, for humans to be overwhelmed by the complexities of technological artifacts around us, and by the attention they demand. While technology provides wonderful support and helpful assistance, it also gives rise to an increased preoccupation with technology itself and with a related fragmentation of attention. But as humans, we would rather attend to a meaningful dialog and interaction with other humans, than to control the operations of machines that serve us. The cause for such complexity and distraction, however, is a natural consequence of the flexibility and choices of functions and features that the technology has to offer. Thus flexibility of choice and the availability of desirable functions are in conflict with ease of use and our very ability to enjoy their benefits. The artifact cannot yet perform autonomously and requires precise specification of the machine's behavior. Standardization, better graphical user interfaces, multimodal human-machine dialog systems, speech, pointing, "mousing" have all contributed to improve the interface, but still force the user to interact with a machine at the detriment of other human-human interaction.

To change the limitations of present day technology, machines must engage implicitly and indirectly in a world of humans, that is we must put Computers in the Human Interaction Loop (CHIL), rather than the other way round. Computers should assist humans engaged in human-human interaction, by providing implicit and

M. Lungarella et al. (Eds.): 50 Years of AI, Festschrift, LNAI 4850, pp. 271–287, 2007.

proactive support. If technology could be "CHIL enabled" in this way, a host of new services could potentially be possible. Could two people be connected with each other at the best moment over the most convenient and best media, without phone tag, embarrassing ring tones and interruptions? Could an attendee in a meeting be reminded of participants' names and affiliations at the right moment without messing with a contact directory? Can meetings be supported, moderated and coached without technology getting in the way? And: Could computers enable speakers of different languages communicate and listen to each other gracefully across the language divide?

Human assistants often provide such services; they work out logistical support, reminders, helpful assistance, and language mediation; they can do it reliably, transparently, tactfully, sensitively and diplomatically. Why not machines? Clearly, a lack of recognition and understanding of human activities, needs and desires are to blame, and an absence of socially adept computing services that mediate rather than intrude. In the following we focus on these two elements: 1.) technologies to track and understand the human context, and 2.) computing services that mediate and support human-human interaction.

2   Understanding the Human Context

In contrast to classical human-machine interfaces, implicit computer support for human-human interaction requires a perceptual user interface with much greater performance, flexibility and robustness, than is available today. This challenge has lead to research aimed at tracking all the communication modalities in realistic recording conditions, rather than individual modalities in idealized recording conditions. CHIL and AMI, both Integrated projects under the 6th Framework Program of the European Commission, as well as CALO, a DARPA program are among the more recent efforts aiming to take on this challenge.

In the following we will discuss computer services that support human-human interaction. To realize this goal, work concentrates on four key areas: The creation of robust perceptual technologies able to acquire rich and detailed knowledge about the human interaction context; the collection and annotation of realistic, audio-visual meeting and seminar data necessary for the development and systematic evaluation of such; the definition of a common software architecture to support reusability and exchangeability of services and technology modules; the implementation of a number of prototypical services offering proactive, implicit assistance based on the gained awareness about human interactions.

2.1 Audio-Visual Perceptual Technologies

2.1.1 Introduction
Multimodal interface technologies "observe" humans and their environments by recruiting signals from multiple AV sensors to detect, track, and recognize human activity. The analysis of all AV signals in the environment (speech, signs, faces, bodies, gestures, attitudes, objects, events, and situations) provides the proper answers

Fig. 1. The "who", "what", "where", "when", "how" and "why" of human interaction

to the basic questions of "who", "what", "where", and "when", that can drive higher-level cognition concerning the "how" and "why", thus allowing computers to engage and interact with humans in a human-like manner using the appropriate communication medium (see Figure 1).

Research work performed and progress made on a number of such technologies is described next. Whereas technological advances for multimodal systems were hard to measure in the past for lack of common benchmarks, recent efforts in the community have led to the creation of international evaluations such as the CLEAR (Classification of Events, Activities and Relationships) [1] and RT (Rich Transcription) [2] evaluations, which offer a platform for large-scale, systematic and objective performance measurements on large audio-visual databases.

2.1.2 Person Tracking

Location and tracking of multiple persons behaving without constraints, unaware of audio/video sensors, in natural, evolving and unconstrained scenarios, still poses significant challenges.

Video-based approaches based on background subtraction are error prone due to varying illumination, shadows and occlusion, whereas those relying on the feature space (e.g. color histograms) are difficult to initialize reliably for every new acquired target. Many approaches that offer higher reliability are simply too computationally expensive to be used in online applications.

Audio-based localization and tracking requires the tracked person to be actively speaking, and have to deal with the variety of acoustic conditions (e.g., room acoustics and reverberation) and, in particular, the undefined number of simultaneous active noise sources and competing speakers found in natural scenarios.

Several strategies are being applied to face the challenges mentioned above. Distributed camera and microphone networks, including microphone arrays placed in different positions in space, provide a better "coverage" of each area of interest. Fusion of sensor data in multi-view approaches overcomes occlusion problems, as in the case of 3D background subtraction techniques combined with shape from silhouette [3]. Probabilistic approaches computing the product of single view

Fig. 2. Audio-visual tracking of multiple persons. Targets are described by an appearance model comprising shape and color information, and tracked in 3D using probabilistic representations [4]. The system tracks 5 people in real-time through multiple persistent occlusions in cluttered environments.

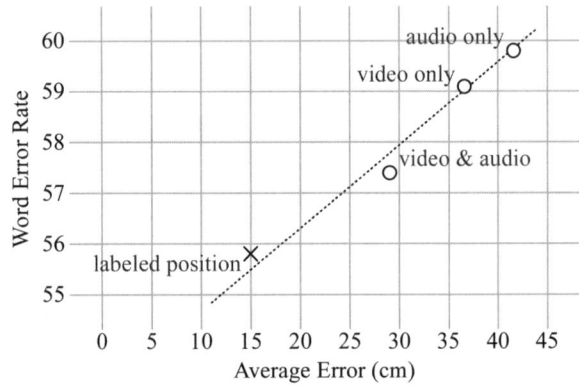

Fig. 3. Acoustic, visual and multimodal 3D person tracking accuracies and resulting word error rate (after beamforming) on the CHIL 2005 dataset

likelihoods using generative models which explicitly model occlusion have proved efficient in managing the trade-off between reliable modeling and computational efficiency [4] (see also Figure 2). Fusion of multimodal data for speaker localization in e.g. particle filtering approaches increases robustness for speaker tracking [5]. Efficient tracking is a useful building block for all subsequent technologies. It has been shown, e.g. that multimodal fusion helps increase localization accuracy, and that this in turn has direct impact on the performance of far-field speech recognition [6,7] (see also Figure 3).

2.1.3 Person Identification

The challenges for audio-visual person identification (ID) in unconstrained natural scenarios are due to far-field, wide-angle, low-resolution sensors, acoustic noise, speech overlap and visual occlusion, unpredictable subject motion, and the lack of position/orientation assumptions to facilitate well-posed signals. Clearly, employing tracking technologies and fusion techniques, either temporal, multi-sensor or multimodal (speaker ID combined with face ID for example) is a viable approach in order to improve robustness.

Identification performance depends on the enabling technologies used for audio, video and their fusion, but also on the accuracy of the extraction of the useful portions from the audio and video streams. The detection process for audio involves finding and extracting the speech segments in the audio stream. The corresponding process for video involves face detection. Developed mono- and multi-modal ID systems within CHIL have been successfully evaluated in the CLEAR'06 and '07 evaluations [1], reaching in many cases near 100% accuracies on databases of more than 25 subjects. Not only was steady progress made on the key technologies over the past years, showing the feasibility of person ID in unconstrained environments, it was also demonstrated that sensor and multimodality fusion help to improve recognition robustness (see Figure 4).

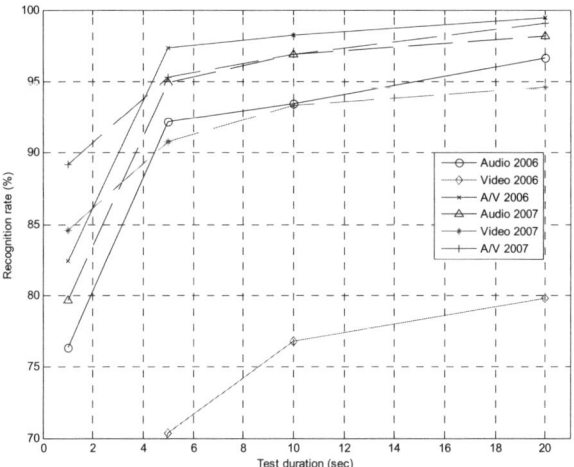

Fig. 4. Acoustic, visual and multimodal identification results for the CLEAR 2006 and 2007 evaluations (only best results shown). Systems were trained on 15 second sequences and tested on 1, 5, 10 and 20 second test sequences. Shown are accuracies for 25 users from 5 sites.

2.1.4 Head Pose, Focus of Attention

Understanding human interaction requires not only to perceive the state of individuals, but also to determine their person or object of interest, the addressees of speech, and so forth. Since people's head orientations are known to be reliable indicators for their direction of attention [8], systems were developed to estimate the head orientations of people in a smart room using multiple fixed cameras (see also Figure 5). In the CLEAR 2006 head pose dry run evaluation, the first formal evaluation for a task of this kind, classification of pan angles into 45° classes was attempted and accuracies of 44.8% were reached [1]. The challenging CHIL database drove the development of more accurate systems and already in 2007, estimation of exact angles was performed and error rates as low as 7° pan, 9° tilt and 4° roll could be achieved.

Once head orientations are estimated, they can be used to automatically determine the foci of attention of people [9].

2.1.5 Activity Analysis, Situation Modeling
Another useful type of information for unobtrusive, context-aware services is the classification of a user's or a group's current activities. In experiments performed in one of the CHIL sites, typical office activities such as "paperwork", "meeting" or "phone call" were distinguished in a multiple-office setup using only one camera and one microphone per room [10]. A hierarchical classification ranging from low level

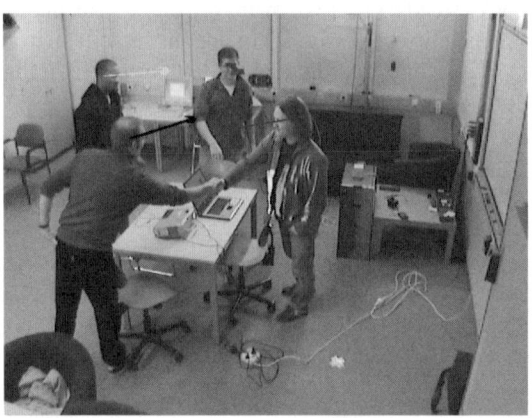

Fig. 5. Estimating Head Pose and Focus of Attention [9]. Head orientations are estimated from four camera views. These are then mapped to likely focus of attention targets, such as room occupants.

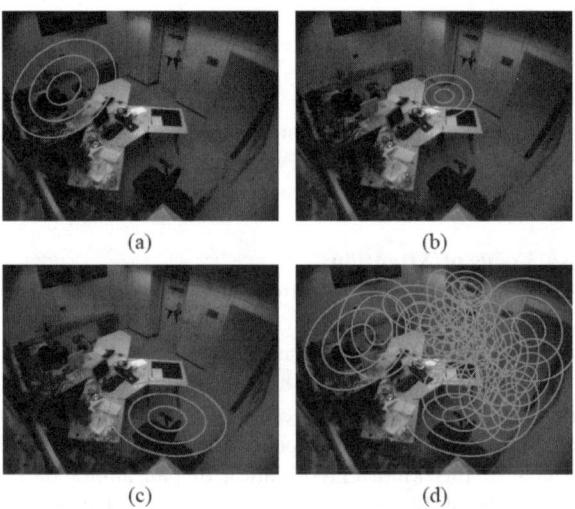

(a)   (b)

(c)   (d)

Fig. 6. Data-driven training of activity regions in an office room[10]. The regions labeled as a), b) and c) represent the learned areas of activities by office workers and their visitors, whereas d) depicts all resulting clusters. Evaluation of an unconstrained one week recording session revealed accuracies of 98% for "nobody in office", 86% for "paperwork", 70% for "phone call" and 60% for "meeting."

isolated events such as desk activity, to complex activities, such as leaving a room and entering another, could be achieved. The event classes were learned by clustering audio-visual data recorded during normal office hours over extended periods of time. Figure 6 depicts an example of data-driven clustering of activity regions within an office.

2.1.6 Speech Activity Detection, Speaker Diarization

These two related technologies are relevant not only for Automatic Speech Recognition (ASR), but also for speech detection and localization and for speaker identification. Speech activity detection (SAD), addresses the "when" of the speech interaction, and speaker diarization, addresses both "who" and "when". Both have been evaluated on the CHIL interactive seminar database in the latest CLEAR and RT evaluations, using primarily far-field microphones.

2.1.7 Recognition of Speech and Acoustic Events

Speech is the most critical human communication modality in seminar and meeting scenarios, and its automatic transcription is of paramount importance to real-time support and off-line indexing of the observed interaction. Although automatic speech recognition (ASR) technology has matured over time, natural unconstrained scenarios present significant challenges to state-of-the-art systems. For example, spontaneous and realistic interaction, with often accented speech and specialized topics of discussion (e.g., technical seminars), as well as overlapping speech, interfering acoustic events, and room reverberation degrade significantly the ASR performance. These factors are further exacerbated by the use of far-field acoustic sensors, which is unavoidable in order to free humans from tethered and obtrusive close-talking microphones.

Various research sites have been developing ASR systems to address these challenges, and have benchmarked their performance, e.g. in the recent RT'06 and '07 evaluations. There, the best far-field ASR system achieved a word error rate (WER) of 44% (52% in 2006), by combining signals from multiple (up to four) table-top microphones. It is interesting to note that this is considerably higher than the 31% (also 31% in 2006) WER achieved on close-talking microphone input – with manual segmentation employed to remove unwanted cross-talk. These results demonstrate the extremely challenging nature of the task at hand.

Various research approaches are being currently investigated to improve far-field ASR. Some employ multi-sensory acoustic input, for example beamforming that aims to efficiently combine acoustic signals from microphone arrays [6], and speech source separation techniques that attempt to improve performance during speech overlap segments. A different multimodal approach considered is to recruit visual speech information from the speaker lips, captured from properly managed pan-tilt-zoom cameras, in order to improve recognition through AV-ASR.

Finally, one should note that speech is only one of the acoustic events occurring during human interaction scenarios. Technology is being developed to detect and classify acoustic events that are informative of human activity, i.e., clapping, keyboard typing, door closing, etc. [1].

2.2 Technology Evaluations, Data Collection and Software Architecture

To drive rapid progress of the presented audio-visual perceptual technologies, their systematic evaluation using large realistic databases and common task definitions and metrics is essential.

Technology evaluations, undertaken on a regular basis, are necessary so that improvements can be measured objectively and different approaches compared. An important aspect is to use real-life data covering the envisioned application scenarios. In CHIL, large numbers of seminars and meetings were collected in five different smart rooms, equipped with a range of cameras and microphones. The recordings were manually enriched with acoustic event and speech transcriptions as well as several visual annotations that allowed to train and evaluate various technology components (see for example [1] for further details). In contrast to many of the evaluation benchmarks that exist for individual technologies such as face recognition, for example, the data from such realistic scenarios is extremely challenging, containing a combination of many difficulties for perceptual technologies, such as varying illumination, viewing angles, head orientations, low resolution images, occlusion, moving people, varying speaking accents, behaviours, room layouts and technical sensor setups.

Starting in 2006, a large effort was undertaken to create an international forum for evaluation of multimodal technologies for the analysis of human activities and interactions. The CLEAR workshop was created in a joint effort between CHIL [11], the US National Institute of Standards and Technology (NIST) and the US Video Analysis Content Extraction (VACE) [12] program. The goal was to provide the needed discussion forums, databases, standards, and benchmarks necessary to drive the development of multimodal perceptual technologies, much like the NIST Rich Transcription Meeting Recognition (RT) workshop for diarization, speech detection and recognition, or the TRECVID [13], PETS [14] and ETISEO [15] programs for visual analysis and surveillance. More than a dozen evaluation tasks were conducted, including face and head tracking, multimodal 3D person tracking, multimodal identification, head pose estimation, acoustic scene analysis, acoustic event detection, etc.

To offer support for the integration of developed technological components, to realize higher level fusion of information and modeling of interaction situations, and to provide well-defined interfaces for the design of useful user services, a proper architectural framework is of great importance. An example of such an infrastructure is the CHIL Architecture [16].

2.3 Human-Human Computer Support Services

Building on the perceptual technologies and compliant to the software architecture, several prototypical services are being developed that instantiate the vision of context-awareness and proactiveness for supporting human-human interaction.

The target domains are lectures and small office meetings. In the following, some example services, relying on the robust perception of human activities and interaction contexts are presented:

2.3.1 The Meeting Browser

The Meeting Browser provides functionality for offline reviewing of recorded meetings, automatic analysis, intelligent summarization or data reduction, generation of minutes, topic segmentation, information querying and retrieval, etc. Although it has been a topic of research for quite some time [17,18], advances in perceptual technologies (such as face detection, speaker separation and far-field speech recognition) have increased its user-friendliness by reducing the constraints on the interaction participants or the need for controlled or scripted scenarios.

2.3.2 The Collaborative Workspace

The Collaborative Workspace (CW) [19] is an infrastructure for fostering cooperation among participants. The system provides a multimodal interface for entering and manipulating contributions from different participants, e.g., by supporting joint discussion of minutes or joint accomplishment of a common task, with people proposing their ideas, and making them available on the shared workspace, where they are discussed by the whole group.

2.3.3 The Connector

The Connector is an adaptive and context-aware service designed for both efficient and socially appropriate communication [20]. It maintains an awareness of users' activities, preoccupations, and social relationships to mediate a proper moment and medium of connection between them.

2.3.4 The Memory Jog

The Memory Jog (MJ) provides background information and memory assistance to its users. It offers "now and here" information by exploiting either external databases: (Who is this person? Where is he/she from?) or own ones (Who was there that day? What did he say?), the latter including information gained from the observation of the interaction context [21]. The MJ can exploit its context-awareness to proactively provide information at the proper time and in the most convenient way given the current situation.

2.3.5 Cross-Lingual Communication Services

Another exciting class of services concern cross-lingual human-human communication. Is it possible to communicate with a fellow human speaking a different language as naturally as if he/she spoke your own? Clearly this would be a worthwhile vision in a globalizing world, when international integration demand limitless communication, while national identity and pride demand recognition and respect for the cultural and linguistic diversity on this planet. How could technology be devised to make this possible? We devote the following section to a discussion of this potentially revolutionary class of human communication support and an area of growing speech, language and interface research.

3   Cross-Lingual Human-Human Communication Services

In the past decade, Speech Translation has grown from an oddity at the fringe of speech and language processing conferences, to one of the main pillars of current research activity. The explosion in interest is driven in part, by considerable market pull from an increasingly globalizing world, where distance is no longer measured in miles but in communication ease and cost. Indeed, effective solutions that overcome the linguistic divide may potentially offer considerable practical and economic benefits. For the research community, the linguistic divide may ultimately prove to be a more formidable challenge than the digital divide as it presents researchers with a number of fascinating new problems. The goal is, of course, good human-to-human communication without interference from technical artifacts, and effective solutions must combine efficient and reliable speech & language processing with effective human factors and interface design.

Early developments provided first prototypes demonstrating the concept and feasibility [22,23]. In the mid '90's a number of projects aiming at spontaneous speech two-way speech translators for limited domains (e.g. JANUS-III, Verbmobil, Nespole) followed suit. The Consortium for Speech Translation Advanced Research (C-STAR) was founded in '91 to promote international cooperation in speech translation research. With the turn of the millennium, activity has proceeded in two directions: The first continues to improve domain-limited two-way translation toward *fieldable*, *robust* deployment where domain limitation is acceptable (humanitarian, health-care, tourism, government, etc.). The second has begun to tackle the open challenge of domain limitation for applications such as broadcast news, speeches and lectures. Large new initiatives (NSF-STR-DUST, EC-IP TC-STAR and DARPA GALE were launched in the US and Europe in '03, '04, and '06, respectively, in response. In the following we review these advances.

3.1 Domain-Limited Portable Speech Translators

Fieldable speech-to-speech translation systems are currently developed around portable platforms (laptops, PDA's) which impose constraints on the ASR, SMT, and TTS components. For PDA's memory limitations and the lack of a floating point unit

Fig. 7. A PDA pocket translator [English-Thai] [1]

---

[1] Courtesy of Mobile Technologies, LLC, Pittsburgh.

require substantial redesign of algorithms and data structures. Thus, a PDA implementation may impose WER increases from 8.8% to 14.6% [24] over laptops. In addition to continued attention to speed, recognition, translation and synthesis performance, however, usability issues such as the user interface, microphone type, place and number, as well as user training and field maintenance must be considered. One of the resulting speech-to-speech graphical user interfaces (GUI) of a PDA pocket translators is shown in Figure 7.

The GUI window is divided into two regions, showing the language pairs. These regions can be populated by recognized speech output (ASR), translation output (SMT), or by a virtual PDA keyboard for backup. A back-translation is provided for verification; a push-to-talk button activates the device and aborts processing for false starts and errors.   Projects (e.g DARPA Transtac) and workshops (e.g. IWSLT, sponsored by C-STAR) provide for collaboration, data exchange and benchmarking that improve performance and coverage in this space.

3.2 Translation of Parliamentary Speeches and Broadcast News

For speech-translation without domain limitation, component technologies first had to be developed that deliver acceptable ASR, SLT (and TTS) performance in face of spontaneous speech, unlimited vocabularies, broad topics, and speaking style characteristic of spoken records. In TC-STAR, speeches from the European Parliament (and their manual transcriptions and translations) were used as data to train and evaluate. Figure 8 shows the improvements over the years in speech recognition and automatic translation within the project. In these experiments it has been seen that there is almost a linear correlation between WER and machine translation quality. We also found that a WER of around 30% is influencing the machine translation quality significantly while a WER of 10% provides for reasonable translation compared to reference transcriptions. The goal of a different ambitious speech translation project, GALE (Global Autonomous Language Exploitation) [25], is to provide relevant information in English, where the input comes from huge amounts of speech in multiple languages (a particular focus is on broadcast news in Arabic and Chinese). However, progress is not measured by WER and BLEU, but how fast a particular goal can be reached.

Figure 9 compares human and computer speech-to-speech translations on five different aspects by human judgment: was the message understandable (understanding), was the output text fluent (fluent speech), how much effort does it take to listen to the translation (effort) and what is the overall quality, where the scale ranges from 1 (very bad) to 5 (very good). The fifth result shows the percent accuracy by which questions of content could be answered by human subjects based on the output from human and machine translators. It can be seen that automatic translation quality still lags behind human translation, but reaches usable and understandable levels already close to human translations. It is interesting to note, that the human translations also fall short of perfection due to the fact that humans translators occasionally omit information.

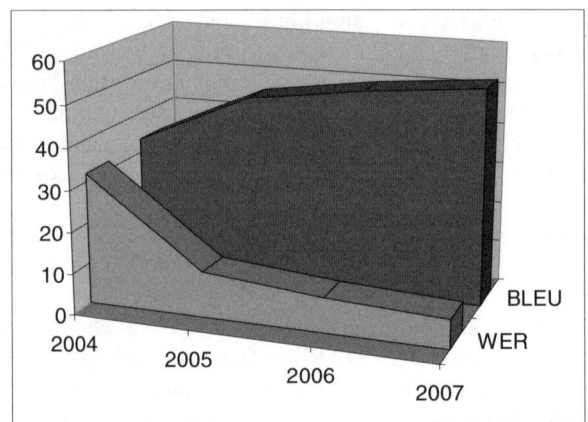

Fig. 8. Improvements in Speech Translation and Automatic speech recognition over the years on English EPPS and translation into Spanish. (source [26,27])

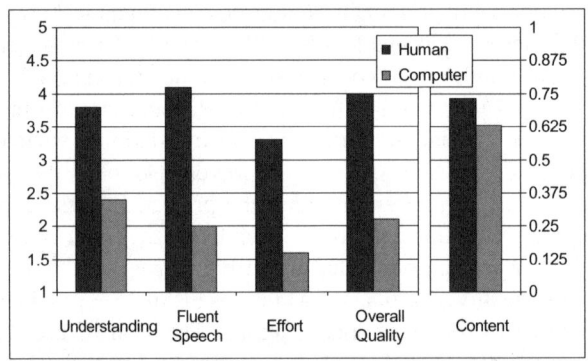

Fig. 9. Human vs. automatic translation performance. (source [28])

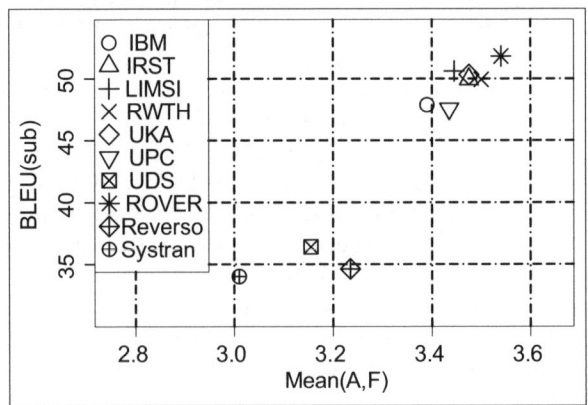

Fig. 10. BLEU scores show good correlation with human judgements (fluency & accuracy) for English to Spanish translations. (source [27])

An important aspect in all automatic evaluations are good metrics that can be evaluated automatically and repetitively. While WER is an established method to measure accuracy of automatic speech transcriptions, automatic MT metrics have only recently been proposed. Figure 10 shows the BLEU score (one of several popular MT scoring metrics) and its good correlation with human judgements (adequacy, fluency) on the European Parliament data.

3.3 Unlimited Domain Simultaneous Translation

The ultimate cross-lingual communication tool would be a simultaneous translator that produces simultaneous real-time translation of spontaneous lectures and presentations. Compared to parliamentary speeches and broadcast news, lectures, seminars, presentations of any kind, present further problems for domain-unlimited speech translation by

- Spontaneity of free speech, the disfluencies, the ill-formed nature of spontaneous natural discourse
- Specialized vocabularies, topics, acronyms, named entities and expressions in typical lectures and presentations (by definition specialized content)
- Real-time and low-latency requirements and on line adaptation to achieve *simultaneous* translation and
- Selection of translatable chunks or segments

3.3.1 The Lecture Translator

To address these problems in ASR and MT engines, changes to an off-line system are introduced as follows:

- To speed up recognition, acoustic models can be adapted to a particular speaker. The size of the acoustic model is restricted (for additional speed up when evaluating the Gaussian mixture model one can use techniques such as Gaussian selection) and the search space is more rigorously pruned.
- To adapt to particular speaker style and domain, the language model is tuned offline on slides and publications provided by the speaker, either by reweighting available text corpora or by retrieving relevant training material through the internet or on previous lectures given by the same speaker.
- As almost all MT systems are trained on data split at sentence boundaries and therefore ideally expect sentence like segments as input, particular care has to be taken for suitable online segmentation. We have observed that extreme deviations from sentence based segmentation can lead to significant decreases in performance. In view of minimizing overall system latency, however, shorter speech segments are preferred. In addition to providing efficient phrase translation on-the-fly, word-to-word alignment is optimally constrained for entire sentence pairs [29].

Figure 11 compares WERs on different domains for English. With a tweaked speaker dependent lecture recognition system we reach a sufficient good performance of 10% WER. On an end-to-end evaluation of the system from English into Spanish we got a BLEU score of 19 while on reference transcripts we got a score of 24 (source [30]).

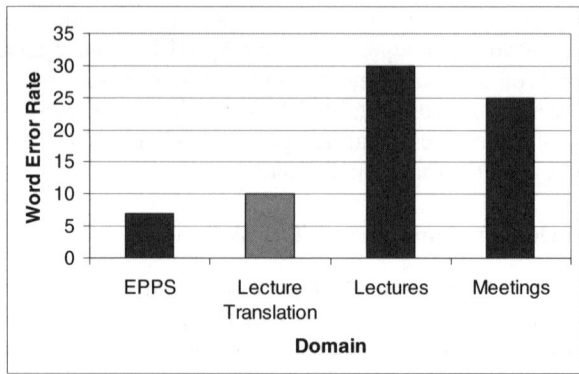

Fig. 11. Current performance of speech recognition systems on different domains (source [28,30,31], black = speaker independent off line system, gray = speaker dependent online system).

3.3.2 Delivering Translation Services (Output Modalities)

Aside from speech and language challenges, lecture translation also presents human factor challenges, as the service should be provided unobtrusively, i.e., with minimal interference or disruption to the human-human communication. Several options are being explored:

- *Subtitles*: Simultaneous translations can be projected to the wall as subtitles. This is suitable if the number of output languages is small.
- *Translation goggles*: Heads-up display goggles that display translation text as captions in a pair of personalized goggles. Such goggles provide unobtrusive translation and exploit the parallelism between the acoustic and visual channel. This is particularly useful, if listeners have partial knowledge of a speaker's language and wish to obtain complementary language assistance.
- *Targeted Audio Speakers*: Under the project CHIL, a set of ultra-sound speakers with high directional characteristics has been explored, that can provide a narrow audio beam to an individual listener or a small area in the audience, where simultaneous translation is required. Since such speakers are only audible in a narrow area, it does not disturb other listeners, or could be complemented by similar translation services into other languages to several other listener areas. [32].
- *PDA's, Display Screens or Head-Phones*: Naturally, output translation can also be delivered through traditional display technology, i.e., displayed on a common screen, a personalized PDA screen or acoustically via head-phones.

3.4 The Long Tail of Language

With promising solutions to the language divide under way, language portability remains the unsolved issue. At current estimates, there are more than 6,000 languages in the world, but language technology is only being developed for the most populous or wealthy languages of the world. Most languages along the long tail of language

(Figure 12) remain unaddressed. Overcoming the language divide thus requires workable solutions to providing solutions to the long tail of language, at reasonable cost. Most current research is focused on improving cross-lingual technology by employing ever larger data, personnel or computational resources. To address the long tail of language, an orthogonal direction should be concerned with making do with less at lower cost.

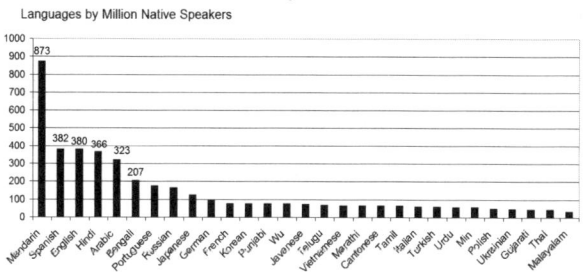

Fig. 12. The long tail of languages

At our center, we are therefore exploring several intriguing possibilities that lower cost that could some day bring this problem within reach as well:

- Language independent or adaptive components (this was demonstrated already for acoustic modeling[33]
- More selective parsimonious use of data and data collection [34]
- Interactive and implicit training by the user [35]
- Training on simultaneously *spoken* translation thereby eliminating the need for parallel text corpora [36]

Acknowledgements

The work presented here was supported in part by the *European Union* (EU) (projects CHIL (Grant number IST-506909) and TC-STAR (Grant number IST-506738)), by NSF (ITR STR-DUST), by DARPA (projects TRANSTAC and GALE). I would also like to thank the CHIL, TC-STAR, GALE, TRANSTAC partners and the InterACT research team at Karlsruhe and Pittsburgh for their collaboration and for data and images reported in this paper. Any opinions, findings, conclusions or recommendations expressed in this paper are those of the author and do not necessarily reflect the views of the funding agencies or the partners.

References

[1] Stiefelhagen, R., Bernardin, K., Bowers, R., Garafolo, J., Mostefa, D., Soundararajan, P.: The CLEAR 2006 Evaluation. In: Stiefelhagen, R., Garofolo, J. (eds.) CLEAR 2006. LNCS, vol. 4122, Springer, Heidelberg (2007)

[2] Fiscus, J., Ajot, J., Michel, M., Garofolo, J.: The rich transcription 2006 spring meeting recognition evaluation. In: Renals, S., Bengio, S., Fiscus, J.G. (eds.) MLMI 2006. LNCS, vol. 4299, Springer, Heidelberg (2006)

[3] Canton-Ferrer, C., Casas, J.R., Pardàs, M.: Human Model and Motion Based 3D Action Recognition in Multiple View Scenarios. In: EUSIPCO, Firenze (September 2006)

[4] Lanz, O.: Approximate Bayesian Multibody Tracking. IEEE Trans. PAMI 28(9) (September 2006)

[5] Stiefelhagen, R., Bernardin, K., Ekenel, H.K., McDonough, J., Nickel, K., Voit, M., Wölfel, M.: Audio-Visual Perception of a Lecturer in a Smart Seminar Room. Signal Processing 86(12) (December 2006)

[6] Wölfel, M., Nickel, K., McDonough, J.: Microphone array driven speech recognition: Influence of localization on the word error rate. In: Renals, S., Bengio, S. (eds.) MLMI 2005. LNCS, vol. 3869, Springer, Heidelberg (2006)

[7] Maganti, H.K., Gatica-Perez, D.: Speaker Localization for Microphone Array-Based ASR: The Effects of Accuracy on Overlapping Speech. In: ICMI, Banff, Canada (November 2006)

[8] Wojek, C., Nickel, K., Stiefelhagen, R.: Activity Recognition and Room-Level Tracking in an Office Environment. In: Proc. of the IEEE Intl. Conference on Multisensor Fusion and Integration for Intelligent Systems, Heidelberg, Germany (2006)

[9] Stiefelhagen, R., Yang, J., Waibel, A.: Modeling Focus of Attention for Meeting Indexing. In: ACM Multimedia, Orlando, Florida (October 1999)

[10] Voit, M., Stiefelhagen, R.: Tracking Head Pose and Focus of Attention with Multiple Far-field Cameras. In: ICMI, Banff, Canada (November 2006)

[11] CHIL – Computers in the Human Interaction Loop, http://chil.server.de

[12] VACE – Video Analysis and Content Extraction, http://www.ic-arda.org

[13] TRECVID – TREC Video Retrieval Evaluation, http://www-nlpir.nist.gov/projects/t01v/

[14] PETS – Performance Evaluation of Tracking and Surveillance, http://www.pets2006.net/

[15] ETISEO – Video Understanding Evaluation, http://www.silogic.fr/etiseo

[16] D2.2 Functional Requirements & CHIL Cooperative Information System Software Design, Part 2, Cooperative Information System Software Design, http://chil.server.de

[17] Waibel, A., Bett, M., Finke, M., Stiefelhagen, R.: Meeting browser: Tracking and summarizing meetings. In: Proceedings of the Broadcast News Transcription and Understanding Workshop, Lansdowne, Virginia, pp. 281–286 (1998)

[18] Bouamrane, M.-M., Luz, S.: Meeting browsing. Multimedia Systems 12(4-5), 439–457 (2006)

[19] Wang, Q.Y., Battocchi, A., Graziola, I., Pianesi, F., Tomasini, D., Zancanaro, M., Nass, C.: The Role of Psychological Ownership and Ownership Markers in Collaborative Working Environment. In: ICMI, Banff, Canada (2006)

[20] Danninger, M., Kluge, T., Stiefelhagen, R.: MyConnector – Analysis of Context Cues to Predict Human Availability for Communication. In: ICMI, Banff, Canada (2006)

[21] Neumann, J., Casas, J.R., Macho, D., Ruiz, J.: Multimodal Integration of Sensor Networks. In: Proc. of AIAI, Athens, Greece, pp. 312–323 (2006)

[22] Waibel, A., Jain, A.N., McNair, A.E., Saito, H., Hauptmann, A.G., Tebelskis, J.: JANUS: A Speech-to-speech Translation Using Connectionist and Symbolic Processing Strategies. In: Proc. of ICASSP 1991, pp. 793–796 (May 1991)

[23] Morimoto, T., Takezawa, T., Yato, F., Sagayama, S., Tashiro, T., Nagata, M., Kurematsu, A.: ATR's speech translation system: ASURA. In: Proc. 3rd European Conf. on Speech Communication and Technology, pp. 1291–1294 (September 1993)

[24] Hsiao, R., Venugopal, A., Köhler, T., Zhang, Y., Charoenpornsawat, P., Zollmann, A., Vogel, S., Black, A.W., Schultz, T., Waibel, A.: Optimizing Components for Handheld Two-way Speech Translation for English-Iraqi Arabic System. In: Proceedings of Interspeech (2006)

[25] GALE – http://www.darpa.mil/ipto/programs/gale

[26] Gauvain, J.L.: Speech transcription: general presentation of existing technologies within TC-Star. In: TC-Star Review Workshop, May 28-30, 2007, Luxembourg (2007)

[27] Ney, H.: TC-Star: Statistical MT of Text and Speech. In: TC-Star Review Workshop, May 28-30, 2007, Luxembourg (2007)

[28] Choukri, K.: Importance of the Evaluation of Human-Language Technologies. In: TC-Star Review Workshop, May 28-30, 2007, Luxembourg (2007)

[29] Kolss, M., Zhao, B., Vogel, S., Hildebrand, A., Niehues, J., Venugopal, A., Zhang, Y.: The ISL Statistical Machine Translation System for the TC-STAR Spring 2006 Evaluation. In: Proc. of the TC-STAR Workshop on Speech-to-Speech Translation, Barcelona, Spain (June 2006)

[30] Fügen, C., Kolss, M., Paulik, M., Waibel, A.: Open Domain Speech Translation: From Seminars and Speeches to Lectures. In: Proc. of the TC-STAR Workshop on Speech-to-Speech Translation, Barcelona, Spain (2006)

[31] Fiscus, J., Ajot, J.: The Rich Transcription 2007 Speech-To-Text (STT) and Speaker Attributed STT (SASTT) Results. In: The Rich Transcription 2007 Meeting Recognition (2007)

[32] Olszewski, D., Prasetyo, F., Linhard, K.: Steerable Highly Directional Audio Beam Louspeaker. In: Proc. of the Interspeech, Lisboa, Portugal (September 2006)

[33] Schultz, T.: Multilinguale Spracherkennung - Kombination akustischer Modelle zur Portierung auf neue Sprachen. PhD thesis, Universität Karlsruhe (June 2000)

[34] Eck, M., Vogel, S., Waibel, A.: Low Cost Portability for Statistical Machine Translation based on N-gram Frequency and TF-IDF. In: Proc. of IWSLT, Pittsburgh, PA (October 2005)

[35] Gavalda, M., Waibel, A.: Growing semantic grammars. In: Proceedings of the COLING/ACL, Montreal, Canada (1998)

[36] Paulik, M., Stüker, S., Fügen, C., Schultz, T., Schaaf, T., Waibel, A.: Speech Translation Enhanced Automatic Speech Recognition. In: ASRU, Cancun, Mexico (December 2005)

A Paradigm Shift in Artificial Intelligence: Why Social Intelligence Matters in the Design and Development of Robots with Human-Like Intelligence

Kerstin Dautenhahn

Adaptive Systems Research Group, School of Computer Science,
University of Hertfordshire,
Hatfield, Herts, AL10 9AB, United Kingdom
K.Dautenhahn@herts.ac.uk

Abstract. The chapter discusses a recent paradigm shift in the field of Artificial Intelligence regarding the nature of human intelligence and its implications for the design and development of intelligent robots. It will be argued that social intelligence is not a mere 'add-on' to intelligent robot behaviour for the practical purpose of enabling the robot to interact smoothly with other robots or people, but that social intelligence might be a stepping stone towards more human-like, embodied artificial intelligence. The argument is supported by discussions in primatology highlighting the social origins of primate intelligence. The chapter also discusses challenges and opportunities provided by socially intelligent robots, with implications for our future.

Keywords: Social Intelligence Hypothesis, Social Robots, Human-Robot Interaction, Paradigm Shift.

1 Introduction

This introductory section provides a brief summary of different approaches towards artificially intelligent systems and paradigm shifts that have occurred during the 50 year history of Artificial Intelligence (AI) in relationship to the nature of human-like intelligence, see Table 1. Section 2 then discusses the social origins of primate intelligence, implications of this for artificially intelligent systems are outlined in section 3. Section 4 highlights some challenges and opportunities of socially intelligent robots in the 21st Century. Section 5 concludes this chapter with remarks on who we are and were we are going in a world shared with artificially intelligent machines.

1.1 When AI Was Born: The Symbolic Era

Since its 'official' origin in 1956 Artificial Intelligence has been a growing area of research which is now been established worldwide in research and education. Several large international conferences are being held regularly, e.g. IJCAI and ECAI, and AI

M. Lungarella et al. (Eds.): 50 Years of AI, Festschrift, LNAI 4850, pp. 288–302, 2007.

courses are part of the curriculum in many undergraduate as well as post-graduate university degree programs.

Artificial Intelligence as a research field has however not lost its controversial nature. At present, we do not find one agreed upon path towards artificially intelligent artifacts, whether they take the form of software agents, robots, or other incarnations of computer technology.

Early approaches to AI emphasized the symbolic nature of human-like intelligence realizing e.g. problem-solving and planning via symbol manipulation. Typically, as seen in expert systems, knowledge has been elicited from human 'experts' and encapsulated in knowledge bases for a particular domain [1]. Rules operating on the knowledge base then try to find answers to problems. Achieving systems that can be applied beyond a limited domain is often seen as requiring increasingly larger knowledge bases. Creating computer programs that could solve problems such as chess, towers of Hanoi, or even model human reasoning in limited domains (e.g. case based reasoning [2]) were exemplary challenges, and a lot of progress has been made in this domain, e.g. over the past 10 years chess playing software has been able to beat human world champions.

However, is chess playing an appropriate benchmark test for human-like artificial intelligence? Who in the first place plays chess? Certainly young children don't, non-human animals don't either, and a large proportion of Earth's population has never been exposed to chess. But even those people growing up in a culture that exposes them to chess and other games don't necessarily like to play chess, and they don't necessarily become good at it, even if they try. Interestingly, algorithms that have been used to model chess-playing, widely based on extensive search algorithms and huge memory capacities, model skills that most humans are not particularly good at: human memory capacities are very limited, the human brain is not a giant database that stores faithfully every detail throughout our lives, but, as we have learnt over the past decades from neurobiology and psychology, the human brain is highly selective, it creates and re-creates *meaningful* information, and forgetting forms an important part of this continuous, dynamic re-organization of experiences and memories structured around narratives [3,4]. To summarize, chess and other games or puzzles often used in AI are not good examples for a bench-mark test of AI. They rather reflect an intellectual interest of a certain proportion of adult members of the Western culture, and they don't tell us much about what it means to be human, and what it could mean to be a human-like robot. Symbolic-based approaches to AI can be applied to many applications e.g. in current software agent applications, or more generally in new communication technologies. Thus, from the point of view of engineering oriented AI research purely symbolic approaches may continue and grow in the research landscape, while at the same time more and more losing touch with the biological realities of naturally intelligent systems.

1.2   When AI Began to Crawl: The Era of Behaviour Based Systems

'Nouvelle AI' emerged in the late 1980s, pioneered by R. A. Brooks in the USA as well as by L. Steels and R. Pfeifer in Europe, soon to be joined by many researchers worldwide [5-7]. Brooks's phrase that 'elephants don't play chess', as explained in

Table 1. A brief history of paradigm shifts that have occurred over 50 years of AI

AI Paradigm	Important principles	AI architectures	Important research themes	Examples of successful results	Primary fields of impact
Symbolic AI	Symbol manipulation	Deliberative / hierarchical architectures	Problem solving, reasoning, navigation	Expert systems	Computation / Cognitive Science
Nouvelle AI	Embodiment, tight sensori-motor couplings	Behaviour-based / evolutionary / neural network architectures	Locomotion, exploration, obstacle avoidance, phototaxis, collective / swarm behaviour	Insect-like walking machines	Ethology, biology, neuroscience
Develop-mental /social robotics	Development of skills, scaffolding/ social context: communication and cooperation	Any: Often based on dynamical systems approaches, emergent computation	Hand-eye coordination, grasping, object manipulation, cognitive development/ coordination and cooperation in teams	Robot hand-eye coordination,/ro bot teams	Developmental psychology/ ethology
Human-Robot interaction	'natural' interaction of people with robots	Any	Human-robot communication, dialogue, cooperation; robots as assistants and companions	Humanoid 'expressive' robots, service robots	Psychology, social sciences

his publications, lead to his research into insect-like, 'behaviour-based' robots, which caused a stir in the AI research landscape. 'Behaviour-based' reflects a certain notion of how to build controllers for autonomous robots, emphasizing the tight connection between sensing and acting and de-empathizing planning, avoiding explicitly represented 'models of the environment' ('the world is its own best model', according to Brooks). Autonomous locomotion, and later *learning* thereof (via applications of neural networks, evolutionary techniques, or other –preferably distributed- machine learning techniques), emerged as new challenges for this new paradigm. Impressive 'fast, cheap and simple' machines roamed around AI labs, e.g. walking insect-like machines in the MIT AI Lab [8], 'self-sufficient' Lego Robots behaving according to biologically inspired principles in the VUB-AI Lab [9], miniature Khepera robots in the Univ. Zürich AI Lab which were learning how to classify different objects based on biologically-inspired principles of sensori-motor coordination, e.g. used to help a robot distinguish between differently sized objects [10], or robots roaming a 'hilly landscape' [11, 12] (cf. Fig. 1). The possibility to build and study (relatively) 'cheap and simple' robots also enabled their use in education: increasingly university courses or summer schools include practical robot building and / or programming laboratories. 'Nouvelle AI' robots typically show a variety of basic behaviours enabling them to explore and interact with and within a certain environment; this usually includes avoiding obstacles, approaching or avoiding lights or using other gradients (odour, heat etc.) to guide behaviour, it may include the ability to pick up and manipulate objects. Simulating or modeling *human-behaviour* has not been a primary interest of Nouvelle AI, inspirations originally derived from insect behaviour (walking, phonotaxis etc.) or rats (navigation), emphasizing the 'bottom-up' approach towards intelligence where simpler systems need to be built and understood before 'moving on' the phylogenetic ladder and targeting 'higher organisms'[1].

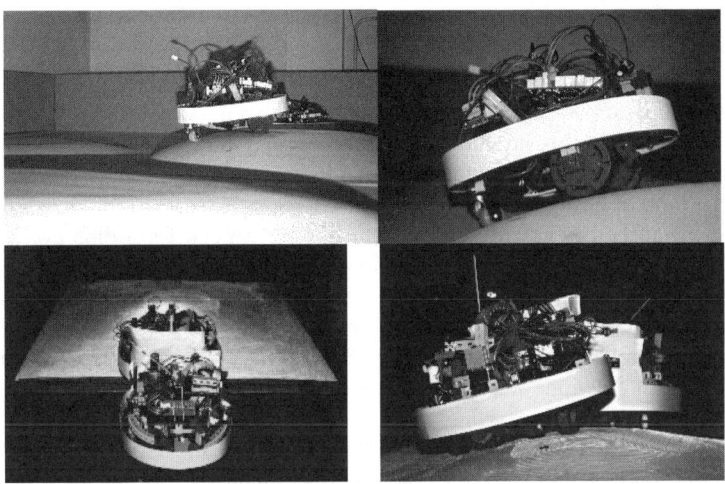

Fig. 1. Early behaviour-based 'social robots', inhabiting a hilly landscape environment, keeping contact with each other via touch sensors, following and learning from each other via imitation (using infra-red sensors and radio communication) [11,12]. In addition the robots had light sensors in order to detect certain areas in the environment, and inclination sensors for perceiving hills. This work was carried out in the mid 1990s at GMD in St. Augustin, Germany and the VUB AI Lab in Brussels, Belgium.

Influenced by principles of behaviour-based robotics, since the early 1990s *swarm robotics/swarm intelligence* [13] has emerged as a research field inspired by interactions among social insects. The goal is to have a large group of relatively simple and, typically, identical robots that can perform tasks on the group level that are impossible to carry out by the individual robot. Typically the robots do not communicate directly, only indirectly via the environment (based on the principles of *stigmergy* and *self-organization*). Related developments were *collective robotics* [14] that could also include direct communication among robots, but still emphasized the bottom-up, *distributed nature of intelligent behaviour.* Other work in behaviour-based robotics on robots interacting with each other included direct interaction and communication, cf. Fig. 1. The inspiration for this work did not come from social

---

[1] I do not imply any hierarchical 'ordering' of animals species; the term 'higher animals' refers to phylogenetically more recent vertebrate, mammalian, primate and human species.

insects, but rather from social behaviour that we can find in individualized societies such as primates [11,15].

1.3 Growing Up Socially: The Era of Developmental and Social Robotics

'Nouvelle AI' has been embracing fundamentally the dynamic nature of agent-environment coupling, influenced by and influencing dynamical systems approaches to programming robots, understanding cognition, or modeling systems that can develop, which has recently lead to fields such as 'developmental robotics' or 'epigenetic' (ontogenetic) robotics. Developmental robotics faces the challenges of building robots that, based on certain basic or 'phylogenetic abilities', can develop more complex behaviour. Different from Nouvelle AI, the main sources of inspiration are taken from child development, e.g. how children learn grasping, hand-eye coordination, manipulating and recognizing objects etc., see [16]. In parallel to, and interconnected with developmental robotics, social robotics emerged, emphasizing direct interactions and communication of robots with other robots (and later giving rise to the field of human-robot interaction (HRI)).

Since the early 1990ies myself and others have argued for a direction of research where robots directly interact with each other or humans, addressing issues inspired by human-human social interaction [11,15]. In the field of HRI, during the past 10 years my team has been investigating robots as therapeutic toys for children [17], and, more recently, robot companions in 'home scenarios' [18,19]. At present, in the year 2007, social interactions between robots and other robots, or between people and robots has become an increasingly active and growing research area. For example, 2006 saw *The 15th IEEE International Symposium on Robot and Human Interactive Communication (RO-MAN06)* with the theme of *Getting to Know Socially Intelligent Robots*, as well as the first *ACM/IEEE Human-Robot Interaction (HRI'06)* conference, just to name two conferences dedicated to Human-Robot Interaction. Human-Robot Interaction is a highly challenging and exciting area of research with potentially many application areas where robots co-exist with people in daily-life environments, such as offices, hospitals, or people's homes. Artificial Intelligence research for such robots poses very different challenges compared to e.g. robotics in manufacturing environments.

Service robotics emphasizes that robots can provide assistance and be useful, which might fulfill the dream of most people, namely to get help with tedious household tasks. Recently the notion of a *Robot Companions* has been investigated, e.g. as part of the IST-FET funded project COGNIRON (Cognitive Robot Companion [20]). There are two important aspects to a robot companion: a) it needs to be able to carry out useful tasks on behalf of or in collaboration with people, and b) it needs to carry out these tasks such that the robot's behaviour is comfortable to and acceptable to its users.

Human-robot interaction is a very inter- and multidisciplinary area that involves not only fields such as robotics, engineering, AI, but also psychology, social sciences, human-computer interaction, and others. Creating robots that people like to 'live with' is demanding; the robot's behaviour, its appearance, its tasks and its (attributed or otherwise designed) 'personality' need to be balanced carefully. Believability and consistency have emerged as a main theme: e.g. people don't want a mechanical looking robot that speaks with a realistic human voice, they don't want a chatty and funny robot in 'serious' (e.g. financial) application domains. The level of a robot's social skills may also much depend on requirements in different application domains, as analysed in [21]. A robot that has repeated, long-term and physically 'close contacts' with people, e.g. in rehabilitation and therapy, will need different social skills compared to a robot that has little and only short-term interactions with people, e.g. a robot cleaning an office building at night. A robot's functionality, environment and context of use will determine its required level of social intelligence and ability to exhibit social skills.

Also, *long-term studies* with repeated exposure are necessary for robot companions that should be around people on a long-term basis: first impressions often change, novelty effects wear out, a robot that might be considered entertaining at first encounter might 'get on one's nerves' quickly. A few long-term studies with people and robots illuminate these issues, e.g. [22-25].

A lot of insights can be gained from studies on human-human interaction and communication, however, results cannot be transferred directly to human-robot interaction. State of the art robots, that are clearly distinguishable from people, are a specific instantiation of interactive technologies that allow new interaction modalities to emerge. Robots are not people, and while robots are given human-like interaction and communication abilities, humans adapt to interacting and communicating with them. Perceptions of and attitudes towards robots are shaped by expectations, based on experience with other machines, computers, as well as inspirations from science fiction movies and novels, but expectations will change with increasing familiarity with robots. Thus, robotic designs that might be suitable now might appear unsuitable in 100 years time. It has become clear that new designs, methods and methodologies are required for the newly emerging research field of Human-Robot Interaction [26-31].

2   The Social Roots of Human Intelligence

Since the early 1990s I have argued for a view of Artificial Intelligence that should acknowledge what is known about the origins of human intelligence [11,15]. A fuller discussion of this argument and its implications for AI and robotics is provided elsewhere [27,32,33]. I can only provide a brief summary here.

While the exact details regarding the evolution of primate and human intelligence are still under investigation, with every new archeological discovery adding a missing piece to the puzzle of 'what we are and where we came from', it has become widely acknowledged that the evolution of primate intelligence cannot be separated from the social context, i.e. from the group-living context of cooperation, collaboration, competition and survival.

The *Social Intelligence Hypothesis*, also called *Machiavellian Intelligence Hypothesis* or *Social Brain Hypothesis*, suggests that primate social intelligence has evolved primarily due to the need to deal with complex social dynamics. In different primate species we find different degrees of 'technical intelligence', while social intelligence seems to be more 'fundamental' (cf. Fig. 2). According to this argument, primates' brains and primate intelligence have evolved in adaptation to the need to

Fig. 2. Ring-tailed lemurs (*Lemur catta*) the subjects of Alison Jolly's studies on primate behaviour. Her seminal article in 1966 made the observation that lemurs are very social, while having less technical skills. Alison Jolly suggested that social intelligence might have been the foundation of later developments of 'technical intelligence' in other primates: "Primate society, thus, could develop without the object-learning capacity or manipulative ingenuity of monkeys. This manipulative, object cleverness, however, evolved only in the context of primate social life. Therefore, I would argue that some social life preceded, and determined the nature of, primate intelligence." ([34], p. 506). (Photo by Kabir Bakie at the Cincinnati Zoo, May 2005 [35], used by permission under the Creative Commons Attribution ShareAlike Licence 2.5).

live in groups, where in order to maintain structure and cohesion an understanding of other group members, their social relationships, and the ability to predict their behaviour became beneficial and accelerated primate brain evolution [36-38], given that maintaining large and 'smart' brains is costly. Increasing social complexity required increasingly sophisticated social skills. Identifying friends and allies, predicting others' behavior, knowing how to form alliances, manipulating group members, making war, love and peace, are important ingredients of primate politics [39]. Thus, there are two interesting aspects to human sociality: it served as an evolutionary constraint which led to an increase of brain size in primates, which in return led to an increased capacity to further develop social complexity. It has been argued that during human evolution intelligent skills gained in the social domain were applied to other domains, e.g. technical domains. Note, for the purpose of the discussion in this chapter it is not important whether sociality has been the primary factor, or was one among other key factors in the evolution of human brain size. It suffices to accept that social intelligence was a driving force behind the evolution of human intelligence, complementing insights into the importance of the social context for development [40].

3   The Social Roots of Artificial Intelligence

The implications of the above argument presented in section 2 for artificially intelligent systems are twofold:

Issues relevant to application oriented AI research:
If human behaviour is fundamentally social, then any artificially intelligent system designed to interact with people needs to be able to deal with social behaviour, e.g. to recognize and respond appropriately to gestures, language or any other movements and expressions people might use in interaction, and be able to generate social behaviour in a timely and situated manner. Note, social intelligence is embodied in a system able to *perceive*, *process* and *express* social intelligence whereby the complexity of each of these three aspects needs to be balanced.

Even for robots that have not been primarily designed to interact with people directly, e.g. a service robot in a restaurant delivering meals from the kitchen to the table where a waiter is serving the customers, even such a robot needs to be *socially aware* of humans in order to work efficiently [41], e.g. it needs to be able to predict where people will go and sit in a crowded place.

If robots operate in an application domain where direct repeated contact with people is necessary, e.g. in care applications, then the success of such robots will depend on its acceptability, not only with regard to its functionality and ability to carry out tasks satisfactorily, but also with regard to *how* it is carrying out the tasks. The robot's presence needs to be acceptable to and comfortable for people.

Issues relevant to foundational AI research oriented towards developing autonomous intelligent machines and understanding intelligence:
If social intelligence, in evolutionary terms, 'came first' in the development of primate intelligence, and then later was applied to other domains, then one may extrapolate and apply this 'evolutionary history' to machines, too. Accordingly, from an evolutionary perspective, then intelligent robots need to be *socially intelligent* robots. Note, a developmental perspective complements this view: children are born into a social network and grow up to become a social being, a 'natural psychologist' *alongside* developing technical skills required for becoming mathematicians, architects, programmers or accountants. Revealing how technical and social intelligence in robots may develop hand-in-hand, mutually benefiting each other, is an interesting area of AI robotics research.

Social intelligence in humans has some universal features, but many that depend on cultural norms and individual differences. Humans develop social intelligence in interaction other humans in a social group. However, who wants to be a full-time caretaker for a robot providing the loving, caring and supporting environment that a 'developing' robot may need? The role of a human as a caretaker for a robot has been explored [42], but not in the sense of interactions over 24 hours / 7 days a week, and for many years. Bringing up a child requires a lot of effort, involving emotional, psychological and physiological investments [27]. Applying this to robots may have unforeseen consequences for people and robots (and ethical implications, for both). The choice of paradigms, caretaker or assistant/companion will certainly influence the way how people interact with and perceive socially intelligent robots of the future.

The paradigm shift towards *socially intelligent* robots opens up exciting possibilities for future research in AI and its applications. The next section raises a few of these issues.

4   Socially Intelligent Robots in the 21st Century: Challenges and Opportunities

Scientifically, for researchers in Artificial Intelligence and other areas, there are a number of exciting future research challenges for socially intelligent robots, for example:

1) Robots that 'have a life' – machines that can remember and re-create meaningful experiences in their 'life-time' and interpret new experiences by relating them to their previous experiences (robots as autobiographic agents, [43,44], robots that possess interaction histories [45]), and using such memories to guide decision-making.

2) Robots that don't pretend to be what they aren't – machines with appearances and behaviour that are consistent with their interaction and other abilities, suitable for a particular task and application domain, able to meet the expectations of people interacting with them. While e.g. Sony's Aibo robot did show some dog-like behaviour, initial encounters with people often could go wrong, for example: a person throwing a ball across the room expecting Aibo to rush and fetch it will be disappointed. A machine closely resembling a living entity will elicit expectations of intelligence and behaviour similar to the biological model. While relying on anthropomorphic and zoomorphic tendencies in people is a possible avenue for engaging people in interactions with robots, often machine-like appearance and behaviour of robots may be more appropriate and reflect more accurately the robots' abilities and level of intelligence and social skills [46]. AI research would benefit from systematically tackling the continuum between (from an AI perspective) truly socially intelligent robots on the one hand, and robots that may be engaging but rely solely on the (psychological) 'attribution' of life-like qualities (socially evocative robots according to [47]) on the other hand. Studies of this kind would help in developing a synthesis of the human-centred and robot cognition (AI) centred viewpoints and may significantly advance the development of socially intelligent robots that people will accept into their lives [27].

3) Robots that are useful companions and increase the quality of life of people – machines that are able to assist, support and entertain humans in a great variety of situations, in the home, at work, in the hospital etc., while *mediating human contact with other people* in order to support the social network of their 'users'.

Robots with human like features that behave socially and may entertain or assist people are expected to continue to enter our daily lives. In the remainder of this section I will make a few remarks based on my experience as a researcher in the field of socially intelligent robots.

Often discussions on the future of (socially) intelligent robots take a turn towards assuming these machines will eventually 'be like us' in any way we might want to evaluate them. In a recent British government-commissioned report scientists suggest that robots might be granted rights by 2056, similar to human rights [48]. Similarly, discussions on robot ethics abound. Popular questions that are being discussed in the public, with almost identical headlines emerging every few years, are <if a robot ever became X, would we be allowed to do Y?>. X can stand for e.g. intelligent, conscious, or sentient, Y may stand for switching it off, selling it, hurt its feelings etc.

While these discussions are certainly entertaining and philosophically interesting, they are in my view beside the point of issues that should be discussed as well, but are less 'sensational', more 'complicated' to explain and thus less likely to reach a large scale public. It is easy to create appealing news stories, let me give you an example: Fig. 3 shows KASPAR, a child-sized minimally expressive humanoid robot that we have developed in our laboratory and has been used to investigate human-robot imitation and social learning. It would be easy to fill in a form and apply for KASPAR to attend the University's nursery. If KASPAR gets accepted to the nursery – great news story. If it doesn't – an even greater news story. Of course it would be completely bizarre and beyond any scientific justification to do so, since after all, like other robots of its kind in the world, KASPAR is a research platform that is more often switched off than switched on, and the time it actually interacts with people in experiments can be counted in minutes or at best hours. But most importantly, KASPAR is a *machine*, it can 'smile' (and we are, from a design perspective, actually quite proud of its beautiful smile), but it only smiles when we tell it to 'smile' according to some algorithmic specification (and it doesn't matter at this point whether the smile is produced by a 'rule' or the firing pattern of an artificial neural network). KASPAR does not 'have a life', it is not a sentient being, if it breaks we'll through it away and build a new one. We designed it to have some human-like expressions in order to provide a 'natural interface' that people like interact with, we did not intend to built a 'robot child'. That's robotics reality in the year 2007.

Of course somebody might get 'attached' to KASPAR, in the way people might get attached to a valuable watch, a precious porcelain figure, or an everyday item with sentimental value. However, this does not make this robot more or less sentient, intelligent, or alive, for this matter. Humans, and probably other sentient species that inhabit this planet, have an enormously rich inner world, a world of imagination and fantasy, memories, stories and real emotions, and we link to the world by projecting some of this to the 'in-animate world'[2], seeing patterns in clouds, becoming attached to a cartoon character or a particular car, anthropomorphizing the world around us and detecting intentionality and goal directed behaviour in other living as well as non-living things. These human capabilities tell us a lot about the nature of being a sentient being, and being human, but they do not necessarily tell us much about the nature of the objects themselves that we are anthropomorphizing.

Discussions about robots that ultimately may be indistinguishable from humans[3] are scientifically interesting, but should not dominate discussions on the future of AI robots in the 21st century. There are many more pressing issues that *should* dominate the discussions, for example:

Robot companions and assistants for the elderly and other vulnerable people are a big topic in Human-Robot Interaction and Service Robotics. Caring for people is labour and thus cost-intensive, and the argument goes that robots can assist people and thus allow e.g. elderly people to stay in their own homes and live independently longer. I personally support this argument but there are some issues that need consideration:

---

[2] For the purpose of this paper I distinguish living things, i.e. biological organisms, from other things, i.e. inanimate objects.
[3] See recent work on androids [49].

The danger of isolation: if a robot can help a person to remain in her own home for longer, keeping her own furniture, within the familiar neighbourhood etc, then this is certainly a worthwhile goal. If interaction with the robot is occasional and focuses on providing assistance e.g. for cooking, going to the bathroom etc, then probably people will primarily view it as a tool, and it will have its place among other machines in the home.

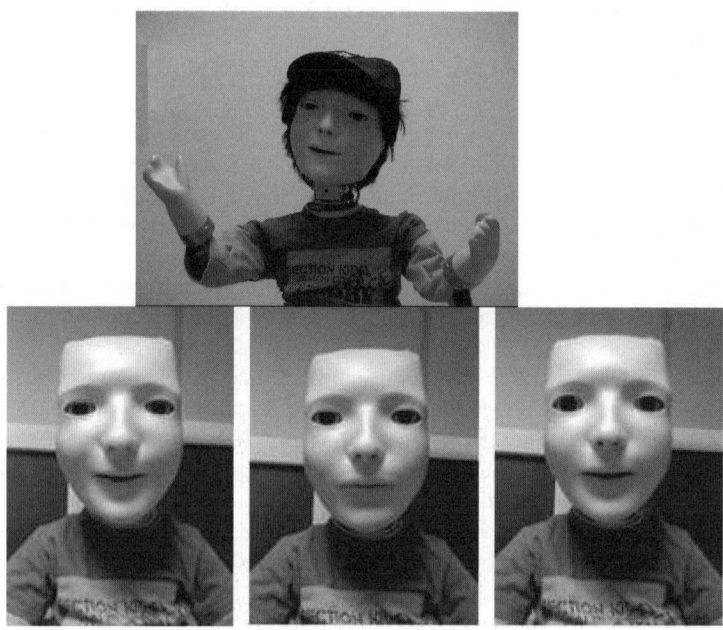

Fig. 3. The child-sized humanoid robot KASPAR [51] used in the European IST-FP6 Robot-Cub project [52] investigating human-robot interaction, specifically issues regarding interaction kinesics, imitation, and computational architectures for a robot to extend its temporal horizon via interaction histories [45]. The aim of designing KASPAR was to study what types of human-robot interactions a minimal set of expressive robot features can afford[4]. The goal is not perfect realism, but sufficient realism for rich interaction. KASPAR has 8 degrees of freedom in the head and neck and 6 in the arms and hands. The face is a silicon-rubber mask, which is supported on an aluminum frame. It has eyelids and 2 DOF eyes fitted with video cameras, and a mouth capable of opening and smiling.

But when the 'mechanical' interaction with the machine is frequent and / or long-lasting, if the person increasingly depends on the machine's abilities, and if at the same time a 'social interface' and artificial social intelligence make the robot more and more human-like, then people are likely to 'bond' with such machines. Bonding

---

[4] This is different from more cartoon-like humanoid face robots, cf. KISMET [42], where exaggerated features are used to trigger a nurturing instinct in people in order to engage them in interactions with the robot. See further discussions on such human-robot relationships in [27].

with people is important, but can we bond with robots? Bonding is a bi-directional relationship, we 'get something back' when bonding with a person, a pet etc. [27]. What are our rewards regarding our emotional needs when bonding with a robot? Questions like these are not purely theoretical, worldwide various humanoid robot interfaces and appearances are currently under development. An investigation is necessary on what the implications of these are, and whether various possible side effects are desirable or undesirable. Pharmaceutical companies need to run extensive clinical testing before releasing a new product: this cannot *guarantee* safety, but at least provides a foundation of knowledge that judgments can be based on. Releasing robots out of the laboratories into the wild of our daily lives, in particular in the context of long-term interaction and when involving vulnerable people (cf. [49]), should require a similarly rigorous procedure. With respect to robots, physical safety for people is not the only concern, this issue can be dealt with using regulations similar to those that e.g. exist for children's toys or other interactive new technologies (e.g. mobile phones etc.). The effects that are usually not considered are psychological: how does interaction with machines effect people's sense of self, sense of autonomy and control, sense of belonging, sense of friendship and love etc. These are *scientific questions* that can be studied experimentally and should be investigated extensively long before actual products are being put on the market.

5 The Future of AI: Who We Are, and What We Can Become?!

Working in the field of social robotics, and more recently human-robot interaction, has alerted me to aspects of AI research that are usually not being considered:

a) being *social is what makes us human*, phylogenetically, when considering the evolution of primate intelligence, and ontogenetically, when considering the development of intelligence and intelligent behaviour in children

b) robotics technology can be scientifically and intellectually exciting, while at the same time leading to the development of *robots that can be useful for people,* as exemplified in many applications in assistive technology and rehabilitation robotics

c) the difficulty of creating artificially autonomous machines shows us how precious life is, how amazingly and wonderfully complex and interesting biological organisms are, how much we depend on nature for our inspirations and ideas from the natural world and living organisms (whether on the level of their 'behaviour', their 'brains', or their 'body/morphology').

If AI creates smart computers and robots, it is my wish that they will also be used to protect our natural environment and the diversity of species, so they will remain our greatest source of scientific inspiration and creativity for still a long time to come. As (socially) intelligent beings we should be able to succeed in this quest.

Acknowledgments. I would like to thank Chrystopher L. Nehaniv for feedback on this chapter. Many thanks to my research team at University of Hertfordshire for discussions and joint research on socially intelligent robots. Some of the work described in this paper was partially conducted within two EU Integrated Projects: COGNIRON ("The Cognitive Robot Companion") funded by the European

Commission Division FP6-IST Future and Emerging Technologies under Contract FP6-002020, and RobotCub ("Robotic Open-architecture Technology for Cognition, Understanding, and Behaviours") funded by the European Commission through Unit E5 (Cognition) of FP6-IST under Contract FP6-004370.

References

[1] Jackson, P.: Introduction to Expert Systems, 3rd edn. Addison Wesley Longman, Harlow, England (1999)

[2] Kolodner, J.L.: Case-Based Reasoning. Morgan Kaufmann, San Francisco, California (1993)

[3] Bartlett, F.C.: Remembering: An Experimental and Social Study. Cambridge University Press, Cambridge (1932)

[4] Bruner, J.: Actual Minds, Possible Worlds. Harvard University Press, Cambridge (1987)

[5] Brooks, R.A.: Cambrian Intelligence. MIT Press, Cambridge, MA (1999)

[6] Pfeifer, R., Scheier, C.: Understanding Intelligence. MIT Press, Cambridge, MA (1996)

[7] Steels, L.: The Artificial Life Roots of Artificial Intelligence. Artificial Life 1, 75–110 (1994)

[8] Brooks, R.A.: A Robot That Walks; Emergent Behaviors from a Carefully Evolved Network. MIT AI Lab Memo 1091 (February 1989)

[9] Steels, L.: A case study in the behavior-oriented design of autonomous agents. In: Cliff, D., Husbands, P., Meyer, J.-A., Wilson, S.W. (eds.) From animals to animats 3 - Proceedings of the 3rd International Conference on Simulation of Adaptive Behavior, pp. 445–452. MIT Press, Cambridge, MA (1994)

[10] Scheier, C., Pfeifer, R.: Classification as sensory-motor coordination. In: Moran, F., Merelo, J.J., Moreno, A., Chacon, P. (eds.) Advances in Artificial Life. LNCS, vol. 929, pp. 656–667. Springer, Heidelberg (1995)

[11] Dautenhahn, K.: Getting to know each other - artificial social intelligence for autonomous robots. Robotics and Autonomous Systems 16, 333–356 (1995)

[12] Dautenhahn, K., Billard, A.: Studying Robot Social Cognition Within a Developmental Psychology Framework. In: Burgard, W., Nehmzow, U., Vestli, S., Schweizer, G. (eds.) Proceedings of the Third European Workshop on Advanced Mobile Robots (EUROBOT), pp. 187–194 (1999)

[13] Bonabeau, E., Dorigo, M., Theraulaz, G.: Swarm Intelligence: From Natural to Artificial Systems. Oxford University Press, Oxford (1999)

[14] Matarić, M.J.: Issues and Approaches in the Design of Collective Autonomous Agents. Robotics and Autonomous Systems 16, 2–4 (1995)

[15] Dautenhahn, K.: Trying to imitate – a step towards releasing robots from social isolation. In: Gaussier, P., Nicoud, J.-D. (eds.) Proc. From Perception to Action Conference, Lausanne, Switzerland, pp. 290–301. IEEE Computer Society Press, Los Alamitos (1994)

[16] Lungarella, M., Metta, G., Pfeifer, R., Sandini, G.: Developmental Robotics: A Survey. Connection Science 15(4), 151–190 (2003)

[17] Robins, B., Dickerson, P., Stribling, P., Dautenhahn, K.: Robot-mediated joint attention in children with autism: A case study in robot-human interaction. Interaction Studies 5(2), 161–198 (2004)

[18] Walters, M.L., Dautenhahn, K., Woods, S.N., Koay, K.L., te Boekorst, R., Lee, D.: Exploratory Studies on Social Spaces between Humans and a Mechanical-looking Robot. Connection Science 18(4), 429–439 (2006)

[19] Dautenhahn, K., Woods, S.N., Kaouri, C., Walters, M.L., Koay, K.L., Werry, I.: What is a Robot Companion - Friend, Assistant or Butler? In: Proc. IROS 2005, IEEE IRS/RSJ International Conference on Intelligent Robots and Systems, Edmonton, Alberta Canada, August 2-6, 2005, pp. 1488–1493. IEEE Computer Society Press, Los Alamitos (2005)

[20] COGNIRON (Last accessed 6th July 2007), http://www.cogniron.org

[21] Dautenhahn, K.: Roles and Functions of Robots in Human Society - Implications from Research in Autism Therapy. Robotica 21(4), 443–452 (2003)

[22] Kanda, T., Hirano, T., Eaton, D., Ishiguro, H.: Interactive Robots as Social Partners and Peer Tutors for Children: A Field Trial. Human Computer Interaction 19(1-2), 61–84 (2004)

[23] Gockley, R., Bruce, A., Forlizzi, J., Michalowski, M., Mundell, A., Rosenthal, S., Selner, B., Simmons, R., Snipes, K., Schultz, A., Wang, J.: Designing Robots for Long-Term Social Interaction. In: IROS 2005. Proc. IEEE/RSJ International Conference on Intelligent Robots and Systems, pp. 1338–1343 (2005)

[24] Robins, B., Dautenhahn, K., te Boekhorst, R., Billard, A.: Robotic Assistants in Therapy and Education of Children with Autism: Can a Small Humanoid Robot Help Encourage Social Interaction Skills? Universal Access in the Information Society (UAIS) 4(2), 105–120 (2005)

[25] Wada, K., Shibata, T.: Robot therapy in a care house – results of case studies. In: Proc. IEEE RO-MAN 2006, pp. 581–586 (2006)

[26] Dautenhahn, K.: Methodology and Themes of Human-Robot Interaction: A Growing Research Field. International Journal of Advanced Robotic Systems (in press, 2007)

[27] Dautenhahn, K.: Socially intelligent robots: dimensions of human–robot interaction. Phil. Trans. R. Soc. B 362(1480), 679–704 (2007)

[28] Thrun, S.: Towards a framework of human-robot interaction. Human Computer Interaction 19(1-2), 9–24 (2004)

[29] Scholtz, J.: Theory and evaluation of human robot interactions. In: Proc. Hawaii International Conference on System Science, vol. 36 (2003)

[30] Yanco, H.A., Drury, J.: A taxonomy for human-robot interaction. In: Proc. AAAI Fall Symposium on Human-Robot Interaction. AAAI, Technical Report FS-02-03, Falmouth, Massachusetts, pp. 111–119 (2002)

[31] Fong, T., Nourbakhsh, I., Dautenhahn, K.: A Survey of Socially Interactive Robots. Robotics and Autonomous Systems 42(3-4), 143–166 (2003)

[32] Dautenhahn, K.: Socially Intelligent Agents in Human Primate Culture. In: Trappl, R., Payr, S. (eds.) Agent Culture: Human-Agent Interaction in a Multicultural World, pp. 45–71. Lawrence Erlbaum Associates, New Jersey (2004)

[33] Dautenhahn, K.: The origins of narrative: In search of the transactional format of narratives in humans and other social animals. In: Gorayska, B., Mey, J.L. (eds.) Cognition and Technology, pp Jolly, A.: Lemur social behavior and primate intelligence. Science 153, 501–506 (1966)

[34] . 127–152. John Benjamins, Amsterdam (2004)

[35] http://commons.wikimedia.org/wiki/Image:Lemur_CinZoo_069.jpg (Last accessed 6th July 2007)

[36] Byrne, R.W., Whiten, A. (eds.): Machiavellian Intelligence. Clarendon Press, Oxford (1988)

[37] Byrne, R.W.: Machiavellian intelligence. Evolutionary Anthropology 5, 172–180 (1997)

[38] Whiten, A., Byrne, R.W. (eds.): Machiavellian Intelligence II: Extensions and Evaluations. Cambridge University Press, Cambridge (1997)

[39] de Waal, F.: Chimpanzee Politics: Power and Sex among Apes. Johns Hopkins University Press, Baltimore (2000)

[40] Lindblom, J., Ziemke, T.: Social situatedness of natural and artificial intelligence: Vygotsky and beyond. Adaptive Behavior 11(2), 79–96 (2003)

[41] Dautenhahn, K., Ogden, B., Quick, T.: From Embodied to Socially Embedded Agents - Implications for Interaction-Aware Robots. Cognitive Systems Research 3(3), 397–428 (2002)

[42] Breazeal, C.: Designing Sociable Robots. The MIT Press, Cambridge, MA (2002)

[43] Dautenhahn, K.: The Art of Designing Socially Intelligent Agents - Science, Fiction, and the Human in the Loop. Applied Artificial Intelligence 12(7-8), 573–617 (1998)

[44] Dautenhahn, K.: Embodiment and Interaction in Socially Intelligent Life-Like Agents. In: Nehaniv, C.L. (ed.) Computation for Metaphors, Analogy, and Agents. LNCS (LNAI), vol. 1562, pp. 102–142. Springer, Heidelberg (1999)

[45] Mirza, N.A., Nehaniv, C.L., Dautenhahn, K., te Boekhorst, R.: Grounded Sensorimotor Interaction Histories in an Information Theoretic Metric Space for Robot Ontogeny. Adaptive Behavior 15(2), 167–187 (2007)

[46] Woods, S., Dautenhahn, K., Schulz, J.: Exploring the design space of robots: Children's perspectives. Interacting with Computers 18, 1390–1418 (2006)

[47] Breazeal, C.: Toward sociable robots. Robotics and Autonomous Systems 42(3-4), 167–175 (2003)

[48] Financial Times (Last accessed 6th July 2007),

[49] http://www.ft.com/cms/s/5ae9b434-8f8e-11db-9ba3-0000779e2340.html

[50] Robins, R., Dautenhahn, K., Dubowski, J.: Robots as isolators or mediators for children with autism? A cautionary tale. In: Dautenhahn, K., te Boekhorst, R. (eds.) Proc. AISB 2005 Symposium on Robot Companions Hard Problems and Open Challenges in Human-Robot Interaction, University of Hertfordshire, UK, 14-15 April 2005, pp. 82–88 (2005) Online papers available:

[51] http://www.aisb.org.uk/publications/proceedings/aisb05/5_RoboComp_final.pdf

[52] MacDorman, K.F., Ishiguro, H.: The uncanny advantage of using androids in cognitive and social science research. Interaction Studies 7(3), 297–337 (2006)

[53] Blow, M.P., Dautenhahn, K., Appleby, A., Nehaniv, C.L., Lee, D.: Perception of Robot Smiles and Dimensions for Human-Robot Interaction Design. In: RO-MAN 2006. Proc. The 15th IEEE International Symposium on Robot and Human Interactive Communication, pp. 469–474. IEEE Press, Los Alamitos (2006)

[54] RobotCub (Last accessed 6th July 2007), http://www.robotcub.org

Intrinsically Motivated Machines

Frédéric Kaplan[1] and Pierre-Yves Oudeyer[2]

[1] Ecole Polytechnique Federale de Lausanne - CRAFT
CE 1628 - Station 1 - CH 1015 Lausanne Switzerland
frederic.kaplan@epfl.ch
[2] Sony Computer Science Laboratory Paris
6 rue Amyot 75005 Paris France
py@csl.sony.fr

Abstract. Children seem intrinsically motivated to manipulate, to explore, to test, to learn and they look for activities and situations that provide such learning opportunities. Inspired by research in developmental psychology and neuroscience, some researchers have started to address the problem of designing intrinsic motivation systems. A robot controlled by such systems is able to autonomously explore its environment not to fulfil predefined tasks but driven by an incentive to search for situations where learning happens efficiently. In this paper, we present the origins of these intrinsically motivated machines, our own research in this novel field and we argue that intrinsic motivation might be a crucial step towards machines capable of life-long learning and open-ended development.

Keywords: Intrinsic motivation, curiosity, exploration, meta-learning, development.

1 Introduction

Have you ever noticed how much fun babies can have by simply touching objects, sticking them into their mouths, or rattling them and discovering new noises? Children seem to engage is such type of activities just for the sake of it. They seem intrinsically motivated to manipulate, to explore, to test - in one word - to learn and therefore they look for activities and situations that provide such learning opportunities. More than 50 years ago, Alan Turing prophetically announced that the child's mind would show us the way to artificial intelligence. "Instead of trying to produce a programme to simulate the adult mind, why not rather try to produce one which simulates the child's?" [Turing, 1950]. We believe it is now time to take this advice seriously. Through hundreds of experiments and models - supervised, unsupervised, reinforced, active, passive, associative, symbolic, connectionist, hybrid, embodied, situated, distributed - we benefit now from a large collection of examples that show how a machine can learn. However, the issue of "why" would a machine learn (or how would it choose what to learn) has not been tackled with the same attention. This is what interests us here.

M. Lungarella et al. (Eds.): 50 Years of AI, Festschrift, LNAI 4850, pp. 303–314, 2007.

During the past five years, we have been working on algorithms that make robots eager to investigate their surroundings. These robots explore their environment in search of new things to learn: they get bored with situations that are already familiar to them, and also avoid situations which are too difficult. In our experiments, we place the robots in a world that is rich in learning opportunities and then just watch how the robots develop by themselves. This research is based on a series of studies showing the importance of *intrinsic motivation* in human development and its neural correlates in the brain. The next sections give a general overview of these findings and discuss the origins of intrinsically motivated machines in artificial intelligence research. We then present our own research in this field through the discussion of a specific architecture and related robotic experiments.

2   What Is Intrinsic Motivation?

In psychology, an activity is characterized as intrinsically motivated when there is no apparent reward except the activity itself. People seek and engage in such activities for their own sake and not because they lead to extrinsic reward. In such cases, the person seems to derive enjoyment directly from the practice of the activity. Following this definition, most children playful or explorative activities can be characterized as being intrinsically motivated. Also, much adult behaviour seem to belong to this category: free problem-solving (solving puzzles, crosswords), creative activities (painting, singing, writing during leisure time), gardening, hiking, etc. At the physiological level, it has been argued that intrinsically motivated activities are directly related to changes in the central nervous system and are quite independant from non-nervous tissues. On the contrary, extrinsic needs (e.g. hunger) are directly related to the state and management of non-nervous-systems tissues [Deci and Ryan, 1985]. Moreover, intrinsically motivated activities are generic in the sense that they can be produced by different kinds of sensory contexts. Finally, at a phenomenological level, a person engages in intrinsically motivating activities to experience particular feelings of competence and self-determination [Deci and Ryan, 1985]. Such situations are characterized by a feeling of effortless control, concentration, enjoyment and a contraction of the sense of time [Csikszenthmihalyi, 1991].

3   Intrinsic Motivation in Psychology and Neuroscience

A first bloom of investigations concerning intrinsic motivation happened in the 1950s. Researchers started by trying to give an account of exploratory activities on the basis of the theory of drives [Hull, 1943], which are non-nervous-system tissue deficits like hunger or pain and that the organisms try to reduce. For example, [Montgomery, 1954] proposed a drive for exploration and [Harlow, 1950] a drive to manipulate. This drive naming approach had many short-comings which were criticized in detail by White in 1959 [White, 1959]: intrinsically motivated exploratory activities have a fundamentally different dynamics. Indeed, they are not homeostatic: the general tendency to explore in is never satiated and is not a consummatory response to a stressful perturbation of the organism's body. Moreover, exploration does not seem to be related to any non-nervous-system tissue deficit.

Some researchers then proposed another conceptualization. Festinger's theory of cognitive dissonance [Festinger, 1957] asserted that organisms are motivated to reduce dissonance, that is the incompatibility between internal cognitive structures and the situations currently perceived. Fifteen years later a related view was articulated by Kagan stating that a primary motivation for human is the reduction of uncertainty in the sense of the "incompatibility between (two or more) cognitive structures, between cognitive structure and experience, or between structures and behaviour" [Kagan, 1972]. However, these theories were criticized on the basis that much human behaviour is also intended to *increase* uncertainty, and not only to reduce it.

Human seem to look for some forms of optimality between completely uncertain and completely certain situations. In 1965, Hunt developed the idea that children and adult look for optimal incongruity [Hunt, 1965]. He regarded children as information-processing systems and stated that interesting stimuli were those where there was a discrepancy between the perceived and standard levels of the stimuli. For, Dember and Earl, the incongruity or discrepancy in intrinsically-motivated behaviours was between a person's expectations and the properties of the stimulus [Dember and Earl, 1957]. Berlyne developed similar notions as he observed that the most rewarding situations were those with an intermediate level of novelty, between already familiar and completely new situations [Berlyne, 1960].

Whereas most of these researchers focused on the notion of optimal incongruity at the level of psychological processes, a parallel trend investigated the notion of optimal arousal at the physiological level [Hebb, 1955]. As over-stimulation and under-stimulation situations induce fear (e.g. dark rooms, noisy rooms), people seem to be motivated to maintain an optimal level of arousal. A complete understanding of intrinsic motivation should certainly include both psychological and physiological levels.

Eventually, a last group of researchers preferred the concept of challenge to the notion of optimal incongruity. These researchers stated that what was driving human behaviour was a motivation for effectance [White, 1959], personal causation [De Charms, 1968], competence and self-determination [Deci and Ryan, 1985]. The difference with optimality theories is mainly a matter of point of view: in one case, human search for some form of optimality as defined by an abstract function, in the other case they look for a particular kind of feelings occurring during challenging situations.

Novel investigations in neuroscience concerning neuromodulation systems have complemented these findings. Although most experiments in this domain focus on the involvement of particular neuromodulators like dopamine for predicting extrinsic reward (e.g. food), some work lends credence to the idea that such neuromodulators might also be involved in the processing of types of intrinsic motivation associated with novelty and exploration (e.g. [Dayan and Belleine, 2002] and [Kakade and Dayan, 2002]). In particular, some studies suggest that dopamine responses could be interpreted as reporting "prediction error" and not only "reward prediction error" [Horvitz, 2000]. At a more global level, Panksepp has compiled a set of evidence suggesting the existence of a SEEKING system responsible for exploratory behaviours. "This harmoniously operating neuroemotional system drives and energizes many mental complexities that humans experience as persistent feelings of interest, curiosity, sensation seeking and, in the presence of a sufficiently complex cortex, the search for higher meaning." [Panksepp, 1998]. However, the gap is still important between neuroscience accounts and research in psychology on intrinsic motivation.

4   The Route to Intrinsically Motivated Machines

During the last ten years, the machine learning and robotics community has begun to investigate architectures that permit incremental and active learning (see for instance [Thrun and Pratt, 1998] or [Cohn et al., 1996]). Interestingly, the mechanisms developed in these papers have strong similarities with mechanisms developed in the field of statistics, where it is called "optimal experiment design" [Fedorov, 1972]. Active learners (or machines that perform optimal experiments) are machines that ask, search and select specific training examples in order to learn efficiently.

A few researchers have started to address the problem of designing intrinsic motivation systems to drive active learning, inspired by research in developmental psychology and neuroscience. The idea is that a robot controlled by such systems would be able to autonomously explore its environment not to fulfil predefined tasks but driven by some form of intrinsic motivation that pushes it to search for situations where learning happens efficiently. One of the first computational system implementing a form of artificial curiosity was described by Schmidhuber in 1991 [Schmidhuber, 1991]. Schmidhuber articulated the idea that in order to learn efficiently a machine should try to reduce prediction error instead of maximizing or minimizing it. More recently, different types of intrinsic motivation systems were explored, mostly in software simulations [Huang and Weng, 2002, Marshall et al., 2004, Steels, 2004]. Most of this research has largely ignored the history of the intrinsic motivation construct as it was elaborated in psychology during the last 50 years and sometimes reinvented concepts that existed several decades before (basically, different forms of optimal incongruity). Technically, such control systems can be viewed as particular types of reinforcement learning architectures [Sutton and Barto, 1998], where rewards are not provided externally by the experimenter but self-generated by the machine itself. The term "intrinsically motivated reinforcement learning" has been used in this context [Barto et al., 2004].

5   Intrinsic Motivation and Development

Our own research in this field aims at showing how such forms of active learning architectures permit to structure the development of a robot. For this issue, Piaget remains a reference. Although he has not written extensively about motivation, his view of motivation is implicit throughout his writings [Piaget, 1952]. According to Piaget, children are intrinsically motivated to encounter activities which involve some assimilation and accommodation. Assimilation is a process whereby children incorporate aspects of the environment into their pre-existing cognitive structure, which are called schemata. This means that the child's cognitive structure influences his perception of the environment. During accommodation, the child adapts his cognitive structures to fit the environment. In Piaget's view, learning is simply an aspect of assimilation and accommodation. When the child encounters an informational input from the environment highly discrepant from existing schemata, the input will most likely be ignored. When inputs are completely predictable, children generally lose interest in them. Therefore, Piaget's theory articulates a concept of intrinsic simulation that bears many resemblance to the notion of optimal incongruity developed by Berlyne and others.

Piaget's work has inspired researchers in artificial intelligence for some time. Many artificial intelligence models make use of internal explicit schema structures under names like *frames* [Minsky, 1975] or *scripts* [Schank and Abelson, 1977]. In such systems, there is a one-to-one mapping between these internal structures and the functional operation that the agent can perform. For instance, Drescher describes a system inspired by Piaget's theories in which a developing agent explicitly creates, modifies and merges schema structures in order to interact with a simple simulated environment [Drescher, 1991]. Using explicit schema structures has several advantages: such structures can be manipulated via symbolic operations, creation of new skills can be easily monitored by following the creation of new schemata, etc. Our work differs notably from this approach first because it does not rely on such explicit representations. We use subsymbolic systems, based on continuous representations of their environment. Nevertheless, as we will see, such systems may display some organized forms of behaviour where clear functional units can be identified. Second and most importantly, it is centrally based the idea of an intrinsic motivation to learn.

6   An Example of Architecture

We have designed a control architecture and performed a series of experiments to investigate how far its intrinsic motivation system, implementing a form of artificial curiosity, can shape the developmental trajectories of a robot [Oudeyer et al., 2007]. The cognitive architecture of our robot can be described as having two modules: 1) one module implements a predictor M which learns to predict the sensorimotor consequences when a given action is executed in a given sensorimotor context; 2) another module is a metapredictor $metaM$ which

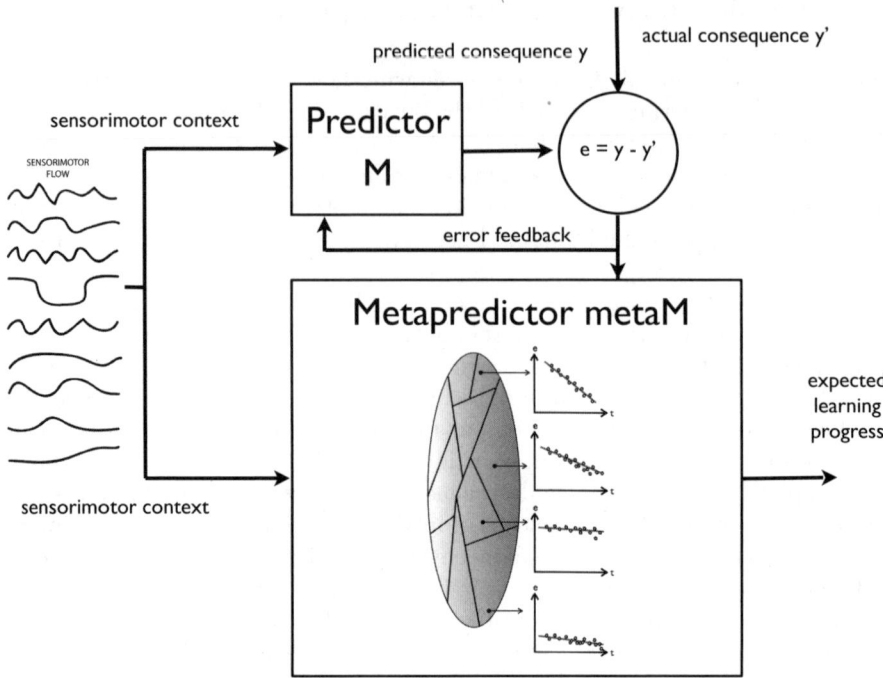

Fig. 1. An intrinsic motivation system including a predictor M that learns to anticipate the consequence **y** of a given sensorimotor context and a metapredictor $metaM$ learning to predict the expected learning progress of M in the same context. Once the actual consequence is known, M and $metaM$ get updated. $MetaM$ re-evaluates the error curve linked with this context and computes an updated measure of the learning progress (local derivative of curve). In order to classify similar contexts, $metaM$ includes a hierarchical self-organizing classifier.

learns to predict the errors that machine M makes in its predictions: these meta-predictions are then used as the basis of a measure of the potential interest of a given situation. The system is designed to be *progress-driven*. It avoids both predictable and unpredictable situations in order to focus on the ones which are expected to maximize the decrease in prediction error. To obtain such a behaviour, the metaprediction system computes the local derivative of the error rate curve of M and generates an estimation of the expected learning progress linked with a particular action in a particular context. In order to really evaluate learning progress, error obtained in one context must be compared with errors obtained in similar contexts (if not the robot may oscillate between hard and easy situations and evaluate these changes as progress). Therefore, the metaprediction system must also be equipped with a self-organized classification system capable of structuring an infinite continuous space of particular situations into higher-level categories (or kinds) of situations. Figure 1 summarizes the key components of such progress-driven systems (see [Oudeyer et al., 2007] for more details).

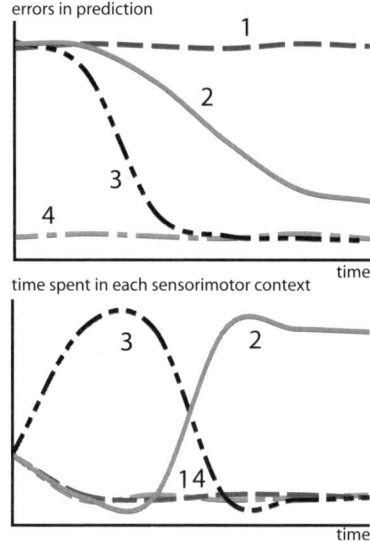

Fig. 2. Confronted with four contexts characterized by different learning profiles, the motivation for maximizing learning progress results in avoiding situations already predictable (context 4) or too difficult to predict (context 1), in order to focus first on the context with the fastest learning curve (context 3) and eventually, when the latter starts to reach a "plateau" to switch to the second most promising learning situation (context 2). This intrinsic motivation system allows the creation of an organized exploratory strategy.

Figure 2 illustrates how progress-driven learning operates on an idealized problem. Confronted with four contexts characterized by different learning profiles, the motivation for maximizing learning progress results in avoiding situations that are already predictable (context 4) or too difficult to predict (context 1), in order to focus first on the context with the fastest learning curve (context 3) and eventually, when the latter starts to reach a "plateau", to switch the second most promising learning situation (context 2). Situations of maximal progress are called "progress niches". Progress niches are not intrinsic properties of the environment. They result from a relationship between a particular environment, a particular embodiment (sensors, actuators, feature detectors and techniques used by the prediction algorithms) and a particular time in the developmental history of the agent. Once discovered, progress niches progressively disappear as they become more predictable.

7   Experiments

We have performed a series of robotic experiments using this architecture. In these experiments, the robot actively seeks out sensorimotor contexts it can learn given its morphological and cognitive constraints. Whereas a passive strategy would lead to very inefficient learning, an active strategy allows the learner to discover and exploit learning situations fitted to its biases. In one of those experiment a four-legged robot is placed on a play mat (for more details, see [Oudeyer and Kaplan, 2006]). The robot can move its arms, its neck and mouth and can produce sounds. Various toys are placed near the robot, as well as a pre-programmed "adult" robot which can respond vocally to the other robot in certain conditions. At the beginning of an experiment, the robot does not know anything about the structure of its sensorimotor space (which actions cause which effects). Given the size of the space, exhaustive exploration would take a very long time and random exploration would be inefficient.

During each robotic experiment, which lasts approximately half a day, the flow of values of the sensorimotor channels are stored, as well as a number of features which help us to characterize the dynamics of the robot's development. The evolution of the relative frequency of the use of the different actuators is measured: the head pan/tilt, the arm, the mouth and the sound speakers (used for vocalizing), as well as the direction in which the robot is turning its head.

Figure 3 shows data obtained during a typical run of the experiment. At the beginning of the experiment, the robot has a short initial phase of random exploration and body babbling. During this stage, the robot's behaviour is equivalent to the one we would obtain using random action selection: we clearly observe that in the vast majority of cases, the robot does not even look at or act on objects; it essentially does not interact with the environment. Then there is a phase during which the robot begins to focus successively on playing with individual actuators, but without knowing the appropriate affordances: first there is a period where it focuses on trying to bite in all directions (and stops bashing or producing sounds), then it focuses on just looking around, then it focuses on trying to bark/vocalize towards all directions (and stops biting and bashing), then on biting, and finally on bashing in all directions (and stops biting and vocalizing). Then, the robot comes to a phase in which it discovers the precise affordances between certain action types and certain particular objects. It is at this point focusing either on trying to bite the biteable object (the elephant ear), or on trying to bash the bashable object (the suspended toy). Eventually, it focuses on vocalizing towards the "adult" robot and listens to the vocal imitations that it triggers. This interest for vocal interactions was not pre-programmed, and results from exactly the same mechanism which allowed the robot to discover the affordances between certain physical actions and certain objects.

The developmental trajectories produced by these experiments can be interpreted as assimilation and accommodation phases if we retain the Piagetian's terminology. For instance, the robot "discovers" the biting and bashing schema by producing repeated sequences of these kinds of behaviour, but initially these actions are not systematically oriented towards the biteable or the bashable object. This stage corresponds to "assimilation". It is only later that "accommodation" occurs as biting and bashing starts to be associated with their respective appropriate context of use. Our experiments show that functional organization

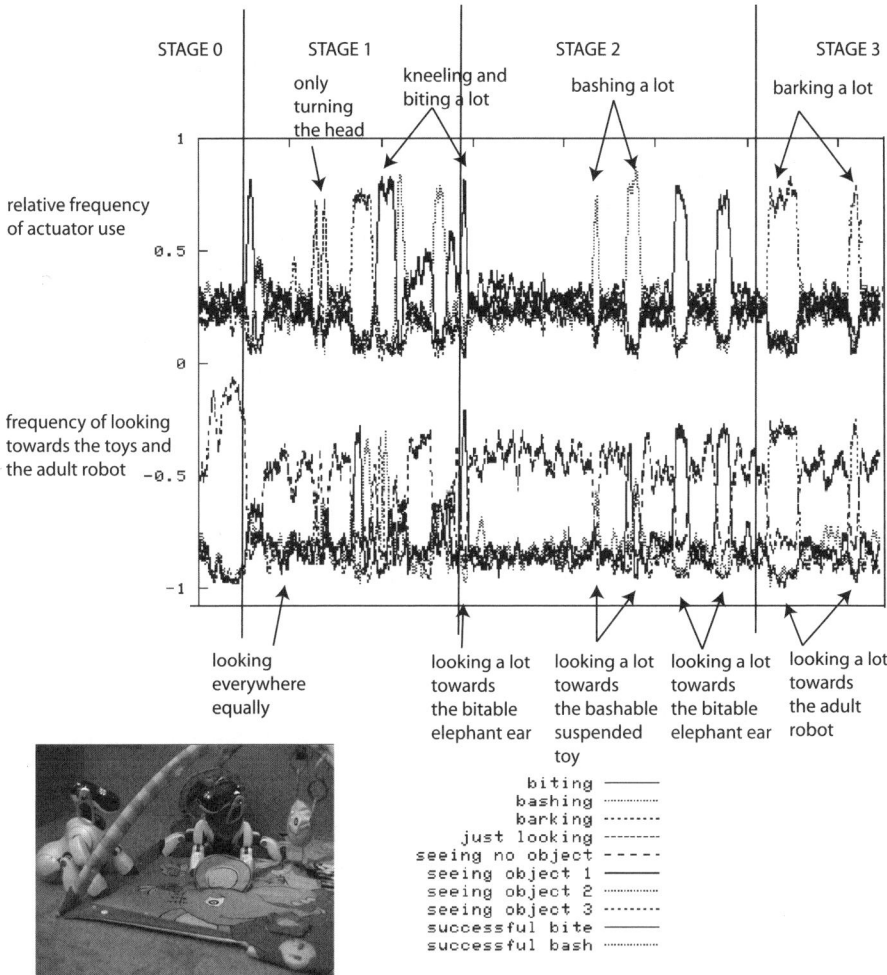

Fig. 3. The robot, placed on a play mat, can move its arms, its neck and mouth and produce sounds. Various toys are placed near the robot, as well as a pre-programmed "adult" robot which can respond vocally to the other robot in certain conditions. Results obtained after a typical run of the experiment are shown. **Top curves**: relative frequency of the use of different actuators (head pan/tilt, arm, mouth, sound speaker). **Bottom curves**: frequency of looking towards each object and in particular towards the "adult" pre-programmed robot.

can emerge even in the absence of explicit internal schema structures and that developmental patterns can spontaneously self-organize, driven by the intrinsic motivation system. Many diverse lines of experimental data can potentially be explained in common terms if we consider that children learn how to focus on what is learnable in the situation they encounter and on what can be efficiently grasped at a given stage of their cognitive and physiological development. For instance, we have discussed elsewhere how progress-driven learning provides an interpretation of developmental sequences in early imitation and sensorimotor development [Kaplan and Oudeyer, 2007a, Kaplan and Oudeyer, 2007b]. What is fundamentally new in these experiments, as compared to what is possible in psychology, is that learning dynamics, embodiment and environmental factors (both social and physical) are *controllable variables*. One experiment can be conducted with the same learning system, but using a different body placed in a different environment. Likewise, the effects of small changes in the intrinsic motivation systems can be studied while keeping the embodiment and environmental aspects similar.

8   The Future of Intrinsically Motivated Machines

To conclude, this novel line of research might also provide radically new techniques for building intelligent robots. Indeed, as opposed to the work in classical artificial intelligence in which engineers impose pre-defined anthropocentric tasks to robots, the techniques we describe endow the robots with the capacity of deciding by themselves which are the activities that are maximally fitted to their current capabilities. Intrinsically motivated machines autonomously and actively choose their learning situations, thus beginning by simple ones and progressively increasing their complexity. Of course, many challenges remain to be solved before we could build intrinsically motivated machines capable to learn like children do. One of them is that children's complex behaviour patterns seem hierarchically organized. This aspect is absent from our current architecture but have started to be tackled by other groups, in particular around the option framework [Sutton et al., 1999, Barto et al., 2004]. This research is very complementary to ours as they experimented the use of a complex reinforcement technique given a simple novelty-based intrinsic motivation system. We believe the future of intrinsically motivated machines lies somewhere between of these two approaches.

References

[Barto et al., 2004] Barto, A., Singh, S., Chentanez, N.: Intrinsically motivated learning of hierarchical collections of skills. In: ICDL 2004. Proceedings of the 3rd International Conference on Development and Learning, Salk Institute, San Diego (2004)

[Berlyne, 1960] Berlyne, D.: Conflict, Arousal and Curiosity. McGraw-Hill, New York (1960)

[Cohn et al., 1996] Cohn, D., Ghahramani, Z., Jordan, M.: Active learning with statistical models. Journal of artificial intelligence research 4, 129–145 (1996)

[Csikszenthmihalyi, 1991] Csikszenthmihalyi, M.: Flow-the psychology of optimal experience. Harper Perennial (1991)

[Dayan and Belleine, 2002] Dayan, P., Belleine, W.: Reward, motivation and reinforcement learning. Neuron 36, 285–298 (2002)

[De Charms, 1968] De Charms, R.: Personal causation: the internal affective determinants of behavior. Academic Press, New York (1968)

[Deci and Ryan, 1985] Deci, E., Ryan, R.: Intrinsic Motivation and Self-Determination in Human Behavior. Plenum Press, New York (1985)

[Dember and Earl, 1957] Dember, W.N., Earl, R.W.: Analysis of exploratory, manipulatory and curiosity behaviors. Psychological Review 64, 91–96 (1957)

[Drescher, 1991] Drescher, G.L.: Made-up minds. MIT Press, Cambridge, MA (1991)

[Fedorov, 1972] Fedorov, V.: Theory of Optimal Experiment. Academic Press, New York (1972)

[Festinger, 1957] Festinger, L.: A theory of cognitive dissonance. Evanston, Row, Peterson (1957)

[Harlow, 1950] Harlow, H.: Learning and satiation of response in intrinsically motivated complex puzzle performances by monkeys. Journal of Comparative and Physiological Psychology 43, 289–294 (1950)

[Hebb, 1955] Hebb, D.O.: Drives and the c.n.s (conceptual nervous system). Psychological review 62, 243–254 (1955)

[Horvitz, 2000] Horvitz, J. C.: Mesolimbocortical and nigrostriatal dopamine responses to salient non-reward events. Neuroscience 96(4), 651–656 (2000)

[Huang and Weng, 2002] Huang, X., Weng, J.: Novelty and reinforcement learning in the value system of developmental robots. In: Prince, C., Demiris, Y., Marom, Y., Kozima, H., Balkenius, C. (eds.) Proceedings of the 2nd international workshop on Epigenetic Robotics: Modeling cognitive development in robotic systems, Lund University Cognitive Studies 94, pp. 47–55 (2002)

[Hull, 1943] Hull, C.L.: Principles of behavior: an introduction to behavior theory. Appleton-Century-Croft, New York (1943)

[Hunt, 1965] Hunt, J.M.: Intrinsic motivation and its role in psychological development. Nebraska symposium on motivation 13, 189–282 (1965)

[Kagan, 1972] Kagan, J.: Motives and development. Journal of Personality and Social Psychology 22, 51–66 (1972)

[Kakade and Dayan, 2002] Kakade, S., Dayan, P.: Dopamine: Generalization and bonuses. Neural Networks 15, 549–559 (2002)

[Kaplan and Oudeyer, 2007a] Kaplan, F., Oudeyer, P.-Y.: The progress-drive hypothesis: an interpretation of early imitation. In: Nehaniv, C., Dautenhahn, K. (eds.) Models and mechanisms of imitation and social learning: Behavioural, social and communication dimensions, pp. 361–377. Cambridge University Press, Cambridge (2007a)

[Kaplan and Oudeyer, 2007b] Kaplan, F., Oudeyer, P.-Y.: Un robot motivé pour apprendre: le role des motivations intrinseques dans le development sensorimoteur. Enfance 59(1), 46–58 (2007b)

[Marshall et al., 2004] Marshall, J., Blank, D., Meeden, L.: An emergent framework for self-motivation in developmental robotics. In: ICDL 2004. Proceedings of the 3rd International Conference on Development and Learning, Salk Institute, San Diego (2004)

[Minsky, 1975] Minsky, M.: A framework for representing knowledge. In: Wiston, P. (ed.) The psychology of computer vision, pp. 211–277. Mc Graw Hill, New York (1975)

[Montgomery, 1954] Montgomery, K.: The role of exploratory drive in learning. Journal of Comparative and Physiological Psychology 47, 60–64 (1954)

[Oudeyer and Kaplan, 2006] Oudeyer, P.-Y., Kaplan, F.: Discovering communication. Connection Science 18(2), 189–206 (2006)

[Oudeyer et al., 2007] Oudeyer, P.-Y., Kaplan, F., Hafner, V.: Intrinsic motivation systems for autonomous mental development. IEEE Transactions on Evolutionary Computation 11(1), 265–286 (2007)

[Panksepp, 1998] Panksepp, J.: Affective neuroscience: the foundations of human and animal emotions. Oxford University Press, Oxford (1998)

[Piaget, 1952] Piaget, J.: The origins of intelligence in children. Norton, New York (1952)

[Schank and Abelson, 1977] Schank, R., Abelson, R.: Scripts, plans, goals and understanding: An inquiry into human knowledge structures. Lawrence Erlbaum Associates, Hillsdale, NJ (1977)

[Schmidhuber, 1991] Schmidhuber, J.: Curious model-building control systems. In: Proceeding International Joint Conference on Neural Networks, vol. 2, pp. 1458–1463. IEEE, Singapore (1991)

[Steels, 2004] Steels, L.: The autotelic principle. In: Iida, F., Pfeifer, R., Steels, L., Kuniyoshi, Y. (eds.) Embodied Artificial Intelligence. LNCS (LNAI), vol. 3139, pp. 231–242. Springer, Heidelberg (2004)

[Sutton and Barto, 1998] Sutton, R., Barto, A.: Reinforcement learning: an introduction. MIT Press, Cambridge, MA (1998)

[Sutton et al., 1999] Sutton, R., Precup, D., Singh, S.: Between mdpss and semi-mdps: A framework for temporal abstraction in reinforcement learning. Artificial Intelligence 112, 181–211 (1999)

[Thrun and Pratt, 1998] Thrun, S., Pratt, L.: Learning to Learn. Kluwer Academic Publishers, Dordrecht (1998)

[Turing, 1950] Turing, A.: Computing machinery and intelligence. Mind 59, 433–460 (1950)

[White, 1959] White, R.: Motivation reconsidered: The concept of competence. Psychological review 66, 297–333 (1959)

Curious and Creative Machines

Hod Lipson

Mechanical and Aerospace Engineering, and Computing & Information Science,
Cornell University, Ithaca NY 14853, USA

1 Introduction

I recently gave a robot demonstration to a class of 1st-grade elementary school children. In the school's gymnasium hall, a few dozen 6-year-olds gathered enthusiastically around a few shiny machines with plenty of sensors and actuators, demonstrating patterns of locomotion. "These robots learned how to move by themselves" – I explained. "Some even developed their own shape", I said, pointing at a set of 3D-printed plastic robots whose morphology and control evolved in simulation.

The kids were not impressed.

One courageous child finally asked the question that was probably on everyone's mind: "But what can they *do*?" I delved into an elaborate discussion of locomotion and manipulation, morphology and control, and machine learning. "But what can they do?" the child persisted. The Emperor's new clothes, I thought, were not that shiny after all. But before long, another child came to my rescue: "Aha! The robot is jogging!" he realized. Yes, that's what the robot was doing. It learned how to exercise, an activity that took western civilization centuries to discover. Perhaps in the not-so-distant future, robots will one day be able to jog an exercise for us. The gymnastics teacher was pleased.

When it comes to intelligence, people are difficult to impress. Children have seen robots that talk, walk, fight and perform a myriad of complex tasks in movies such as Star Wars™ and Terminator™, but they know very well that robots in movies are not real. Artificial Intelligence is almost an oxymoron: Whenever breakthroughs are achieved – from Deep Blue's mastery of chess to Stanley's autonomous traversal of the Mojave desert – something is still missing. If it is just doing what it was *designed* to do, is it *truly* intelligent?

It is fascinating to watch how teachers and parents, when asked about signs of intelligence, quickly point out: Curiosity and creativity are hallmarks of a gifted child. Can we make such curious and creative machines? Will we relinquish some control over what they discover and create?[1] Are we ready to give up on our human-centric claim to curiosity and creativity?

2 The Second Half of AI

One of the hallmarks of human intelligence is the ability to design: To synthesize a set of elementary building blocks in order to achieve some novel, high-level and

M. Lungarella et al. (Eds.): 50 Years of AI, Festschrift, LNAI 4850, pp. 315–319, 2007.

open-ended functionality. Imagine a Lego set at your disposal: Bricks, rods, wheels, motors, sensors and logic components are your "atomic" building blocks, and you must find a way to put them together to achieve a given high-level functionality: A machine that can move[2], say. You know the physics of the individual components' behaviors; you know the repertoire of pieces available, and you know how they are allowed to connect. But how do you determine the combination that gives you the desired functionality? This is the problem of *Synthesis*.

In the last two centuries, engineering sciences have made remarkable progress in their ability to analyze and predict physical phenomena. We understand the governing equations of thermodynamics, electromagnetics, and fluid flow, to name but a few. Numerical methods such as finite elements allow us to solve these constitutive equations with good approximation for many practical situation. We can use these methods to investigate and explain observations, as well as to predict the behavior of products and systems long before they are ever physically realized.

But progress in systematic *synthesis* has been frustratingly slow. Robert Willis, a professor of natural and experimental philosophy at Cambridge, wrote back in 1841:

> *[A rational approach is needed] to obtain, by direct and certain methods, all the forms and arrangements that are applicable to the desired purpose. At present, questions of this kind can only be solved by that species of intuition that which long familiarity with the subject usually confers upon experienced persons, but which they are totally unable to communicate to others. When the mind of a mechanician is occupied with the contrivance of a machine, he must wait until, in the midst of his meditations, some happy combination presents itself to his mind which may answer his purpose."* [3]

Almost two centuries later, a rational method for general open-ended synthesis is still not at hand. Design is still taught today largely through apprenticeship: Engineering students learn about existing solutions and techniques for well-defined, relatively simple problems, and then – through practice – are expected to improve and combine these to create larger, more complex systems.

How is this synthesis process done? We do not know, but we cloak it with the term "creativity". Fields such as humanities and arts, share the same conundrum: You can learn to appreciate good poetry, music, and sculpture, but how do you systematically create it?

The field of Artificial Intelligence has not escaped this inevitable course either. Over the last fifty years, AI – and its modern incarnation as machine learning in particular – has been primarily occupied with deduction, modeling and prediction, but not synthesis of *new things*. Learning from examples, combining logical facts, and propagating constraints, leave us interpolating inside the convex hull of our existing knowledge. I am not claiming that this is either easy or that it is not useful, nor that it has been fully mastered. But it is a fundamentally different direction than the quest for open-ended creativity, where the results are unbounded in their complexity and performance.

3   On Creativity

While computers can compute – and now analyze – almost anything, open ended creativity is still the unconquered Holy Grail still seen as distinctively human. Human intelligence is ultimately a natural biological phenomenon, and like any other biological phenomenon, it is a product of evolution. Many theses have been written about the evolutionary origin of intelligence[4], and one argument is that intelligence was driven by the need to create and use new tools. Not blindly execute an innate recipe for building a nest, a dam, or a hive – but a true *adaptive* ability to construct new things that rapidly exploit current resources, strengths and weaknesses of others.

Indeed the two standing examples of systematic synthesis we have to inspire us are both evolutionary: One is natural evolution, governed by Darwinian natural selection and variation. The other example is engineering design – not by the mythical maverick designer, but by a slow evolutionary progress, accumulating successive small variations and recombination of exiting technologies made by millions of ordinary designers, subject to the natural selection of the market[5]. These evolutionary processes are admittedly slow, inefficient, and provide no guarantees of optimality or even success, but perhaps there are fundamental limits on the conversion of energy into new information – a kind of thermodynamic law[6].

Over the last few decades, a number of results have appeared showing how evolutionary search is able to generate new solutions to open-ended synthesis problems. Whether or not these solutions are deemed "creative" is a matter of opinion, but they certainly satisfy some objective criteria of innovation such as patentability and publishability in their own right. The number of such inventions is growing, and Fig 1 shows two examples.

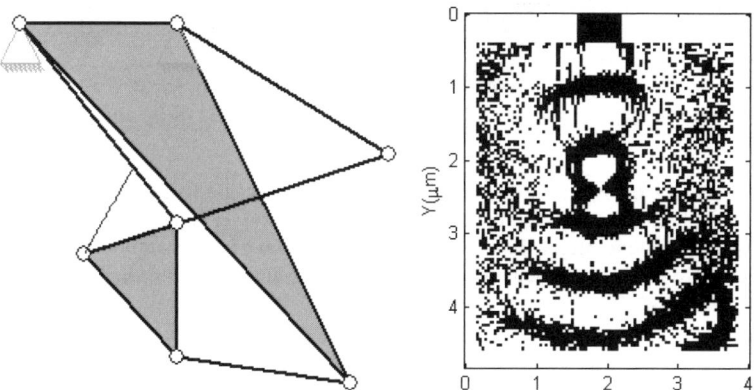

Fig. 1. Machine Creativity. Two examples of open-ended synthesis. Left: A kinematic machine automatically designed to trace a straight line without requiring any straight line in the mechanisms[7]. Right: An automatically designed light-confining nanophotonic structure with a novel hourglass-shaped element[8].

4   On Curiosity

Perhaps the most fascinating form of intelligence is the one that combines open-ended synthesis with open-ended analysis. *Curiosity* is the pursuit of new knowledge: Not only passively searching for patterns in data, but actively probing and perturbing the world to extract new information – like a child asking questions. Asking the right question is again an open-ended synthesis problem, involving creation of new predictive hypotheses and generation of actions to best test their consequence[9]. Though the field of artificial curiosity is in its infancy, it is rooted in the principles of active learning. Fig 2 shows one recent examples.

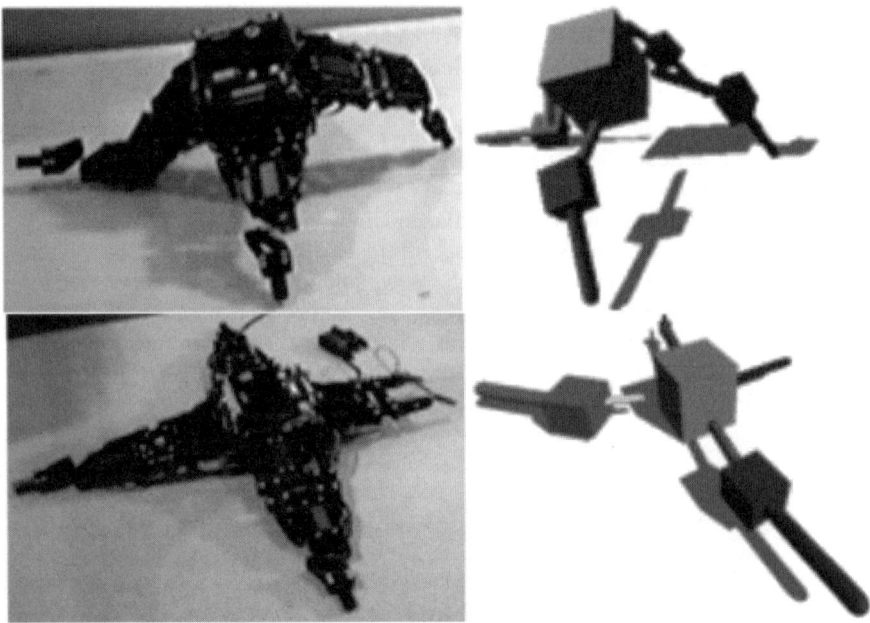

Fig. 2. Machine Curiosity: Through a series of self-directed actions, a robot explores itself and creates an explicit internal model of its topology. The self-model is then adapted after damage. Left: Physical machine; Right: Emerged internal abstraction of that morphology, that the robot has developed to explain its actuation-sensation relationships. Top: Intact; Bottom: Damaged (from Bongard *et al*[10]).

5   Conclusion

I am not alone in this quest for a new AI that can creatively generate new things[11] and ask new questions[12], nor am I unique in my view that natures' evolutionary processes provide the key; but *open-ended* evolutionary computation and active learning[13] have existed on the periphery of mainstream AI for decades. In this fiftieth anniversary of AI, I seek a new thrust – from analysis to synthesis, and from learning machines, to curious and creative machines.

References

[1] Kelly, K.: Out of Control: The New Biology of Machines, Social Systems and the Economic World. Perseus Books Group (1995)

[2] Lipson, H., Pollack, J.B.: Automatic Design and Manufacture of Artificial Lifeforms. Nature 406, 974–978 (2000)

[3] Willis, R.: Principles of Mechanism (1841)

[4] Geary, D.: Origin of Mind: Evolution of Brain, Cognition, and General Intelligence. American Psychological Association (2004)

[5] Ziman, J. (ed.): Technological Innovation as an Evolutionary Process. Cambridge University Press, Cambridge (2000)

[6] Kauffman, S.: The Origins of Order: Self-Organization and Selection in Evolution. Oxford University Press, Oxford (1993)

[7] Lipson, H.: Evolutionary Synthesis of Kinematic Mechanisms. Journal of Computer Aided Design (in press)

[8] Gondarenko, A., Preble, S., Robinson, J., Chen, L., Lipson, H., Lipson, M.: Spontaneous emergence of periodical patterns in a biologically-inspired simulation of photonic structures. Physical Review Letters 96, 143904 (2006)

[9] Bongard, J., Lipson, H.: Nonlinear system identification using coevolution of models and tests. IEEE Transactions on Evolutionary Computation 9(4), 361–384 (2005)

[10] Bongard, J., Zykov, V., Lipson, H.: Resilient Machines Through Continuous Self-Modeling. Science 314(5802), 1118–1121 (2006)

[11] Koza, J.: Genetic Programming: Routine Human-Competitive Machine Intelligence. Springer, Heidelberg (2005)

[12] Schmidhuber, J.: Developmental Robotics, Optimal Artificial Curiosity, Creativity, Music, and the Fine Arts. Connection Science 18(2), 173–187 (2006)

[13] Angluin, D.: Learning regular sets from queries and counter examples. Information and Computation 75, 87–106 (1987)

Applying Data Fusion in a Rational Decision Making with Emotional Regulation

Benjamin Fonooni[1], Behzad Moshiri[2], and Caro Lucas[2]

[1] Young Researchers Club, Tehran, Iran
bfonooni@hotmail.com
[2] Control and Intelligent Processing, Center of Excellence, School of Electrical and Computer Engineering
University of Tehran, Tehran, Iran
moshiri@ut.ac.ir, lucas@ipm.ir

Abstract. This paper focuses on designing a goal based rational component of a believable agent which has to interact with facial expressions with humans in communicative scenarios like teaching. One of the main concerns of the proposed model is to define interactions among rationality, personality and emotion in order to fulfill the idea of making rational decisions with emotional regulation. Our research aims are directed towards improving decision making process by means of applying Data Fusion techniques, especially Ordered Weighted Averaging (OWA) operator as a goal selection mechanism. Also the issue of obtaining weights for OWA aggregation is discussed. Finally the suggested algorithm is tested and results are provided with a real benchmark.

Keywords: Data Fusion, OWA, Rationality, Artificial Emotions, Decision Making.

1 Introduction

The research on believable agents focuses on creating interactive agents that give users the illusion of being human. Application domains include, among others, human-computer interaction, interactive entertainment, and education. Believability is accomplished by convincing the humans interacting with the agents express their emotions in their behavior and equip the agents with clearly distinguishable personalities. This has consequences for the agent's internal model of deliberation. It has to have knowledge of both emotions and know how these can be expressed.

As we look inside the general architecture [5], it's obvious that agents which are equipped with emotions will deliberate processes that assumed to be their best choices. The rationality means acting appropriately in various situations. However, it enables applications to have more believable interactions between man and machine which is the most important consideration of these agents. Combination of emotions, rationality and personality will yield believable agents. A well-formed agent is an agent which makes decisions according to its perceptions, rational and emotional states. Because they simulate men's rationality, it would be necessary to learn from their own experiences, manifest personality and eventually modifying their features.

M. Lungarella et al. (Eds.): 50 Years of AI, Festschrift, LNAI 4850, pp. 320–331, 2007.

In agents of this type rationality, emotions, personality and behavior are inherent characteristics of agency. This job, then, investigates the simplest relations that have place between these various aspects and also on the base of the reading of previous studies and approaches [14].

The fundamental objective of this research is the improvement of interaction between man and machine mainly in the domains applied to humans in which does not centralized on the oral communication. For instance, instructor robot which presents courses remotely and has to change its behaviors according to student's reactions. In this kind of projects the visual mediums are the main channels of interaction between the man and robot.

In other words, this research project is headed for developing a fundamental understanding of the role and usefulness of the notions of emotions and personality in designing rational artificial agents that are to operate within complex uncertain environments populated by humans. This paper draws on emerging technology of rational agent design from artificial intelligence on the one hand, with briefly research on human emotions in cognitive science and psychology on the other hand to consider personality for the agent in order to express consecutive emotions according to its environment. We have to consider the decision-theoretic paradigm of rationality for the agent to enhance the goal and action selection process [3]. Enhancement is occurred due to improving decision making algorithm by obtaining Data Fusion techniques, especially Ordered Weighted Averaging (OWA) [15].

2 Rational-Emotional Architecture

The Figure 1 shows the architecture of the rational component and its subcomponents. The heart of the architecture constitute from a production system with forward chaining approach. Usually, system contains a series of rules to make conditions-actions and use the sensory input of the agent to detect conditions and verifies them in relation of the rules and decides which actions to choose towards the outside world or other agents [6].

In case we use production system exclusively data driven, therefore it does not expect explicitly the presence of goal. These should be managed with mechanisms timely prepared to the outside of the production component of the same system. To accomplish such a task, we form goal management so that it has the function to choose the objectives of the whole system and verifies its state periodically. The presence of goals also needs to be controlled inside the production system, so the rules are structured for obtaining the requested operation.

The purpose of creating Action Selection component is make selection of actions to be performed which is affected on the base of the current emotional state of the agent. In this case, emotional state will determine which action to complete among those that production system has judged applicable and equivalent from the rational point of view. In the following paragraphs different parts of architecture are explained.

The rational input component picks up the information coming from the other components of the agent's architecture and transforms them so that is usable for the

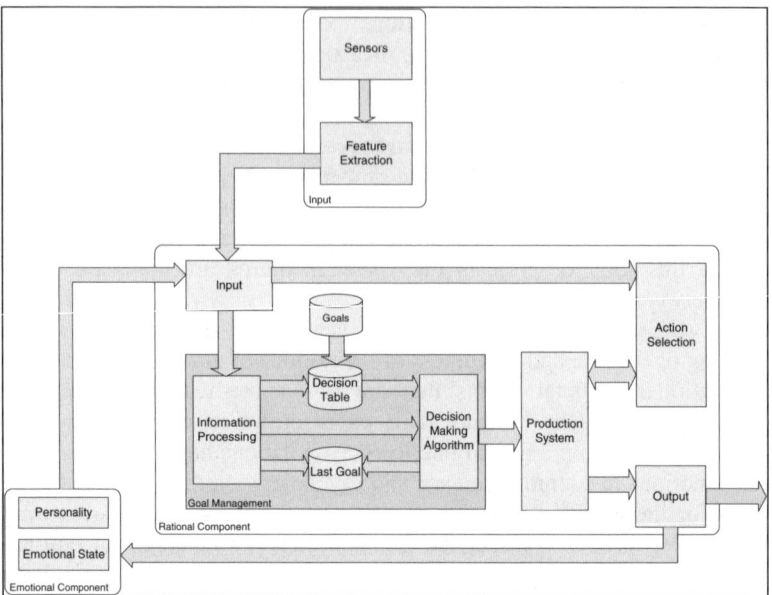

Fig. 1. Overall Architecture

production system, goal manager and action selector. The rational output component produces the information to send to the other components of the same or another agent.

Goal Management has a fundamental role in the rational architecture by handling goal selection and managing goals list. The selection will be carried out on the bases of the goal's origin information percept from the environment and of the subjective appearance relevant to the personality and to the current emotional state of the agent. In this manner, the goal management component realizes the possible interactions between rationality and emotions. It has access to the database of goals with technological information from the author of the application. Even though it does not be explicitly expected, it is included inside of the rational component with the mechanism that conforms to the agent learning rules which come from the actual experience of further goals that going to add themselves to the list. It also allows database of goals to be integrated with new goals which built from other components of the agent's architecture in result of interactions that agent has with the world or other agents. In this research, goals are emotions that has to be expressed according to the situation that agent encounters. Also actions are emotions which have to be expressed by facial expressions.

Goal Management component is divided into two principal parts:

1) Module of information processing which starts from environment and emotional information, deals with input data and update decisional parameters.

2) Module of decision making comprises the function of choosing the goal that best satisfies agent's needs.

The information processing module receives the emotional information as well as the environmental information from the rational input and calculates the new parameters which shall be used to make the decisions. One can therefore, assume that this module develops a good enough part of managing the goals as well as the current goal. In particular, with the new values calculated, it will be able to verify whether the current goal is still prioritized. It shall also determine whether the current goal was a success or a failure. This can be done by utilizing the production system whose rules can be more adaptable to the rules of the goal's success or failure. The internal functions of the information processing module that calculate the new values of the decision parameters can be implemented in more disparate manner. In particular, learning mechanisms can be taken into consideration [6].

The decision making module could use one of the algorithms for the multi-attribute decision making in the scope of which different goals are the alternatives to choose. What we used in this research, was OWA operator which will be explained in the next section. In this prospective, each goal characterized by a number of attributes which its value must be evaluated according to environment and emotional information, which enter the goal management component. So a table of decisions will be made of output of the mentioned algorithm. It is also important to track the last chosen goal in such a way to change the priority periodically and not to choose same goal again in the next request of selection.

The production system is responsible for maintaining and updating rational state of the system and determining on the base of states what actions are eligible to be undertaken in order to pursue the current active goal's objectives. The production system also will be able to supply the goal management component with the necessary support to check success or failure of the current goal. To obtain such results, the application designer should define four set of rules:

- A set of rules for the evolution of the rational state.
- A set o rules to verify the state of the current goal, if the goal management component utilizes them. Generally there will be a success and failure test for each goal.
- A set of rules to determine what actions are applicable in the current situation. The production system sends the applicable actions to the action selector component that chooses one of them. Generally there will be a subset of such rules for each goal.
- A set of rules to verify if the action chosen by the action selector is immediately executable. In fact, if the selected action is too complex, it can be decomposed in sub goals with a top down method. After that, the rational component repeats the action selection process until action selector chooses an executable action and sends it to the rational component in order to generate the instructions to execute it.

The functionality of the action selection component, under certain appearance, is like goal management component described in previous lately. In fact, the task of the action selection component is to choose which action to undertake between those received from the production system according to the current environmental conditions. Our approach uses an emotional space to find nearest action to the current emotional state of an agent. In other words, the role of emotions in this architecture would be choosing the most desirable action among those which have chosen rationally. One sample of emotional space is depicted in Figure 2 which is used in kismet [4]. Nevertheless, the decision of the goal management component is a higher level with respect to the action selection component. In fact, we have to consider that decisions made by the goal manager are at higher level than the decisions made by action selector. The goal manager selects between several long term goals, while the action selector chooses an action that should be applied in particular and limited situations. Another point is the fundamental interactions between rationality and emotions which will be discussed: The undertaken actions, in some manner, have to reflect the emotional state and the personality that the agent understands and demonstrates. Then possibility of connoting some actions with their emotional implications would be considered as subordinate to the emotional state that their execution communicates with those who interacts with the agent. Compared to the production system, action selection component does not receive the environmental information on the bases of the input information twisted with applicable actions which are rationally equivalent among other choices.

One of the important components of the architecture is decision making module which comprises multi-attribute decision making algorithm. In order to completely understand the mechanism, in the next section Data Fusion and OWA operator will be briefly discussed.

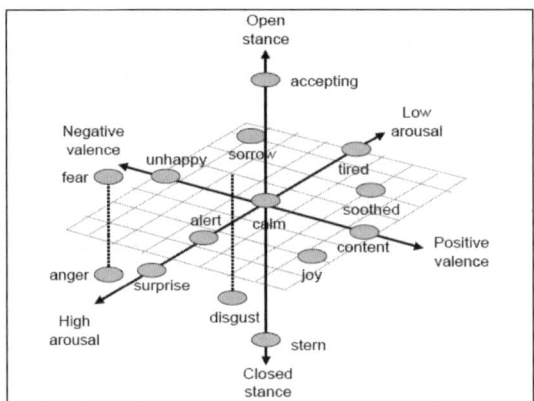

Fig. 2. Sample of Emotional Space

3 Overview on Data Fusion and Ordered Weighted Averaging (OWA)

Data Fusion is the process of combining data and knowledge from different sources with the aim of maximizing the useful information content [1]. It improves reliability or discriminate capability while offering the opportunity to minimize the data retained. Data Fusion algorithms has been categorized into several categories [11]. Among all these approaches, we chose OWA which discussed in the following paragraph.

Ordered weighted aggregation (OWA) operators were introduced by Yager [15].

An OWA operator of dimension n is a mapping $f : [0,1]^n \rightarrow [0,1]$, which has an associated weighting vector

$W = (w_1, \ldots, w_n)^t$, s.t.

$$\sum_i w_i = 1, \ w_i \in [0,1] \tag{1}$$

and where

$$f(x_1,\ldots,x_n) = \sum_i w_i x_{k_i} \tag{2}$$

The vector $K = (k_1, \ldots, k_n)^t$ is such permutation of $(1, 2, \ldots, n)^t$ that x_{k_i} is the ith largest element in $(x_1, \ldots, x_n)^t$. The fundamental aspect of the OWA operator is that a particular weight w_i is associated with a particular ordered position i of the arguments. OWA operators include min, max, and arithmetic mean for the appropriate choice of vector W.

Yager introduced a measure to characterize the type of aggregation performed by OWA operators. He calls it the **orness measure**. It is defined as

$$\text{Orness(w)} = \frac{1}{n-1} \sum_{i=1}^{n} (n-i)w_i \tag{3}$$

It can be shown that orness of max operator is 1, orness of min operator is 0, and orness of the arithmetic mean is 0.5. Orness of other OWA operators lies in the unit interval. The measure of orness is used frequently as an additional constraint when determining weights of the operator.

One of the main concerns in using OWA operator is how to obtain weights vector. There are several approaches introduced by Yager and other people [7], [8], [9], [10]. In the next section we will go through this issue in details.

4 Decision Making Algorithm

Our algorithm needs to specify a set of attributes for each goal which has to be assigned by a user. Set of attributes includes priority, importance and other items listed in Table 1 which considered as collection of aggregated objects in the unit interval, a_i, which has to be ordered and stored in decision table. Agent's personality is used as coefficient which determines its degree of pessimism. In case of using OWA operator in decision making algorithm, a weighting vector W should be defined and initialized. The main question would be obtaining the weights associated with OWA, because it models process of aggregation used on data set.

We used a back propagation method to learn from agent's observations [2], [7]. Suggested algorithm is described below:

1) Each aggregated value will be calculated by classic Hurwicz's multi-attribute method and is considered as desired value:

$$\rho \operatorname{Max}_i a_i + (1-\rho) \operatorname{Min}_i a_i = d \qquad\qquad \rho : \text{Agent's personality} \qquad (4)$$

2) Following learning algorithm should be applied to estimate the corresponding weights:

$$\lambda_i(l+1) = \lambda_i(l) - \beta w_i(b_{k_i} - \hat{d}_k)(\hat{d}_k - d_k) \qquad \beta : \text{Learning rate} \qquad (5)$$

$$w_i = \frac{e^{\lambda_i(l)}}{\sum_{j=1}^{n} e^{\lambda_j(l)}} \qquad (6)$$

$$\hat{d}_k = b_{k_1} w_1 + b_{k_2} w_2 + \dots + b_{k_n} w_n \qquad \hat{d}_k : \text{Current estimation of } d_k \qquad (7)$$

Parameters λ_i determine the OWA weights and are updated with back propagation of the error ($\hat{d}_k - d_k$).

Finally, after 10000 iterations, the best $\hat{d}_k$ with maximum value will be selected as current agent's goal and will be delegated to the production system. Also priority of each goal will be decreased and checked out with a threshold so that it would be decayed after a while.

The value of ρ could be interpreted as a measure of the agent's "pessimism". In fact, if $\rho \rightarrow 1$, the agent should tend to pay greater attention to the minimum value of the attributes, whereas if $\rho \rightarrow 0$, agent should consider mainly the maximum value of the attributes [6]. Because of having only two types of personality in this research (Introversive and Extroversive), we have to set a proper value for ρ according to the definition mentioned above. In general, introversive person is usually pessimistic, so we likely use values greater than 0.5 for it. Hence, for extroversive person who is more optimistic, we use values less than 0.5. Also the weights w_i reflect in some way the agent's personality, too.

Among possible attributes that we can associate with a goal, we need to consider those, which contain the emotion and rational aspects based on personality of an agent. In this case, first we prepared a psychological questionnaire and distribute it to about 60 people whose personality has been determined by psychological tests and proved to be introversive or extroversive [12].

Table 1 is a sample questionnaire that illuminates all the needed attributes for each goal which most of them are inputs of MIT's Kismet [4]. This indicates that goals in this research are expressing nine different emotions with facial expressions. Because our main objective is to show how data fusion can affect decision making process of rational-emotional agent, attributes listed in Table 1 are used to explain the algorithm. Also new set of attributes can be replaced anytime in order to have more believable agents and accurate decision making.

Table 1. A sample questionnaire

Emotions	Imp.	Environment		Skin	Motion		Tone		Color			Sentence	
		Calm	Noisy		Fast	Slow	Low	Loud	R	G	B	Kind	Threaten
Joy	x_{11}	x_{12}	x_{13}	x_{14}	x_{15}	x_{16}	x_{17}	x_{18}	x_{19}	x_{110}	x_{111}	x_{112}	x_{113}
Surprise	x_{21}	x_{22}	x_{23}	x_{24}	x_{25}	x_{26}	x_{27}	x_{28}	x_{29}	x_{210}	x_{211}	x_{212}	x_{213}
Anger	x_{31}	x_{32}	x_{33}	x_{34}	x_{35}	x_{36}	x_{37}	x_{38}	x_{39}	x_{310}	x_{311}	x_{312}	x_{313}
Fear	x_{41}	x_{42}	x_{43}	x_{44}	x_{45}	x_{46}	x_{47}	x_{48}	x_{49}	x_{410}	x_{411}	x_{412}	x_{413}
Interest	x_{51}	x_{52}	x_{53}	x_{54}	x_{55}	x_{56}	x_{57}	x_{58}	x_{59}	x_{510}	x_{511}	x_{512}	x_{513}
Disgust	x_{61}	x_{62}	x_{63}	x_{64}	x_{65}	x_{66}	x_{67}	x_{68}	x_{69}	x_{610}	x_{611}	x_{612}	x_{613}
Sorrow	x_{71}	x_{72}	x_{73}	x_{74}	x_{75}	x_{76}	x_{77}	x_{78}	x_{79}	x_{710}	x_{711}	x_{712}	x_{713}
Boredom	x_{81}	x_{82}	x_{83}	x_{84}	x_{85}	x_{86}	x_{87}	x_{88}	x_{89}	x_{810}	x_{811}	x_{812}	x_{813}
Calm	x_{91}	x_{92}	x_{93}	x_{94}	x_{95}	x_{96}	x_{97}	x_{98}	x_{99}	x_{910}	x_{911}	x_{912}	x_{913}

Priority of a goal is calculated on the basis of how much it can be determinant the realization of goal referring to the current external conditions, it is in connection with the actual state of the environment in which the agent works. Importance of a goal is calculated on the basis of how much is determinant the realization of the goal in relationship to the inner state of the agent and so in connection with its personality to its emotional state [6].

Other attributes can be sensed via visual and audio sensory devices and calculated with high level feature extraction techniques to improve the decisional process.

Initial x_{ij} values, are taken from the people mentioned above or set by an expert (Although this will not yield a valid data) and for including in OWA fusion algorithm, they must be normalized between 0 and 1.

During the process, as mentioned before, x_{ij} are updated by the information processing module based on arrived emotional and environment information.

5 Experimental Results

In order to understand the whole mechanism, an example of decision making for an extroversive agent is discussed. First, input component gathers required data from the environment with its visual and audio sensory systems. After feature extraction, raw data yields to the information which indicates how agent cognizes its environment.

For instance, suppose personality type is *extroversive*, environment is *without noise*, initial emotion state is *Calm* and we show the agent a *Red* object. Other values are calculated by means of high level feature extraction. Final perceptions are shown in Table 2.

Table 2. Agent's perceptions

Importance	Environment	Skin Detection	Motion	Tone	Color	Sentence
1	0.3	0.7	0.35	0.2	0.8	0.5

Next step is calculating desired values for all of the states. As we mentioned before, values of each attribute is available from the questionnaire and stored in database. Table 3 contains all the information provided in the questionnaire and Table 5 indicates initial values for each weight.

Table 3. A sample of stored values in database to calculate desired values

Emotions	Imp.	Environment		Skin	Motion		Tone		Color			Sentence	
		Calm	Noisy		Fast	Slow	Low	Loud	R	G	B	Kind	Threaten
Calm	0.4	0.1	0.9	0.5	1	0.2	0.1	1	0.2	0.2	0.3	0.9	0.2

According to Table 3 and equation (4), if ρ =0.15 and β =0.35, desire values are calculated and shown in Table 4. By using equations (5) and (6), λ_i and w_i are calculated. Table 6 and 7 indicates their values.

Table 4. Desired values of nine emotions

d_1	d_2	d_3	d_4	d_5	d_6	d_7	d_8	d_9
(Joy)	(Sorrow)	(Anger)	(Calm)	(Disgust)	(Fear)	(Surprise)	(Interest)	(Boredom)
0.32	0.205	0.235	0.22	0.21	0.19	0.232	0.31	0.34

Table 5. Initial values of Weights

w_1	w_2	w_3	w_4	w_5	w_6	w_7
0.14	0.14	0.14	0.14	0.14	0.14	0.16

Table 6. Calculated λ_i values

λ_1	λ_2	λ_3	λ_4	λ_5	λ_6	λ_7
1.244	-0.654	-0.737	-1.35	-1.144	-0.428	3.141

Table 7. Calculated w_i values

w_1	w_2	w_3	w_4	w_5	w_6	w_7
0.12	0.018	0.016	0.008	0.011	0.022	0.802

As it can be understood, $\sum_i w_i = 1$, $w_i \in [0,1]$.

Finally, according to equation (7), value of each $\hat{d}_k$ after 10000 iterations calculated and shown in Table 8.

Table 8. Estimated desired values of nine emotions

$\hat{d}_1$	$\hat{d}_2$	$\hat{d}_3$	$\hat{d}_4$	$\hat{d}_5$	$\hat{d}_6$	$\hat{d}_7$	$\hat{d}_8$	$\hat{d}_9$
(Joy)	*(Sorrow)*	*(Anger)*	*(Calm)*	*(Disgust)*	*(Fear)*	*(Surprise)*	*(Interest)*	*(Boredom)*
0.321	*0.208*	*0.233*	*0.210*	*0.219*	*0.187*	*0.234*	*0.313*	*0.334*

Figure 3 depicts learning curve of w_i.

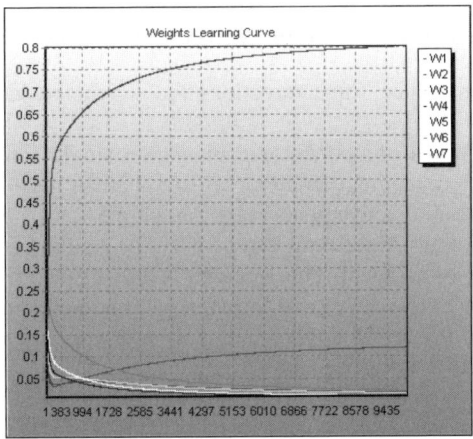

Fig. 3. Weights Learning Curve

According to equation (2), the output of decision making algorithm is:

$$f(x_1,...,x_n) = \sum_i w_i x_{k_i} = 0.322$$

The selected goals regarding to threshold of 0.05 are *Joy*, *Interest* and *Boredom* with values of 0.32, 0.31 and 0.34 respectively.

Now this is a time to regulate the decision with current emotional state of an agent. Since it is *Calm*, regarding emotional space depicted in Figure 2 and algorithm mentioned lately, final selected goal is *Joy*. The selected goal will be delegated to the Output System in order to be executed with facial expressions.

6 Conclusion and Future Works

A main goal of the mentioned architecture is to experiment the interaction between rationality, personality and emotions in the framework of rational-emotional agents in the application domain which communication with humans is based on facial expressions. In particular, the rational component supports a rational state evolving on the basis of both rational and emotional knowledge. Such evolutions take place in two ways: implicitly, the decision parameters are updated depending on the current emotional state; explicitly, the emotional knowledge is encoded as facts in the production system working memory. The model supports a rational choice of a set of possible actions and an emotional and personality based selection of the current goal and action. The rational component has been implemented and under testing with different decision algorithms based on Data Fusion techniques especially OWA operators. In other words, what actually more in attention, is to show OWA and its extensions are making an agent capable to decide more like human and therefore creating more believable agents. These additions to similar previous implementations [4], [5], [6], we believe, in no way diminish the embeddedness of the proposed architecture. Even virtual agents can be considered profoundly embodied and situated. On the other hand, many real embodied and situated robots lack profound embeddedness. Their bodies and situational circumstances are considered not affordances to be taken advantage of, but further problems to be solved by the proposed AI methodology [13].

At the end, we introduce some directions that might be interesting as a future works:

This prototype of rational component can manage a single active goal at the same time. In applications containing complex real-time human-machine interaction, there might be emerge of several active goals at the same time.

Other Decision Making Algorithms can be used for goal or action selections which improve believability of an agent. This research is only headed for two major types of personality which are Extroversive and Introversive, but there are more types that can be modeled and used inside the application. Also utilizing any learning algorithm in order to learn personality would be another interesting future research in this area.

References

1. Abidi, M.A., Gonzalez, R.C.: Data Fusion in robotics and machine intelligence. Academic Press, San Diego (1992)
2. Beliakov, G.: How to Build Aggregation Operators from Data. International Journal of Intelligent Systems 18, 903–923 (2003)
3. Boutilier, C., Dean, T., Hanks, S.: Decision-theoretic planning: Structural assumptions and computational leverage. Journal of Artificial Intelligence Research 11, 1–94 (1999)
4. Breazeal, C.L.: Emotion and Sociable Humanoid Robots. International Journal of Human-Computer Studies 59, 119–155 (2003)
5. Camurri, A., Caglio, A.: An Architecture for Emotional Agents. IEEE Multimedia Journal (1998)
6. Camurri, A., Volpe, G.: A Goal Directed Rational Component for Emotional Agent. Systems, Man, and Cybernetics. In: IEEE SMC 1999 Conference Proceedings, IEEE Computer Society Press, Los Alamitos (1999)
7. Filev, D.P., Yager, R.R.: Learning OWA Operator Weights From Data. In: Proceedings of the Third IEEE International Conference on Fuzzy Systems, Orlando, pp. 468–473. IEEE Computer Society Press, Los Alamitos (1994)

8. Filev, D., Yager, R.: On the issue of obtaining OWA operator weights. Fuzzy Set Systems 94, 157–169 (1998)
9. Fuller, R., Majlender, P.: An analytic approach for obtaining maximal entropy OWA operator weights. Fuzzy Set Systems 124, 53–57 (2001)
10. Fuller, R., Majlender, P.: On obtaining minimal variability OWA operator weights. Fuzzy Sets and Systems 136(2), 203–215 (2003)
11. Hall, D.L., LLinas, J.: Handbook of Multisensor Data Fusion. CRC Press, Boca Raton (2001)
12. Millon, T., Lerner, M.J.: Handbook of Psychology. Personality and social psychology 5 (2003)
13. Sartre, J.P.: Being and Nothingness. Trans. Hazel E. Barnes. Quokka Books, New York (1978)
14. Scassellati, B.M.: Foundations for a Theory of Mind for a Humanoid Robot. MIT University (2001)
15. Yager, R.R.: On Ordered Weighted Averaging Operators in Multicriteria Decision Making. IEEE Transaction on Systems, Man and Cybernetics (1988)

How to Build Consciousness into a Robot: The Sensorimotor Approach

J. Kevin O'Regan

Laboratoire Psychologie de la Perception, CNRS Université Paris Descartes, Centre
Universitaire des Saints Pères, 45 rue des Saints Pères, Paris 75270 Cedex 06, France
kevin.oregan@psycho.univ-paris5.fr

Abstract. The problem of consciousness has been divided by philosophers into
the problem of Access Consciousness and the problem of Phenomenal
Consciousness or "raw feel". In this chapter it is suggested that Access
Consciousness is something that we can logically envisage building into a robot
because it is a cognitive capacity giving rise to behaviors or behavioral
tendencies or potentials. A few examples are given of how this is being done in
current research. On the other hand, Phenomenal Consciousness or "raw feel" is
problematic, since we do not know what we really mean by "feel". It is
suggested that three main properties are what characterize feel: the fact that
feels are different from each other, that there is structure in these differences,
and that feels have sensory presence. It is then shown how, by taking the
sensorimotor approach [24], [27] it is possible to account for these properties in a
natural way and furthermore to make counter-intuitive empirical predictions
which have recently been confirmed. In conclusion it is claimed that when we
take the sensorimotor approach to feel, building raw feel into a robot becomes a
theoretical possibility, even if we are a long way from actually attaining it.

1 Introduction

Consider a robot programmed so that it *acts* in every way as though it is conscious.
For example when injured, it screams and shows avoidance behavior, imitating in all
respects what a human would do when in pain. The robot is able to talk about its pain,
and it reasons and acts like it has the pain. The philosopher Ned Block would say that
the robot has *Access Consciousness* to the pain [6].

However all this would not guarantee that to the robot, there was actually
something it was like to have the pain. The robot might simply be going through the
motions of manifesting its pain: perhaps it actually feels nothing at all. Something
extra might be required for the robot to *actually experience* the pain, and that extra
thing is *raw feel*, or what Ned Block calls *Phenomenal Consciousness*.

2 Access Consciousness

From a theoretical standpoint (although currently no one has actually done it), there
would appear to be no logical obstacle to implementing Access Consciousness in a

M. Lungarella et al. (Eds.): 50 Years of AI, Festschrift, LNAI 4850, pp. 332–346, 2007.
© Springer-Verlag Berlin Heidelberg 2007

robot: the reason is that Access Consciousness ultimately corresponds to a behavioral capacity. What we mean when we say someone has Access Consciousness to something is that the person currently knows that he (considered as a person with a self) is poised to make use of that thing in his ongoing rational decisions, in his planning, intentions and linguistic behavior [6]. Agreed, the notions of "self", "rational", "decision", "planning", "intention", and "language" required to have access consciousness are all difficult notions. We are far from understanding these notions, and once we do, building them into a robot may require as yet undiscovered principles. But the important point is that there is no logical impossibility preventing this from being done: it has been termed the "easy" problem of consciousness [8]. Indeed, as the following illustrations show, cognitive scientists and artificial intelligence researchers are busy analyzing the components and prerequisites necessary to achieve this goal.

2.1 The Self, Intentions, and Theory of Mind

One critical aspect of Access Consciousness that has to be understood is the notion of self. Studies in cognitive science reveal that the notion is not a unitary notion, but is an umbrella term, covering capacities going from the individual to the social, and going from knowledge about only the organism itself, to knowledge about other organisms and their motivations [13], [22], [32], [33], [47]. Different aspects of the self become established at different times as humans grow up, with social pressures and individual experience contributing to their development in complicated ways. The notion of self is related to "intentions" and to "Theory of Mind", that is, the ability to understand other agents' thoughts and goals. The following are just a few illustrations where current robotics research is attempting to implement some very simple aspects of the self.

Self-discrimination has been investigated with Domo, a robot constructed at the Humanoid Robotics Group at the MIT Computer Science and Artificial Intelligence Laboratory (CSAIL). The robot consists of an upper torso equipped with moveable eyes, head, arms and grippers. It uses vision-based movement detection algorithms to determine whether something is moving in its visual field. It checks whether by commanding movements of its own body, the movements it sees are correlated with the movements it commands. If such a correlation occurs, it assumes that what it is seeing is part of its own body. In this way it is able to figure out what its own hand looks like, and later, what its own fingers look like [11].

Work on the higher notions of self, namely *self-knowledge* and *knowledge of self-knowledge* is being done using the COG platform, also developed at CSAIL. COG is actually one of the first robotic platforms that was built at the Humanoid Robotics group, and one might say it is approaching "retirement". It is an upper-torso humanoid robot equiped with visual, auditory, tactile, vestibular, and kinesthetic sensors, and which can move its waist, spine, eye, head, arms and primitive hands. COG has been used by a variety of groups at MIT to do experiments in object recognition, tactile manipulation, and human-robot interaction. Since the development of COG, many groups throughout the world have been constructing similar devices to study embodied cognition.

The ideas being used to study the emergence of the higher notions of self in COG and similar robots are based on analyses of what psychologists consider to be the most basic capacities postulated to underlie this notion in humans [39]. One such basic capacity is the ability to locate and follow an agent's gaze direction. For this, skin, face and eye detection algorithms in the robot's visual system allow it to locate eyes and infer where a human is looking. By extending this gaze following capacity, the researchers hope to implement algorithms for joint attention, that is, algorithms that allow the robot to attend to an object that another agent is also attending to.

A further example of a capacity that might be involved in the genesis of the self is the ability to distinguish mechanical motion due to inanimate objects and animate motion due to living agents like animals and humans. This is perhaps the basis of the notion of the ability to ascribe intentions and goals to other agents. To test this idea with COG, an algorithm has been used that estimates the variability in the velocity of moving objects. Presumably an object whose velocity does not follow simple laws of physics probably has a "will of its own" and is likely to be animate.

Another robotic platform that is being used to investigate the emergence of the self is Sony's domestic dog robot, the AIBO. At Sony Computer Science Laboratory in Paris the AIBO has for example been used to study joint attention and pointing, except this time in a "social" context, that is, with another AIBO or with a human [18].

The Domo, COG and AIBO projects are just three samples of work in progress. They are only painfully preliminary steps towards implementation of different self notions in a robot. But such studies in developmental robotics are a growing research field in which researchers attempt to show how from a few basic capacities, robots can acquire the social skills that we know humans acquire over the first few years of life, skills that are at the root of humans' notions of intention and "Theory of Mind" [1], [12], [19], [21]. Much work needs to be done, but the vitality of this and related research projects shows that researchers are confident that providing robots with a realistic notion of self and accompanying Theory of Mind is an achievable goal. Even if it takes many more years, the problem of building a robot with a self seems *in principle* solvable.

2.2 Language

Whether beings without language are conscious will probably have to remain a matter of debate. But it is obvious to us humans that insofar as we possess the faculty of language, it is an important component for Access Consciousness: after all among the things we *mean* by having conscious access to something is being able to talk about it.

However the goal of providing artificial agents with human-like natural language understanding is still far from being attained [10], [17]. The main problem seems to be in anchoring the symbols used by machines in the real world. One attempt to do this is progressing by accumulating vast amounts of "common sense" knowledge from large natural language databases like the web [20] (see http://www.cyc.com). More recently researchers are trying to physically embed artificial agents in the real world in order to facilitate proper human-like use of language [34], [35]. As was the case for the notion of the self, work at MIT and Sony CSL is also illustrative of this.

An example of how immersion in the real world can help solve problems is Ripley, a kind of robot dog from the MIT Media Lab. Ripley can move its neck and pick things up with its mouth. Because Ripley is embedded in the real world, it does not need to do any very complicated reasoning concerning how it is physically placed with respect to the objects it is dealing with, and how they are placed with respect to the person it is talking to: this kind of information is available at any moment in front of its eyes, so when someone says "pick up the one on your left", it can just look over on the left and find what is being referred to. Furthermore, when it learns words like push, pull, move, shove, light, heavy, red, hard, soft, it can make use of information it obtains from interacting with objects in order to ground the meaning of the words in physical reality, imitating what probably happens when real infants interact with their caretakers [36], [37].

A similar project is being undertaken at Sony CSL, where Sony's robot dog AIBO learns the meanings of simple words by interacting with a human [44], [45] (see also http://www.csl.sony.fr and http://playground.csl.sony.fr/). Other work at Sony CSL is investigating how word meaning and syntax can emerge when humans or robotic agents play language-oriented games together in order to achieve common purposes [42], [43]. Of course in these examples the interactions between robots and humans is much more focussed and the number of utterances involved is much more limited than in normal human interactions. But this work suggests that we may be starting to model human language acquisition in a plausible way.

2.3 Conclusion on Access Consciousness

Though the illustrations in the preceding paragraphs are obviously ridiculously simple, and are clearly only the very first steps towards implementation of Access Consciousness, they nonetheless suggest that Access Consciousness, though a difficult problem, can be decomposed into a collection of simpler problems which are logically not beyond the bounds of robotic implementation. Access Consciousness is ultimately an aggregate of behavioral capacities, and the necessary ingredients, while out of reach today, can conceivably be achieved in the future. Perhaps the most tricky problems are on the one hand the notion of "self", with its accompanying concepts of intention and Theory of Mind, and on the other hand natural language understanding. The hope today is that these problems will be successfully dealt with when researchers start working more with actual physically embodied agents in real world settings. Indeed, one cannot neglect the fact that humans live, move, and interact in particular ways with the objects that they use and with other humans in the real world. Human language and thought are not just raw symbol manipulation: the concepts that are manipulated are constrained by the physical world and the particular way humans interact with it. People have bodies and interact with other people who themselves also have bodies. People live in a shared social environment and have desires, emotions and motivations that play an important role in conditioning communication. It may be that only machines that have human-like immersion in the world will be able to have notions of self and use language like humans [10], [29].

3 Phenomenal Consciousness or "Feel"

We have seen that Access Consciousness, though clearly difficult, is not a logically insoluble problem. We can hope that we will gradually progress towards its implementation in robots. With *Phenomenal Consciousness* however, we are in quite a different situation. People are convinced that they *feel* things, but it is hard to say exactly what feel is.

To try to understand better what is meant by feel, consider what happens when I look at a red patch of color: I see red. What exactly is this feel of red? What do I *experience* when I feel the feel of red?

One aspect of the feel of red is the **mental associations** that I have with red: Redness is associated in my mind with, among other things: the word "red", with roses, ketchup, blood, red traffic lights, stopping, anger, and certain red cough-drops... But these mental associations are *in addition to* the raw sensation: they are added over and above the experience of red itself. Having these mental associations might of course produce *additional* experiences: for example the association with anger may make me more likely to get angry. But such effects, if we want to call them experiences, would appear to be *in addition* to the basic, core, raw experience of red itself.

Another aspect of the feel of red may be the automatic **physiological states or tendencies** it creates. For example, red may be a color that has the direct effect on my nervous system of making me more excited, whereas blue may calm me down. The existence of such effects is controversial, but if they do exist, they are surely *over and above* the actual raw feel of the redness of red: in this example they correspond to excitement, not to red.

Yet another aspect of the feel of red is the **learnt bodily reactions** that redness may engender, caused for example by habits that are associated with red: for example pressing on the brake at red traffic lights. But again, such bodily reactions are add-ons to the actual raw experience of red.

To summarize: experiencing red may be accompanied by various mental associations, physiological tendencies and bodily reactions. These are extra behaviors that *come with* the feel of red, and they may in turn produce their own, additional experiences. But at the root of the feel of red there surely must be more than simply mental associations, physiological tendencies and bodily reactions. This extra component, which we could call the "raw feel" of red itself, is presumably what makes red quite different from green, or from the sound of a bell, or any other sensation.

Now it seems clear that if we wanted to build a robot that experienced sensations as humans do, then at least conceptually, building in the *additional components* accompanying the sensations poses no particular problem. This is because these components are behaviors or behavioral tendencies or capacities. One could fairly easily build into the robot a higher probability of saying "red" when it sees red; one could have the robot be more active or agressive in red rooms, and even have it make subliminal brake-pressing movements.

But what could be done to provide the robot with the "core" component of the red sensation, namely the raw red feel? This seems to be a much harder problem. What "circuit" should be added into the robot to provide it with the *raw feel* of red?

The robot played by Arnold Schwarzenegger in the film "Terminator I" is a good example of this problem. The designers of the robot could incorporate circuits that make the robot wince, say "ouch", and otherwise manifest its disapproval in cases where for example it gets its arm chopped off. The robot could know that this kind of injury is a *bad thing* for it, and it could be programmed to avoid getting into such nasty situations in the future. But what extra circuit would the designers have to build into the robot so that it actually felt the *raw pain itself*, instead of just going through the motions of feeling the pain?

3.1 Three Properties of Raw Feel

Let us look more closely at the raw, core component that is at the basis of feel. In this exercise we are purposefully leaving aside all the "extra" components like the mental associations and bodily manifestations that might come with feel.

There are three important aspects to note about raw feel.

First, *raw feels are different from each other*. For example there is red, green, pink, black. There is the sound of a tractor, of a violin, of middle C, of the wind in the willows. There is the smell of lemon, the taste of onion, the touch of a feather, the cold of ice, among innumerable others.

Second, *there is structure in the differences*. Sensations can be grouped together according to their similarity. For example sensations of light form a collection which is separate from sensations of sound, which are in turn different from sensations of touch, etc. Within each such collection or "modality" there may be further structure. Tones for example can be compared and contrasted, and they form a linear order going from low pitched to high pitched. Color is more complex, since one can distinguish the hue or tint of a color, and its "saturation", that is, the intenseness of the color it contains (a color is less saturated when it contains a lot of grey). Furthermore the dimension of hue is circular rather than linear: you can arrange colors in a closed circle of similarity going from red to orange to yellow to green to blue to purple and back to red again.

Sounds are another complicated case. Clearly loudness is something that can be defined along a linear dimension, but then sounds also have "timbre", which seems not to be describable in terms of dimensions that can easily be agreed upon. Smells also are complicated, and no consensus has been reached on a set of dimensions to describe them. A recent study suggests that a minimum of 30 independent dimensions are needed to account for smell judgments.

But whereas sensations form sensory orders [9], [48] or "modalities" within which they can be compared and contrasted in this way, across such modalities they cannot. For example, how is red different from middle C? Or how is cold different from onion flavor? It seems to make no sense to try and compare Red and middle C, since they have nothing to do with each other, and can't really be compared at all. Perhaps there is one attribute which is common even across different modalities, namely intensity: we can talk about intense colors, intense sounds. But apart from that sounds and colors are incommensurate (except perhaps to synesthetes!). The same goes for cold and onion flavor.

In summary, sensations form more or less separate modalities across which making comparisons is impossible. But within each such modality there may be a structure, which may be of varying complexity, depending on the modality.

Third, *raw feels have a quality rather than no quality, and are perceptually present.*

To understand this statement, consider the fact that the brain continually monitors blood oxygen and carbon dioxide, keeps the heartbeat steady and controls a variety of other bodily functions. All these activities involve sensors signalling their measurements via neural circuits and are processed by the brain. Yet one does not feel them, whereas one does feel the redness of the light or the prick of the needle.

Why should brain processes involved in processing input from certain sensors (namely the eyes, the ears, etc.), give rise to a felt sensation, whereas other brain processes, deriving from other senses (namely those measuring blood oxygen levels etc.) do not give rise to a felt sensation?

A related, but not identical case is thinking. Thinking obviously involves brain processing like controlling the oxygen level in the blood, and like analyzing inputs from the sensory systems. But does thinking have a feel?

Clearly, like the situation for sensory inputs, one is *aware* of one's thoughts. One knows what one is thinking about, and one can, to a large degree, control one's thoughts. But being *aware* of something is not the same as feeling something. Indeed, thoughts are more like blood oxygen levels than like sensory inputs: thoughts are not associated with any kind of sensory presence. Thoughts may be *about* things like blood and red traffic lights and red cough drops, or even about the raw feel of red, but such thoughts do not *themselves* have a red quality or indeed any sensory quality at all. Thoughts may of course be accompanied by feels: the thought of an injection makes me almost feel the pain and almost makes me pass out. But the pain I feel is the sensory pain normally associated with the injection, not the sensory quality of thinking. The thought *itself* has no sensory quality.

3.2 Neurophysiological Explanations for Feel

We have concluded that there are three important aspects of raw feel: feels are different from each other, there is structure in the differences, and feels have a quality and sensory presence, rather than no quality.

One's first impulse in seeking for an explanation for these facts is to look in the brain.

Neuroscientists have certainly localised different brain areas which seem to be involved in consciousness, but to date no explanation of how any such areas contribute is in view. Many hypotheses are entertained and discussed in the literature, such as the possibility that consciousness is generated by recurrent activation in corticothalamic networks, or by widespread synchrony of oscillations in the gamma band, or even that consciousness could be linked to quantum gravity effects in neuron microtubules. Such mechanisms might account for the behavioral capacities involved in Access Consciousness, but how any such mechanisms could explain why feels have the properties that they do is never addressed.

It would seem that there is a logical problem: Whatever mechanism is invoked to generate consciousness, additional "linking hypotheses" will always have to be made: a linking hypothesis is a hypothesis that establishes a link that justifies, for example, why different neurons or neural mechanisms or firing patterns or quantum mechanisms should produce the particular different sensory qualities, with the particular structure of similarities and differences that is found, and with the experienced sensory presence. The problem is that there would appear logically to be no non-arbitrary way of making such links between the neural states or physical characteristics of the firing patterns of neurons, and the experienced sensations.

4   The Sensorimotor Approach

A possible alternative way to understand the problem of Phenomenal Consciousness of feel is the sensorimotor approach [24], [25], [27]. This starts from the postulate that looking for a circuit or mechanism that generates Phenomenal Consciousness is to make what the philosopher Gilbert Ryle called a "category mistake" [38]: Phenomenal Consciousness is simply not the kind of thing that can be generated at all. Just as it makes no sense to search for the meaning of a word in the shapes of the particular letters that compose it, it makes no sense to search for a circuit that generates Phenomenal Consciousness in the brain.

Instead, the sensorimotor approach suggests that what it really means for a person to have Phenomenal Consciousness or feel is that the person:

1. is currently engaged in exercising a certain sensorimotor skill, and
2. is attending to this engagement and the skill's qualities.

Under this approach, the *quality* of a feel is constituted by the particular laws that govern an individual's sensorimotor interaction when he is experiencing the feel.

To understand the idea, one can take as analogy the feel of driving a Porsche as compared to driving a Volkswagen. Where lies the essential difference between the feel of Porsche driving and the feel of Volkswagen driving? It comes from the mode of sensorimotor interaction you have with the cars. It comes from *the things you can do* and the *way the car reacts when you do them*. When you press on the accelerator, the Porsche *whooshes rapidly forward*, whereas nothing very much happens in a Volkswagen. When you so much as slightly touch the steering wheel the Porsche *swerves immediately* whereas the Volkswagen only lumbers slightly to the side. Thus:

1. The Porsche driving feel comes from being engaged in exercising a certain sensorimotor skill, namely the Porsche driving skill. What provides the Porsche driving feel with its distinctive quality is the different mode of interaction you have with the Porsche as compared to other cars.

2. Furthermore to actually feel the Porsche driving feel, you have to be paying attention to the fact that you are doing Porsche driving things. If while you drive you get very involved in a discussion with a friend, you might no longer be noticing that you were driving the Porsche, and you would rather be experiencing the fact that you are conversing with your friend even though your body was actually doing the same Porsche driving things as before.

Taking the Porsche driving analogy seriously and applying it to sensory feels in general provides a way of accounting for feel in which feel is not something which can be located in some circuit, or which is generated by some mechanism. Instead, feel is a way of doing things. Taking this stance allows one to escape from many of the conundrums connected with phenomenal consciousness, and provides a principled way of explaining the three main properties of feel.

Thus, *why are feels different from one another*, and how exactly are they different? If we were to take the neurophysiologist's view that feels are different because they correspond to different input channels or different areas or mechanisms in the brain, we would be left with the question: What makes this channel or brain area or mechanism generate an experience of seeing, and that channel or brain area or mechanism generate an experience of hearing?

But the sensorimotor approach suggests that we should look for the differences between sensations in the *things that we can potentially do* when we have sensations: Thus, take the example of seeing and hearing. Seeing is a form of interaction in which blinks, movements of the eyes, of the body and of outside objects provoke very particular types of change in sensory input. The laws governing these changes are quite different from the laws governing sensory input in the auditory modality. For example, when one sees, moving forward potentially produces an expanding flow-field on the retina, whereas when one hears, the change in sensory input is now mainly an increase in amplitude of the signal. The claim is now that the sum total of these differences constitute precisely what differentiates the sensations of seeing and hearing. The same would be true for differences between experiences across other sensory modalities. This explanation for differences in sensory modalities escapes from the arbitrariness inherent in neurophysiological explanations appealing to brain channels, areas or mechanisms. No "linking hypothesis" need be made, because the quality of feel is considered to be constituted by what one does when one engages in a particular sensorimotor interaction.

The second main question one can ask about feel is: *What determines the structure of the differences between feels* within a sensory modality? The sensorimotor approach suggests that these differences correspond to differences in the laws that govern one's interaction with the world when one is experiencing different sensations. Contrary to a neural correlate explanation where we have no natural metric linking neural firing rates or other brain phenomena with differences in sensation, in the sensorimotor approach, there is a natural metric allowing sensations to be compared, namely the same metric used by subjects to compare sensorimotor skills in everyday parlance. Quite naturally, because the laws governing sensorimotor interactions are complex and vary from modality to modality, the structure of the differences between feels will be complex. Across two modalities, the sensorimotor interactions are so different that little comparison is possible. Within a modality, each change in one's mode of interaction determines the change in the quality of the experience involved. Later sections will discuss two examples, namely the sensation of touch and the sensation of color.

And finally, what can be said about the third question we asked about sensations, namely: *Why do they have sensory presence, that is, Why do they have a feel at all, rather than having no feel?* We will devote a few paragraphs to this question here.

If having a feel consists in attending to the fact that one is engaged in exercising a sensorimotor skill, and if the quality of the feel is constituted by the laws of sensorimotor interaction that the skill involves, then by the very definition of feel, the feel must have a quality, namely the quality constituted by exercising the particular sensorimotor law involved. Thus feels have a quality rather than no quality.

Then, just as the sensorimotor approach invokes differences in skills to account for differences between sensations, the approach will also invoke differences in skills to account for the difference between experiences involved in perceptual acts and the experiences associated with other brain activities. In particular, two facts about perceptual skills distinguish them from other brain activities.

First, whereas perceptual acts invariably involve, at least potentially, changes caused by motor behavior, this is not true either of internal physiological states or "mental" activities. What we call sensory experience can always potentially be modified by a voluntary motion of the body: Sensory input to the eyes, ears, or any other sensory system is immediately changed in a systematic and lawful way by body motions. On the other hand, people cannot reliably control their internal physiological states by moving their bodies (although of course states like heartbeat and blood oxygen will be affected indirectly by body motions). Likewise, "mental" activities like thoughts, memories, and decisions, to the extent that these can be considered as skills, are not skills that intrinsically involve voluntary body motions. This then is one thing that makes the skills constituting sensory experiences special as compared to other brain processes: they are by nature sensorimotor. Even if at any particular moment there need be no motion, they have what the sensorimotor approach calls "corporality" or "bodiliness" [24],[25],[26]. This strong potential effect of body motions on sensory input is what distinguishes sensor states deriving from the world from sensor states internal to the body or brain.

A second characteristic that distinguishes the skills involved in sensory experience from those of mental functions is what is termed "alerting capacity" or "grabbiness" [24],[25],[26]: this is the fact that sensory systems are genetically endowed with the capacity to deflect our cognitive processing. A loud noise or bright flash will automatically, incontrovertibly, attract our attention to the locus of the event. We are thus, in some sense, "cognitively at the mercy" of sensory input. This is not generally the case for either internal states or mental activities. If a change occurs in the visual field, like a mouse flitting across the floor, one's attention will immediately be caught by it. But variations in heart beat, for example, generally provoke no attentional orienting. Only exceptionnally, and then only indirectly through the pounding it produces on the chest, for example, is one aware that one's heart is beating very fast. Like visceral states, memory, and in general other mental activities, possess no alerting capacity or grabbiness: If one forgets a fact, one only discovers this if one actively tries to recover the fact from memory (an exception might however be, for example, obsessive thoughts).

Thus: a characterisation of the differences in skills associated with sensory acts, as compared to those involved in internal physiological states and mental acts, reveals differences which naturally account for the difference in felt quality of sensations as compared to other brain processes. What has been called the "presence" of sensations seems precisely to consist in the fact that they are both under our control (in that we can modify sensory input by our bodily actions — they have corporality or bodiliness), and also not under our control (they can cause uncontrollable alerting reactions that interfere with our normal cognitive processing: they have alerting capacity or grabbiness). The sensorimotor approach suggests that this captures the idea that there is "something it is like" to have a sensory experience provoked by an outside sensory event, as opposed to it feeling like nothing to have one's heart beat or to have a thought.

In summary for this section: the sensorimotor approach, unlike appeals to neurophysiological mechanisms that generate consciousness, provides a principled account of the three main properties of raw feel, namely

1. the fact that raw feels are different from each other
2. there is structure in the differences
3. raw feels have a quality rather than no quality, and are perceptually present.

The next sections will consider applications of the sensorimotor approach that provide interesting new predictions and results.

4.1 Sensory Substitution

The sensorimotor approach claims that what determines whether one has the feel of seeing, rather than say, the feel of hearing or the feel of smelling, is not the particular sensory input channel, but the laws that characterize the sensorimotor interactions that are involved when one sees, hears or smells. If this is true, then one ought to be able to create conditions where for example one "sees" through one's ears, or through stimulation of the skin: In order to do this, things would have to be arranged so that the laws that govern the sensory input through the ears or skin corresponded to visual-type laws, rather than the usual auditory-type laws.

Indeed the possibility of such "sensory substitution" has been known since Bach y Rita equipped blind subjects with an array of tactile stimulators positioned on their abdomen or back, connected to a video camera that produced a tactile "image" of the world on the observer's skin. Bach y Rita reports in his book that whereas passive stimulation was inconclusive, active camera manipulation by subjects very rapidly provided them with a sensation that they qualified as "seeing" [5], [14], [15].

Recently, technical advances have facilitated further development of sensory substitution devices, and an active community is investigating different types of substitution, ranging from transposing vision to audition, vision to tongue stimulation, vestibular to tongue, among others. For reviews see [3], [4].

4.2 The Localisation of Touch Sensation

Another prediction of the sensorimotor approach is as follows. If the quality of sensory feel is provided, not by the particular nervous pathways, but by the particular mode of sensorimotor interaction that is involved, then one should predict for example that the perceived localisation of a touch, say, on the arm, is not caused by the activation of a particular brain region, but by the particular sensorimotor laws that are involved when that location is touched. More precisely, a touch is felt to be on one's arm, rather than, say, on one's leg, when the touch can be modified by moving one's arm rather than one's leg; when the touch is accompanied by a visual stimulation in the region of the arm, rather than in the region of the leg.

Conversely, if one were presented with a tactile stimulation on the arm that is systematically accompanied by a visual stimulation on some outside object, like say a fake arm put on the table in front of one, then the prediction is that one should come to feel the sensation on the fake arm.

This is precisely the situation that has been investigated in an extensive literature on the "rubber hand illusion", which confirms this counter-intuitive prediction [7], [46].

The finding is also compatible with a large body of work showing that people can rapidly adapt when their body is "extended" by the use of tools or other artefacts. For example, you "feel" the paper under the tip of your pencil, not in your fingers. When you park your car, you "feel" the curb on the wheels of the car, not on the steering wheel.

4.3 The Structure of Color Sensations

Psychophysicists since Hering and Helmholtz have been trying to understand the structure of color space. Certain colors are considered to be "special" or "unique" in the sense that other colors are perceived to be composed of them. For example red, yellow, blue and green are seen as "pure", and not containing other colors, whereas orange is not pure because it is seen to contain red and yellow. Though some success is obtained by the classic opponent process theory of color vision, color scientists today agree that the finer details of these phenomena have not up till now been adequately accounted for by any neurophysiological findings.

The sensorimotor approach claims that in fact the structure of color space is to be sought not in sensory channels per se, but in the laws of interaction that characterize color perception. Thus when one moves a coloured piece of paper under different illuminants, or when one moves one's eyes on or off the paper, there are precise laws that govern the changes in photon catches made by the three photoreceptor types that humans possess. A recent attempt to apply this idea has come up with surprising success [30]. In the case of red, for example, it is found that the changes in photon catches are confined to a single dimension of variation, suggesting why red is a special colour as compared to, say, orange, where three dimensions of variation are observed. Unique hues are accurately predicted in this way. Furthermore, the approach also accurately correlates with well-known anthropological data concerning the way people name colors [30].

4.4 Change Blindness

An interesting point about the sensorimotor approach is the way it explains humans' experience of a very rich and continually present visual world. Instead of supposing, as does the classic approach to vision, that the perceived richness of visual experience requires continuous activation of a rich internal representation of the world, the theory says that richness and continuity are due to the fact that a perceiver has immediate access, via a flick of attention or an eye movement, to any information about the outside world that the perceiver wishes to investigate. The analogy is made of the light in the refrigerator: every time you open the fridge, the light is on, so you assume it is continually on. Similarly, the reason you feel the visual world as being continually present is that whenever you attend to any portion of it, information is available about that portion [23].

According to the sensorimotor approach, a further fact that buttresses the illusion of continual presence lies in the "grabbiness" of visual stimuli. The low level visual system is equipped with "transient-detectors" that register sudden motion or fast changes in luminance or color. These automatically provoke attentional orienting: when a flash of light occurs in the visual periphery, you cannot help moving your eye in that direction. One therefore has the illusion that one is continually seeing everything, because if anything should suddenly change, one's attention is automatically directed to it, and one sees the change [24], [26].

An exception to this however would occur if the transient detectors were somehow rendered inoperative. This can be done by swamping them with extraneous luminance transients. "Change blindness" is a phenomenon which is coherent with this prediction: large changes in pictures can go unnoticed if the change occurs simultaneously with a global white flash [31] or with several "mudsplashes" [28] distributed all over the visual field. Another way of rendering transient detectors inoperative is to make the changes so slow that they are no longer "grabby". This is what happens in experiments with progressive changes [2], [41], where a large region of a picture changes color or appears or disappears without this being noticed. Although the interpretation has been contested (cf. [40]), because the phenomenon of change blindness is so striking and counter-intuitive, it serves as a convincing confirmation of the sensorimotor approach.

5 Conclusion: Building Consciousness into a Robot

Can we build Access Consciousness into a robot?

The first part of this chapter argued that Access Consciousness is a behavioral capacity, which, though outside the bounds of current work in AI and robotics, presents no fundamental *logical* problem to a robotic implementation. Future work, particularly with embodied systems, bears the hope of gradually approaching the notions of self, intentions, and Theory of Mind, as well as natural language understanding that are undoubtedly prerequisites to Access Consciousness.

Can we build Phenomenal Consciousness or "feel" into a robot?

Although Phenomenal Consciousness or feel is generally considered the "hard" problem of consciousness, this chapter has argued that feel may in fact turn out to be the easier problem (for a similar view see [16]). If we take the stance suggested by the sensorimotor approach, according to which having a feel is, first, engaging in a sensorimotor skill, and second, having access consciousness to the skill, then clearly once a robot has access consciousness, it will suffice for the robot to engage in an embodied interaction with the environment for it to have feel.

References

1. Adolphs, R.: Could a robot have emotions? Theoretical perspectives from social cognitive neuroscience. In: Fellous, J.M., Arbib, M. (eds.) Who Needs Emotions: The Brain Meets the Robot, pp. 9–28. Oxford University Press, Oxford (2005)
2. Auvray, M., O'Regan, J.K.: l'Influence des facteurs sémantiques sur la cécité aux changements progressifs dans les scènes visuelles. Année Psychologique 103, 9–32 (2003)
3. Auvray, M., O'Regan, J.K.: voir avec les oreilles: enjeux de la substitution sensorielle. Pour la Science 39, 30–35 (2003)
4. Bach-y-Rita, P., Kercel, S.W.: Sensory Substitution and the Human-Machine Interface. Trends in Cognitive Sciences 7, 541–546 (2003)

5. Bach-y-Rita, P.: Brain mechanisms in sensory substitution. Academic Press, New York (1972)
6. Block, N.: On a Confusion about a Function of Consciousness. Behav. Brain Sci. 18, 227–247 (1995)
7. Botvinick, M., Cohen, J.: Rubber Hands 'Feel' Touch that Eyes See [Letter]. Nature 391, 756 (1998)
8. Chalmers, D.J.: Facing Up to the Problem of Consciousness. Journal of Consciousness Studies 2, 200–219 (1995)
9. Clark, A.: Sensory qualities. Clarendon Press, Oxford (1993)
10. Dreyfus, H.L.: What computers still can't do: A critique of artificial reason. MIT Press, Cambridge, Mass. (1992)
11. Edsinger, A., Kemp, C.C.: What can I Control? A Framework for Robot Self-Discovery (2006)
12. Fellous, J., Arbib, M.: Who needs emotions?: The brain meets the robot. Oxford University Press, Oxford (2005)
13. Gallagher, S., Shear, J.: Models of the self. Imprint Academic (1999)
14. Guarniero, G.: Tactile Vision: A Personal View. Journal of Visual Impairment and Blindness 71, 125–130 (1977)
15. Guarniero, G.: Experience of Tactile Vision. Perception 3, 101–104 (1974)
16. Harvey, I.: Evolving robot consciousness: The easy problems and the rest. In: Fetzer, J.H. (ed.) Evolving Consciousness. Advances in Consciousness Research Series, pp. 205–219. John Benjamins, Amsterdam (2002)
17. Hutchins, J.: Has Machine Translation Improved? In: MT Summit IX: proceedings of the Ninth Machine Translation Summit, New Orleans, USA, September 23-27, 2003, pp. 181–188 (2003)
18. Kaplan, F., Hafner, V.: The challenges of joint attention. In: Berthouze, L., Kozima, H., Prince, C.G., et al. (eds.) Proceedings of the Fourth International Workshop on Epigenetic Robotics, pp. 67–74 (2004)
19. Lee, M.H., Meng, Q.: Psychologically Insired Sensory-Motor Development in Early Robot Learning. International Journal of Advanced Robotic Systems 2, 325–334 (2005)
20. Lenat, D.B.: CYC: A Large-Scale Investment in Knowledge Infrastructure. Commun. ACM 38, 33–38 (1995)
21. Lungarella, M., Metta, G., Pfeifer, R., et al.: Developmental Robotics: A Survey. Connect. Sci. 15, 151–190 (2003)
22. Martin, R., Barresi, J.: The rise and fall of soul and self: An intellectual history of personal identity. Columbia University Press, Vancouver (2006)
23. O'Regan, J.K.: Solving the real Mysteries of Visual Perception: The World as an Outside Memory. Can. J. Psychol. 46, 461–488 (1992)
24. O'Regan, J.K., Myin, E., Noë, A.: Skill, Corporality and Alerting Capacity in an Account of Sensory Consciousness. Prog. Brain Res. 150, 55–68 (2006)
25. O'Regan, J.K., Myin, E., Noë, A.: Phenomenal Consciousness Explained (Better) in Terms of Bodiliness and Grabbiness. Phenomenology and the Cognitive Sciences 4, 369–387 (2005)
26. O'Regan, J.K., Myin, E., Noë, A.: Towards an analytic phenomenology: The concepts of "bodiliness" and "grabbiness". In: Carsetti, A. (ed.) Proceedings of the International Colloquium: Seeing and Thinking. Reflections on Kanizsa's Studies in Visual Cognition, University Tor Vergata, Rome, June 8-9, 2001, pp. 103–114. Kluwer Academic Publishers, Dordrecht (2004)

27. O'Regan, J.K., Noe, A.: A Sensorimotor Account of Vision and Visual Consciousness. Behav. Brain Sci. 24, 939–973 discussion 973-1031 (2001)
28. O'Regan, J.K., Rensink, R.A., Clark, J.J.: Change-Blindness as a Result of Mudsplashes. Nature 398, 34 (1999)
29. Pfeifer, R., Bongard, J.: How the body shapes the way we think: A new view of intelligence. MIT Press, Cambridge (2006)
30. Philipona, D.L., O'Regan, J.K.: Color Naming, Unique Hues, and Hue Cancellation Predicted from Singularities in Reflection Properties. Vis. Neurosci. 23, 331–339 (2006)
31. Rensink, R.A., O'Regan, J.K., Clark, J.J.: On the Failure to Detect Changes in Scenes Across Brief Interruptions. Visual Cognition 7, 127–145 (2000)
32. Rochat, P.: Five Levels of Self-Awareness as they Unfold Early in Life. Conscious. Cogn. 12, 717–731 (2003)
33. Rochat, P.: The self in infancy: Theory and research. Advances in Psychology. Elsevier, Amsterdam, New York (1995)
34. Roy, D.: Grounding Words in Perception and Action: Computational Insights. Trends Cogn. Sci. 9, 389–396 (2005)
35. Roy, D.: Semiotic Schemas: A Framework for Grounding Language in Action and Perception. Artif. Intell. 167, 170–205 (2005)
36. Roy, D.: Grounded Spoken Language Acquisition: Experiments in Word Learning. Multimedia, IEEE Transactions on 5, 197–209 (2003)
37. Roy, D., Pentland, A.: Learning Words from Sights and Sounds: A Computational Model. Cognitive Science 26, 113–146 (2002)
38. Ryle, G.: The concept of mind. Hutchinson, London (1949)
39. Scassellati, B.: Theory of Mind for a Humanoid Robot. Autonomous Robots 12, 13–24 (2002)
40. Simons, D.J., Rensink, R.A.: Change Blindness: Past, Present, and Future. Trends Cogn. Sci. 9, 16–20 (2005)
41. Simons, D.J., Franconeri, S.L., Reimer, R.L.: Change Blindness in the Absence of a Visual Disruption. Perception 29, 1143–1154 (2000)
42. Steels, L.: The Emergence and Evolution of Linguistic Structure: From Lexical to Grammatical Communication Systems. Connect. Sci. 17, 213–230 (2005)
43. Steels, L.: Grounding Symbols through Evolutionary Language Games. Simulating the evolution of language table of contents, 211–226 (2002)
44. Steels, L., Kaplan, F., McIntyre, A., et al.: Crucial Factors in the Origins of Word-Meaning. The Transition to Language, 252–271 (2002)
45. Steels, L., Kaplan, F.: AIBO's First Words: The Social Learning of Language and Meaning. Evol. Commun. 4, 3–32 (2000)
46. Tsakiris, M., Haggard, P.: The Rubber Hand Illusion Revisited: Visuotactile Integration and Self-Attribution. J. Exp. Psychol. Hum. Percept. Perform. 31, 80–91 (2005)
47. Vierkant, T.: Is the self real? LIT Verlag, Münster (2003)
48. von Hayek, F.A.: Routledge & Kegan Paul, London (1952)

A Human-Like Robot Torso ZAR5 with Fluidic Muscles: Toward a Common Platform for Embodied AI

Ivo Boblan[1], Rudolf Bannasch[2], Andreas Schulz[2], and Hartmut Schwenk[1]

[1] Technische Universität Berlin, FG Bionik und Evolutionstechnik,
[2] EvoLogics GmbH, F and E Labor Bionik,
Ackerstr. 76, 13355 Berlin, Germany
boblan@bionik.tu-berlin.de, www.zar-x.de

Abstract. "Without embodiment artificial intelligence is nothing." Algorithms in the field of artificial intelligence are mostly tested on a computer instead of testing on a real platform. Our anthropomorphic robot ZAR5 (in German Zwei-Arm-Roboter in the 5th version) is the first biologically inspired and completely artificial muscle driven robot torso that can be fully controlled by a data suit and two five finger data gloves. The underlying biological principles of sensor technology, signal processing, control architecture und actuator technology of our robot platform meet the requirements of biological based technical realization and support a distributed programming and control as well as an online self-adaptation and relearning processing. The following elaboration focuses on biological inspiration for the embodiment of artificial intelligence, gives a short insight into technical realisation of a humanoid robot, which is of high importance in this context, and accentuates highlights relating to a possible paradigm shift in artificial intelligence.

Keywords: embodied artificial intelligence, biological archetype, humanoid robot, biological inspired construction, fluidic muscle, muscle-tendon system, weight saving construction, common platform.

1 Introduction in the Biological Inspiration of the Robot ZAR5

ZARx is a joint project of the Technische Universität Berlin department Bionik und Evolutionstechnik, the company EvoLogics and the company Festo[1].

The aim of this project using the fluidic muscle of Festo [1] is to show the current possibilities of biologically inspired construction in embodiment, muscle-tendon system, control architecture, radius of action, and weight saving.

The robot ZAR5 is a human-like torso with two arms and two five-finger hands which are strictly developed according to bionical considerations. The combination of biology and robotics leads to smoother and compliant movement which is more pleasant for us as people. Biologically inspired robots embody non-rigid movements which are made possible by special joints and actuators that give way and can both actively and passively adapt stiffness in different situations. The more the technical

---

[1] This project was supported by Markus Fischer TC-D, FESTO AG & Co. KG.

M. Lungarella et al. (Eds.): 50 Years of AI, Festschrift, LNAI 4850, pp. 347–357, 2007.

realisation corresponds with the biological role model the successful is the reflection of the true reality. If we want to learn more about the control architecture and their functionality in the human being, we have to build an exact copy of the natural role model as much as possible to improve our conceivability of artificial intelligence.

Biological inspiration is not only the morphology – size, proportions and load-bearing inner structures – but also the physiology – moving mechanical parts and muscle tendon systems – as well as parts of the all driven control architecture. The better the morphology is understood and transferred to the artificial body the better the physiological parts can act and thus finally the controlled software. Morphology, physiology and control are an entity and have to be considered always together.

The next two chapters engage with the importance of a biological inspired embodiment for missions in artificial intelligence (AI) and the rest of the paper gives a deeper look in selected issues of the humanoid robot ZAR5.

2 Why Is an Embodiment in AI Necessary?

Intelligence is the Latin word for cognition, perception or comprehension. Based on the natural intelligence of man or animal the essential intelligent criteria can be abstracted to:

- The ability of processing any symbols (not only data),
- The constitution of an inner model of the outer world,
- The ability of an adequate use of the knowledge, and more minor features like
- Reasoning, generalising and specialising.

As archetype of intelligence the human brain is named nearly exclusive. The matter of AI is to understand and reproduce the ability of the human brain in technical applications. Current areas of AI are pattern recognition, speech synthesis/ recognition, programmable machines and expert systems. The fundamental idea of AI is to analyse under which conditions computers can reproduce the behaviour pattern of the intelligence-based creatures.

The intelligence of creatures is evolved by interaction with their environment over millions of years. The peculiarity of the carrier – the embodiment – of the brain has an essential hand in whose development status and intellectual level. The embodiment is the interface between natural intelligence and environment. The complexity or the intellectual height of the brain is determined by the complexity or miscellaneousness of the embodiment. We think always in the complexity of our doing. The more (complex) we can do the more intellectual we are generally.

According to this a detach of an algorithm of AI from their environmentally connecting embodiment leads to an incorrect simulation condition and finally to an insufficient solution by definition.

3 Why an Embodiment Close to a Biological Archetype?

If we accept the above agreement which type of embodiment we should use? That depends on what do you want to do. If we are investigate in special bat skills we have to validate the results on an embodiment which fulfils the requirements of a real bat, concerning the asked features and interfaces. A test on an other mechanical platform seems to be hardly meaningful.

If we test an AI algorithm in the field of muscle control, we have to build an appropriate application which fits the requirements on a muscle-tendon system. The better the respective technical solution meets the underlying methods of the biological role model, the better works the AI algorithm and reflect the real conditions.

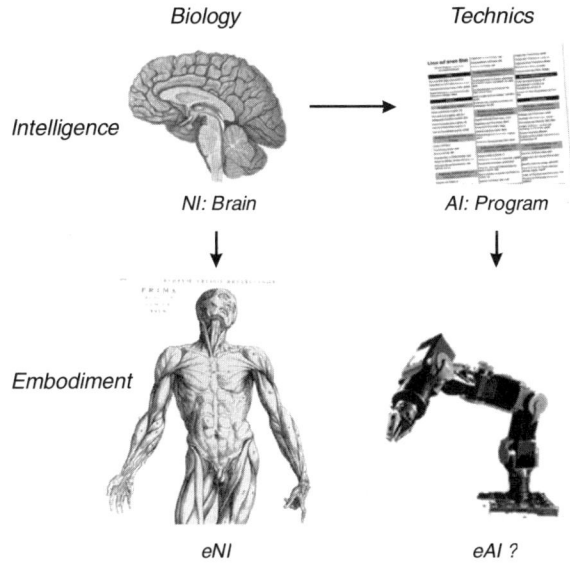

Fig. 1. If the human brain and thinking are archetype to an algorithm or program of AI, how should an appropriate body structure be to prove the functional efficiency?

The possible limits have to be considered depending of what we want to do with our AI algorithm and where shall the algorithm run. Surely we do not want to steer a man but rather reproduce a human skill on a technical application. But if we want to improve a robot with a man's skill, we have to build a human-like robot as far as possible (Fig. 1). Only such an embodiment has the needed requirements to fulfil adequate the posed task.

The AI in terms of a computer algorithm is the technical realisation of the natural intelligence – of the thinking brain. This small piece of reality should be validated gainfully. To come closer to the reality we have to build an embodiment which produces an adequate feedback compared to the archetype. It is not important that the embodiment looks like a man but it is highly important that it copes with the task and the descriptive functionality.

The scientific discipline which deals with the analysis of biological systems and transferring the underlying principles into technical implementations, is called Bionics in German Bionik. Bionics is concerned with decoding 'inventions' made by living organisms and utilising them in innovative engineering techniques. Bionics is a made-up word that links biology and technology. However, nature does not simply supply blueprints which can merely be copied. Findings from functional biology have to be translated into materials and dimensions applicable in practical engineering. It is less the form here but rather the functional coherences which have to translated in a proper way [2].

What can we learn from nature about morphology and physiology for the design of humanoid robots? If we concur with the law of survival of the fittest, then we believe that only optimised individuals can exist in nature in their respective surrounding conditions. Bionics initial task is to search for individuals in nature which have the same characteristics as the object to be developed. In our case, we are searching for a model of a humanoid robot arm and hand. We are thus looking for animals which are able to hold and/or carry several kilograms and which have human-like proportions with respect to weight and inherent compliance. When looking at the problem more closely, the intrinsic problem is how we can produce a multiple of force that is able to hold objects that are heavier than the embodiments own weight. This is the so-called power-weight ratio; this ratio is about one to one for electric motors. We have found other solutions for actuators in nature, particularly linear actuators that produce tractive force. The power-weight ratio of these actuators is multiplicatively higher than those known for technical actuators. Thus, it seems that nature has a better solution for our technical problem under the given terms and conditions.

We will not look at industrial robots here, as they carry out rigid tasks among themselves, or in contact with a technical environment. This field, called contact stability [3-5], has been widely investigated and has presented large problems for robotic manipulation tasks till date.

We will instead focus on human-like robots and their interaction with humans and the environment. This contact or physical touching between robot and human is subject with special requirements regarding softness and compliance of motion [6, 7]. The aim of humanoids is not to assemble printed circuit boards that is also hard for humans, but to master soft and energy-optimised movement in different situations of life.

The question of the appropriate embodiment – morphology and physiology – cannot be answered generally. Certainly it is true that for various questions also a more technical embodiment fulfils the given task. A specific analysis of the question concerning to sensor input, signal processing, actuator output, control loop and interaction with the environment should provide the solution.

Other humanoid robot projects suitable for AI without claim of completeness are e.g. Cog [8], iCub [9], Kismet [10], the Shadow hand [11] or the well-known ASIMO.

4   Humanoid Robot ZAR5 in the Face of the Embodied AI

The current humanoid robot project ZAR5 located in the department *Bionik und Evolutionstechnik* of the *Technische Universität Berlin* is the development and construction of a two-arm robot with two five-finger hands attached to a rigid spinal column (Fig. 2).

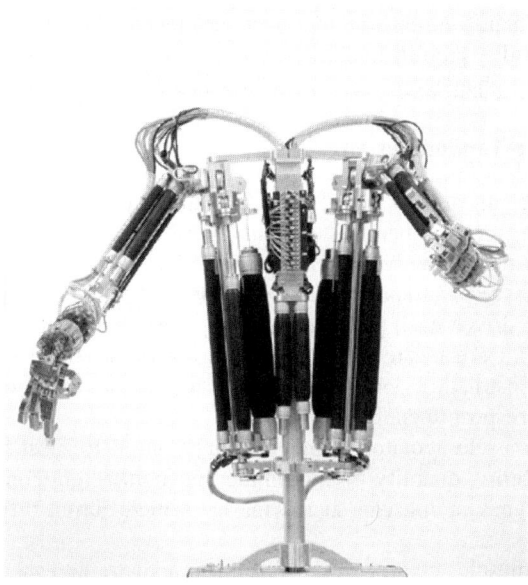

Fig. 2. The humanoid robot torso ZAR5 with two arms and two five-finger hands

The whole robot is 190 cm tall, the torso – the upper part of the robot – has a human shape and a weight of about 45 kg and is thus similar to humans of this size. The humanoid robot torso is developed according to a biologically inspired approach as far as possible. Not only the shape, proportions and radii of action but also the deeper and major qualities like skeleton, joints, muscle-tendon systems and data processing of the archetype man are implemented. The company Festo has provided the linear actuators of the fluidic muscles [1]. Tendons of Dyneema® filaments are used to convey the tractive force to the joints regarding tensile strength, lightweight and little bending radius.

The robot ZAR5 can be operated by a batch file, by teach-in and by a data suit and two five-finger data gloves. All joint angles of the data suit wearing man are read and transferred via a main PC located in the base to the controlling microcontrollers and then to the corresponding robot joints. All angular data are read every 20 ms and transferred to the CAN-bus connected microcontrollers of the robot body. This is the path planning stage. Each main body part of the whole robot: right and left hand as well as right and left side of the body is controlled by a system of two microcontrollers. One microcontroller organises the control loops of the connecting joints of this body part and the other one is responsible for the generation of the PWM signals for the fast switching valves. These inner controllers try to follow the given path. Before the tasks are finished they are normally overwritten by the next datagram's from the suit or gloves and thus lead to complex movement trajectories.

The goal of this project in the face of embodied AI is to provide a biologically inspired and manlike platform for the most different algorithms in AI. This platform could get a common interface and base for the latest developments in the algorithmic

AI. The requirements on realistic approach, efficiency, openness, decentralisation and free programmability are fulfilled [12].

The next three chapters give a deeper look in selected issues of the robot ZAR5.

4.1 The Muscle-Tendon System

The study of the physiology of the muscle-tendon system [13-16] of a man and its activation by the central nervous system gives us insight into the functions and activities of the human body. A tendon transmits the tractive power of the human muscle across tissues and special parts of bones. A pair of muscles called 'agonist' and 'antagonist' drives each joint and pulls against each other to build a tonus. All muscles of a joint are located always on the top or proximal side to the centre of rotation. This construction detail leads to less torque and the ability to carry out fast movement with respect to energy need.

The fluidic muscle actuator from Festo [1] is used to meet the requirements on dimensional stability, quantity of shortening and lightweight construction. There are three different types of muscles at the market which output different tensile forces dependent on diameter.

This fluidic muscle actuator works as a linear actuator and shortens in length by an inside pressure above atmospheric. The advantages are a high power-weight ratio, no stick-slip behaviour, works as closed system, no maintenance and no needed retention forces. In the robot application the muscle is used with air pressure of 7 bar. The muscle shortens as an ideal cylinder and is modelled in [17, 18]. The greater the affected force by a constant air pressure is, the smaller is the shortening referred to as base length of the muscle rubber tube. Moreover, the higher the air pressure by a constant force is, the greater is the shortening.

A muscle pair in an antagonistic setup drives each joint of the robot ZAR5 except of the fingers. Tendons of Dyneema® are guided via Bowden cables and pulleys to the joint and transmit the tractive force of the artificial muscle.

The dimensioning of the muscle type, the length and the deflection pulley are the most important tasks in order to fulfil the requirements regarding radius of action, velocity of movement and, in the end, weight to be lifted. Due to being scaled to human proportions, the type and the length of the muscle is limited. The relationship between muscle length and radius of the deflection pulley has been well defined and is calculated beforehand. The smaller the pulley, the smaller the length of the muscle can be, however the muscle have to be more powerful.

Two of the new muscle actuators in combination with an artificial tendon build a completely new actuator system which allows both soft, elastic and compliant as well as force-guided and exact-positioned movements depending on the tonus. This compliance is not actively caused but inherent by the medium air and the material of the muscle, opened the possibility of energy storage and conversion from kinetic energy to potential energy and vice versa and is more pleasant in contact with humans e.g. in the field of assistant robotics. The challenge is the mastering of the nonlinearities and out of it the utilizing of the advantages in the face of energy-optimization, material saving and finally efficiency toward to a more natural movement.

4.2 The Joints

The joints of a humanoid robot are essential for the later capability of its movement shown in different approaches [19-24].

The human shoulder is a ball and socket joint. A technical replica has proven to be a bold venture; this is because the construction involves a group of muscles which covers the shoulder joint and helps to keep the shoulder in the socket and enable the movement of the arm. A surface muscle or the placing of muscles around the joint to imitate the human shoulder muscle-tendon system is awkward to construct and susceptible in operation. A better way to build a complex shoulder joint is to split the multi-freedom joint into separate rotational joints each of which have one degree of freedom. These single joints are easier to construct, can be attached directly to the muscle-tendon system and are more rugged in use. Each of the three rotational joints spans a 2D vector plane around an axis of the Cartesian coordinate system.

The elbow joint – biceps-triceps system – is constructed according to the human system. It is technical a hinge and allows bending and straightening but does not rotate.

The human twist behaviour of the ulna-radius system is a rotary motion of the wrist which can be simplified by a joint with pulley and vertical rotation axis. The challenge of the wrist joint is to duplicate full functionality of the human wrist with a simultaneously simple and durable construction. All tendons of the finger joints have to be concentrated in the middle of the rotation axes. The mechanical resistance in the joint arise from the guidance of the tendons to the sockets of the fingers. In particular, the tilt and lift muscle works against this rising mechanical resistance. For this reason, we have to limit the maximum range of the joint in each direction. Two muscles – flexor und extensor respectively – are used to tilt and lift the joint and are arranged as pairs of antagonists. In the technical sense one speaks of an ellipsoid joint which is a less flexible version of the shoulder's ball-and-socket joint.

The robot's hand has 12 DOF without the wrist. Only the flexor muscle is attached to each finger limb and lays on the extensor as the pullback spring. This construction does not constrict the task of grasping, but only active releasing. This leads to a decrease in size and mass and, due to this, to a smaller inertia of masses and control effort. A disadvantage of this concurrence is the unnecessary additional expenses of providing tractive force via the muscles to overcome the resilience of the springs.

Not only the appearance but rather the function have to be reproduced to build a human-like robot. Thus we have to turn attention to human skills which are determined by the construction of the concerned joints. The better the joints achieve the desired radii of action of man the better the whole robot acts humanoid. Often the archetype of a joint is too complex to reproduce it in detail – we have to do Bionics. This means we reduce the degrees of freedom of e.g. the shoulder joint and separate it in its reference axis. Only by this way we have the chance to control the joint in a proper way. The combined activation of the grounding axis leads to the original capability of movement of the considered joint. The use of simplified joints allow the application of standard bearings and ensure good friction and abrasion properties. More construction details of the joints of our application ZAR5 can be found in [12].

4.3   The Control Architecture

The challenge of the control architecture of an anthropomorphic robot is the design of the electronic components concerning their decentralised tasks and the connecting communication pathways [8, 25-28]. In techniques data cannot really be processed in parallel in opposite to the human brain and the central nervous system of a man. Engineers till date have not been able to reproduce this data flow and communication network in vitro. The task will be to assemble, place and manage electronic parts in the same way as to achieve results similar to that of the human. Many small activities and reactions are not controlled by the brain, but rather initiated by the spinal cord or local reflexes. The advantage of this is shorter reaction time; specialised distributed units can be used as a paradigm to design decentralised control architecture. This approach applied to a technical system is tolerant of failure, enables short distances in the sensor-control-actuator loop and provides a control and command hierarchy.

The robot is divided into four units, completely separately assembled and controlled, one unit for each five-finger hand and one for each arm and shoulder. Each unit has identical circuit devices, functional ranges and consists of two communication directions which can be addressed both separately and independent of each other. All units are connected among each other via CAN-bus. A barebone PC in the base is on the one side connected with the data suit and the two data gloves to read the data of the path planning and on the other side connected to the CAN-bus to address the units and to monitor various values of the whole robot system.

The strict separation of different components and data directions enables speedier troubleshooting and is a first step towards decentralisation. The distribution of responsibilities and the break down of information handling reduces data activities on the bus and the complexity of the units. The fast response time of an unit in a control loop in case of emergency cannot be affected by a fewer crucial task of monitoring or finger play. The remote unit receives a command from the control PC or from another unit via CAN-bus and decides which operations to be done. Without any errors, the unit will initiate the appropriated control loop to reach the demanded goal angle. This stand-alone execution can be interrupted by the control PC or by an exceeded sensor limit value.

The control architecture consists of PC (technical brain), the CAN-bus (technical spinal cord) and the sensory and motor units including controlling electronics (sensory-motor units) is in simple words the connecting system between command (intended action) and action (executive embodiment). The way of doing and extent of the signal transmission from the technical brain to the executive embodiment determine on the one hand the parallelisation, decentralisation and finally the variety of the possible simultaneous and mutually independent movements on the other hand the complexity of the common interface for the users. Actually we have four independently operating units which can be further subdivided in one independent programmable unit per joint in the future. By now the actual software of a unit allows independent quasi-parallel software snippet per joint. We use the CAN-bus layer 2 for the mutual communication of the units [12]. This CAN-bus is also used as a common user interface too. There exists a dedicated interface and protocol description to integrate software parts on top of or into the control architecture to test different algorithms with the humanoid platform ZAR5.

5 Novelty of the Approach and Future Challenges

The used fluidic muscle as a pure pulling actuator seems to be an applicable alternative to the popular electric motor. Through its advantageous properties it is more suitable for humanoid robotics than other drive concepts. The main disadvantage is the use of a second energy form: compressed air. The electric motor obtains the power from the electric current and the fluidic muscle from the attached fluid. The air pipes occupy a bigger volume compared to electric cables but allow the direct quantification of the compliance of each muscle using pressure sensors.

The muscle actuator is suited to locate away from the point of force utilisation. The actuator mass can be easily located in the centre of rotation whereas the produced pulling force is guided via tendons to the point of force taking. Attention has to be turned to the joints which have to house the necessary cables and pipes in its centre of rotation to prevent the forming of loops or kinks during movements. A reduced to the reference axes joint simplifies the measuring of its position and thus the amount of electronic and control effort.

One local electronic control unit per joint allows not only the management of the connecting sensor data, signal processing, control and actuator triggering but also the implementation of special local functions like reflexes, online learning strategies and exception handlings. Local functions are joint dependent, preferably not interruptible from the higher levels, independent of the path planning and thus applicable to real-time use.

A proper control architecture connects all lower level units and achieves the higher level path planning from the main controller in our application an usual personal computer.

The next steps in the development process of the humanoid robot project ZAR5 are to divide the actual four units into one smaller and low-end unit for each joint for basal functions, to provide the possibility of changing and updating the software code of all CAN-bus connected units, to implement higher and lower level learning strategies and to make the sensor technology and the cable connection points more robust.

Learning strategies enable the optimisation of control parameters or the whole control structures during operation. An individual joint or a chain of joints can adapt its parameters depending on different requirements like overshoot, transient response or simply the speed of movement. Depending on the priority of the measured values, that have to be processed, the optimising algorithms can be located on the different control levels. The main processor – the brain – is responsible for holding, up-to-dateness, replacement and finally management of the different kinds of learning mechanisms.

Such kind of open platform is always limited by the used components both hardware and software. Are they designed too open no common rules and interfaces emerge. Are limited to few features it does not meet the complexity of the real situation. A reasonable platform has to be restricted as possible and extensive as necessary.

6 Conclusion

An adequate embodiment as reasonable interface to the environmental condition seems to be necessary for a testbed of an AI algorithm. A method, which links together software and a physical object, is only as good as the flimsiest element. The better the used embodiment represents the reality the better the expected solution for a characterised task in AI will be. Each task may require its own particular embodiment. An AI algorithm for the reproduction of a human motion pattern and reflex demands an anthropomorphic representation of at least one joint of a man. The respective embodiment has absolutely not look like a man but must reproduce the essential requirements on structure and function. The underlying principles of the biological archetype have to be implemented.

With the briefly introduced anthropomorphic and man-like torso a worldwide leadoff platform can emerge which facilitates a common working and testing under same conditions. Repeated tasks are prevented and the solutions are comparable among each other.

Acknowledgement. The company FESTO AG & Co. KG supports the work on the various versions of the humanoid robot ZAR.

References

1. anon., Fluidic Muscle DMSP/MAS, FESTO AG & Co. KG (2005), https://xdki.festo.com/xdki/data/doc_de/pdf/de/mas_de.pdf
2. BIOKON e.V., BIOKON - Competence in Bionics, What is Bionics? (2006), http://www.biokon.net/bionik/bionik.html.en
3. Šurdilović, D.T.: Synthesis of Robust Compliance Control Algorithms for Industrial Robots and Advanced Interaction Systems. In: Mechanical Engineering Faculty, University in Niš, p. 475 (2002)
4. Okada, M., Nakamura, Y., Ban, S.: Design of Programmable Passive Compliance Shoulder Mechanism. In: Proceedings of the 2001 IEEE International Conference on Robotics & Automation, Seoul, Korea (2001)
5. Rocco, P., Ferretti, G., Magnani, G.: Implicit Force Control for Industrial Robots in Contact with Stiff Surfaces. Automatica 33(11), 2041–2047 (1997)
6. Pratt, G.A., et al.: Stiffness Isn't Everything. Fourth International Symposium on Experimental Robotics. In: ISER 1995, p. 6 (1995)
7. Williamson, M.: Series Elastic Actuators. In: MIT Department of Electrical Engineering and Computer Science, p. 83 (1995)
8. Brooks, R., et al.: The Cog Project: Building a Humanoid Robot. In: Nehaniv, C.L. (ed.) Computation for Metaphors, Analogy, and Agents. LNCS (LNAI), vol. 1562, pp. 52–87. Springer, Heidelberg (1999)
9. Beira, R., et al.: Design of the Robot-Cub (iCub) Head. In: IEEE International Conference on Robotics and Automation, ICRA (2006)
10. Bar-Cohen, Y., Breazeal, C.: Cognitive Modeling for Biomimetic Robots. In: Biologically Inspired Intellilgent Robots, pp. 253–283. SPIE Press, Washington USA (2003)

11. anon., Shadow Dexterous Hand C5, Technical Specification, Shadow Robot Company (2006), http://www.shadowrobot.com/downloads/shadow_dextrous_hand_technical_specification_C5.pdf
12. Boblan, I., et al.: A Human-like Robot Hand and Arm with Fluidic Muscles: Biologically Inspired Construction and Functionality. In: Iida, F., et al. (eds.) Embodied Artificial Intelligence, pp. 160–179. Springer, Heidelberg (2004)
13. Fenn, W.O., Marsh, B.S.: Muscular force at different speeds of shortening. Journal of physiology 85(3), 277–298 (1935)
14. Gordon, A.M., Huxley, A.F., Julian, F.J.: The variation in isometric tension with sarcomere length in vertebrate muscle fibres. Journal of Physiology 184, 170–192 (1966)
15. Carlson, F.D., Wilkie, D.R.: Muscle Physiology. Prentice-Hall, Inc., Englewood Cliffs (1974)
16. Huxley, A.F.: Muscular Contraction. Journal of Physiology (London) 243, 1–43 (1974)
17. Neumann, R., Bretz, C., Volzer, J.: Ein Positionierantrieb mit hoher Kraft: Positions- und Druckregelung eines künstlichen pneumatischen Muskels. In: 4. Internationalen Fluidtechnischen Kolloquium, Dresden, p. 12 (2004)
18. Boblan, I., et al.: A Human-like Robot Hand and Arm with Fluidic Muscles: Modelling of a Muscle Driven Joint with an Antagonistic Setup. In: 3rd Int. Symposium on Adaptive Motion in Animals and Machines, Technische Universität Ilmenau, Germany (2005)
19. Caffaz, A., et al.: The DIST-Hand, an Anthropomorphic, Fully Sensorized Dexterous Gripper. In: IEEE Humanoids 2000. MIT, Boston, USA (2000)
20. Folgheraiter, M., Gini, G.: Blackfingers an artificial hand that copies human hand in structure, size, and function. In: Proc. IEEE Humanoids 2000, p. 4. MIT, Cambridge (2000)
21. Fukaya, N., et al.: Design of the TUAT/Karlsruhe Humanoid Hand, Karlsruhe, p. 6 (2000)
22. Casalino, G., et al.: Dexterous Object Manipulation via Integrated Hand-Arm Systems. In: IEEE Humanoids 2001, Tokyo, Japan (2001)
23. Kawasaki, H., Komatsu, T., Uchiyama, K.: Dexterous Anthropomorphic Robot Hand With Distributed Tactile Sensor: Gifu Hand II. IEEE/ASME Transactions on Mechatronics 7(3), 296–303 (2002)
24. Osswald, D., et al.: Integrating a Flexible Anthropomorphic Robot Hand into the Control System of a Humanoid Robot. In: Proc. of the Int. Conf. on Humanoid Robots, p. 12 (2003)
25. Northrup, S., Sarkar, N., Kawamura, K.: Biologically-Inspired Control Architecture for a Humanoid Robot, Center for Intelligent Systems, Nashville, Vanderbilt University, TN 37235. p. 6
26. Folgheraiter, M., Gini, G.: A bio-inspired control system and a VRML Simulator for an Autonomous Humanoid Arm. p. 16
27. Folgheraiter, M., Gini, G.: Human-Like Hierarchical Reflex Control for an Artificial Hand. In: Proceedings of the IEEE-RAS International Conference on Humanoid Robots, p. 8 (2001)
28. Williamson, M.: Robot Arm Control Exploiting Natural Dynamics. In: MIT Department of Electrical Engineering and Computer Science, p. 192 (1999)

The *iCub* Cognitive Humanoid Robot: An Open-System Research Platform for Enactive Cognition

Giulio Sandini[1], Giorgio Metta[1,2], and David Vernon[3]

[1] Italian Institute of Technology (IIT), Italy
[2] University of Genoa, Italy
[3] Etisalat University College, UAE

Abstract. This paper describes a multi-disciplinary initiative to promote collaborative research in enactive artificial cognitive systems by developing the *iCub* : a open-systems 53 degree-of-freedom cognitive humanoid robot. At 94 cm tall, the *iCub* is the same size as a three year-old child. It will be able to crawl on all fours and sit up, its hands will allow dexterous manipulation, and its head and eyes are fully articulated. It has visual, vestibular, auditory, and haptic sensory capabilities. As an open system, the design and documentation of all hardware and software is licensed under the Free Software Foundation GNU licences so that the system can be freely replicated and customized. We begin this paper by outlining the enactive approach to cognition, drawing out the implications for phylogenetic configuration, the necessity for ontogenetic development, and the importance of humanoid embodiment. This is followed by a short discussion of our motivation for adopting an open-systems approach. We proceed to describe the *iCub*'s mechanical and electronic specifications, its software architecture, its cognitive architecture. We conclude by discussing the *iCub* phylogeny, *i.e.* the robot's intended innate abilities, and an scenario for ontogenesis based on human neo-natal development.

1 Enactive Cognition: Why Create a Cognitive Humanoid Robot?

Until recently, the study of cognition and the neuro-physiological basis of human behaviour was the subject of quite separate disciplines such as psychology, neurophysiology, cognitive science, computer science, and philosophy, among others. Cognitive processes were mainly studied in the framework of abstract theories, mathematical models, and disembodied artificial intelligence. It has now become clear that cognitive processes are strongly entwined with the physical structure of the body and its interaction with the environment. Intelligence and mental processes are deeply influenced by the structure of the body, by motor abilities and especially skillful manipulation, by the elastic properties of the muscles, and the morphology of the retina and the sensory system. The physical body and its actions together play as much of a role in cognition as do neural processes, and

M. Lungarella et al. (Eds.): 50 Years of AI, Festschrift, LNAI 4850, pp. 358–369, 2007.

human intelligence develops through interaction with objects in the environment and it is shaped profoundly by its interactions with other human beings.

This new view of artificial intelligence represents a shift away from the functionalism and dualism of cognitivism and classical AI towards an alternative position that re-asserts the primacy of embodiment, development, and interaction in a cognitive system [1]. Cognitivism and classical physical symbol systems AI are dualist in the sense that they make a fundamental distinction between the computational processes of the mind and the computational infrastructure of the body, and they are functionalist in the sense that the computational infrastructure is inconsequential: any instantiation that supports the symbolic processing is sufficient. They are also positivist in the sense that they assert a unique and absolute empirically-accessible external reality that is apprehended by the senses and reasoned about by the cognitive processes.

This contrasts with the emergent embodied approach which is based to a greater or lesser extent on principles of self-organization [2,3] and best epitomized by enactive approaches originally formulated in the work of Maturana and Varela [4,5,6,7,2,8,9]. The enactive stance asserts that cognition is the process whereby an autonomous system becomes viable and effective in its environment. In this, there are two complementary processes operating: one being the co-determination of the system and environment (through action and perception and contingent self-organization) and the second being the co-development of the system as it adapts, anticipates, and assimilates new modes of interacting.

Co-determination implies that the cognitive agent is specified by its environment and at the same time that the cognitive process determines what is real or meaningful for the agent. Co-determination means that the agent constructs its reality (its world) as a result of its operation in that world. Perception provides the requisite sensory data to enable effective action [9] but it does so as a consequence of the system's actions. Thus, cognition and perception are functionally-dependent on the richness of the system's action interface [10].

Co-development is the exploratory cognitive process of establishing the possible space of mutually-consistent interaction between the system and its environment. The space of perceptual possibilities is predicated not on an objective environment, but on the space of possible actions that the system can engage in whilst still maintaining the consistency of the coupling with the environment. Through this ontogenetic development — through interaction — the cognitive system develops its own epistemology, *i.e.* its own system-specific history- and context-dependent knowledge of its world. This knowledge that has meaning exactly because it captures the consistency and invariance that emerges from the dynamic self-organization in the face of environmental coupling.

It is important to understand what exactly we mean here by the term interaction. It is a shared activity in which the actions of each agent influence the actions of the other agents engaged in the same interaction, resulting in a mutually constructed pattern of shared behavior [11]. This aspect of mutually constructed patterns of complementary behaviour is also emphasized in Clark's notion of joint action [12]. According to this definition, explicit meaning is not necessary for anything to be communicated in an interaction: it is simply important that the agents are mutually engaged in a sequence of actions. Meaning then emerges through shared consensual experience mediated by interaction.

Enactive approaches assert that the primary model for cognitive learning is anticipative skill construction rather than knowledge acquisition and that processes that both guide action and improve the capacity to guide action while doing so are taken to be the root capacity for all intelligent systems [13]. While cognitivism entails a self-contained abstract model that is disembodied in principle because the physical instantiation of the systems plays no part in the model of cognition [14,15]. In contrast, enactive approaches are intrinsically embodied and the physical instantiation plays a pivotal constitutive role in cognition [14,16,17]. A strong consequence of this is that one cannot short-circuit the ontogenetic development because it is the agent's own experience that defines its cognitive understanding of the world in which it is embedded. Furthermore, since cognition is dependent on the richness of the system's action interface and since the system's understanding of its world is dependent on its history of interaction, a further consequence of enactive AI is that, if the system is to develop an understanding of the world that is compatible with humans, the system requires a morphology that that is compatible with a human. It is for this reason that a robot which is to be used in the research of human-centred natural and artificial cognition should be humanoid and should possess as rich a set of potential actions as possible.

2   Why Open-Systems?

The *iCub* is a freely-available open system. This openness is guaranteed by releasing the mechanical and electronic design under a GNU Free Document Licence (FDL) and all embedded software (controller software, interface software, and cognition software) under a GNU General Public Licence. Thus, the scientific community can use it, copy it, and alter it, provided that all alterations to the humanoid design and the embedded software are also made available under a FDL/GPL.

We have two goals in making the *iCub* so open. First, we hope that it will become the research platform of choice for the scientific community. This will help establish a *de facto* standard and therefore increase the likelihood of collaboration among research groups and, consequently, the amount of resources that can be shared among these groups. The nature of the GNU licences helps greatly in this. Second, we hope that by removing the very significant cost of system specification, design, and validation, it will lower the barrier to entry in humanoid research both for people who are expert in humanoid robotics and also for those who simply wish to carry out empirical research in cognitive neuroscience science and developmental psychology.

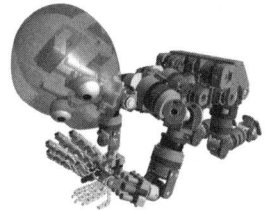

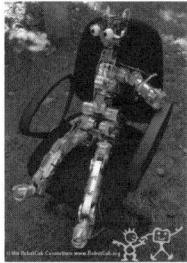

Fig. 1. Details of the *iCub* design and construction

3 The *iCub* Robot: Mechanical and Electronic Specifications

To ensure that the *iCub's* interaction is compatible with humans, for the reasons outlined above, the design is aimed at maximizing the number of degrees of freedom of the upper part of the body, *i.e.* the head, torso, arms, and hands. The lower body, *i.e.* the legs and feet, has been designed to support crawling and sitting on the ground in a stable position with smooth autonomous transition from crawling to sitting. The *iCub* has 53 degrees of freedom in total: six in the head (two for azimuth & vergence, one for coupled eye-tilt, and three for the neck) [18], seven degrees of freedom in each of the arms (three in the shoulder, one in the elbow, and three in the wrist), nine degrees of freedom in each of the hands to effect under-actuated control the 17 joints comprising the five fingers), six degrees of freedom in each of the legs (three for the hip joints, one for the knee, and two for the ankle), with the waist also having three degrees of freedom.

The sensory system includes a binocular vision system, touch, audition, and inertial sensors to allow it to coordinate the movement of the eyes and hands, grasp and manipulate lightweight objects of reasonable size and appearance, crawl, and sit up.

Figure 1 shows some details of the current status of design and construction of the *iCub* .

Although we are focussing for the present on locomotion by crawling, the torque capabilities of the feet, leg, and hip joints have been specified to be sufficient to support bi-pedal locomotion. The development of a bi-pedal gait controller is something we expect will be contributed to the *iCub* software repository under its GNU licence at some point by a third-party developer.

All of the motors and sensors are controlled by a suite of DSP chips which channel data over a CAN bus to an on-board PC-104 hub computer. This hub then interfaces over a Gbit ethernet cable to an off-board computer system which takes responsibility for the *iCub's* high-level behavioural control. Because the *iCub* has so many joints to be configured and such a wealth of sensor data to be processed, to achieve real-time control it is almost inevitable that the *iCub* software has be configured to run in parallel on a distributed system of computers. This in turn creates a need for a suite of interface and communications libraries — the *iCub* middleware — that will run on this distributed system, effectively hiding the device-specific details of motor controllers and sensors and facilitating inter-process and inter-processor communication. We discuss this middleware briefly in the next section.

4   The *iCub* Software Architecture

We decided to adopt YARP as the *iCub* middleware [19]. YARP (Yet Another Robot Platform) is a multi-platform open-source framework that supports distributed computation with an focus on robot control and efficiency. Yarp comprises a set of libraries which can be embedded in many different systems and robots, and the *iCub* is just one of the systems in which YARP is embedded.

YARP provides a set of protocols and a C++ implementation for inter-process communication on a local network (thereby enabling parallel multi-processor computation), for standardization of the hardware interface through run-time dynamically loadable modules, for providing data types for images, vectors, buffers, *etc.*, and for providing various interfaces to commonly used open-source packages (*e.g.* openCV).

Typically, when writing the *iCub* software, each module will spawn a set of YARP processes and threads whose complexity will be hidden within the module. The lowest level of the software architecture consists of the level-0 API which provides the basic control of the *iCub* hardware by formatting and unformatting IP packets into appropriate classes and data structures. IP packets are sent to the robot via the Gbit Ethernet connection. For software to be compliant to the *iCub* the only requirement is to use this and only this API. The API will be provided for both Linux and Windows operating systems. It is then possible to consider multiple levels of software development and level-n APIs that re-use the underlying levels to create even more sophisticated modules. The same rationale of level-0 APIs clearly applies to higher levels.

Higher-level behaviour-oriented application sofware will typically comprise several coarse-grained Yarp processes. This means that to run *iCub* applications, you only need to invoke each process and instantiate the communication between them. The YARP philosophy is to decouple the process functionality from the specification of the inter-process connections. This encourages modular software with reusable processes that can be used in a variety of configurations that are not dependent on the functionality of the process or embedded code.

We plan on implementing the *iCub* cognitive architecture (see next section) as a set of YARP processes. That is, we expect that each of the *iCub* phylogenetic abilities as well as the modules for their modulation, for prospection and anticipation, and for self-modification, will be implemented as distinct YARP processes.

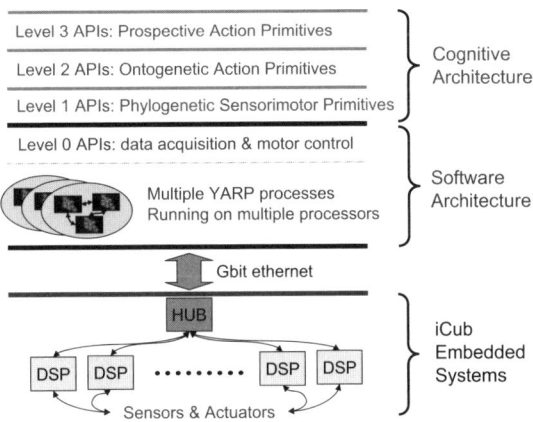

Fig. 2. The layers of the *iCub* architecture

5   The *iCub* Cognitive Architecture: An Infrastructure for Developmental Learning and Cognition

The *iCub* cognitive architecture is based on a survey of cognitivist, emergent, and hybrid cognitive architectures [20], an analysis of the phylogeny and ontogeny of human neonates [21,22], and a review of design principles for developmental systems [23,24,16]. The cognitive architecture comprises a network of competing and cooperating distributed multi-functional perceptuo-motor circuits, a modulation circuit which effects homeostatic action selection by disinhibition of the perceptuo-motor circuits, and a system to effect anticipation through perception-action simulation. The modulation circuit comprises three components: auto-associative memory, action selection, and motivation, based loosely on the hippocampus, basal ganglia, and amygdala, respectively, while the anticipatory circuit comprises paired motor-sensor and sensor-motor heteroassociative memories [25,26,27,28,29,30]. The anticipatory system allows the cognitive agent to rehearse hypothetical scenarios and in turn to influence the modulation of the network of perception-action circuits. Each perception-action circuit has its own limited representational framework and together they constitute the phylogenetic abilities of the system. The crucial issue of self-modification is catered for in two ways, the first through parameter adjustment of the phylogenetic skills through learning, and the second through the developmental adjustment of the structure and organization of the system so that it is capable

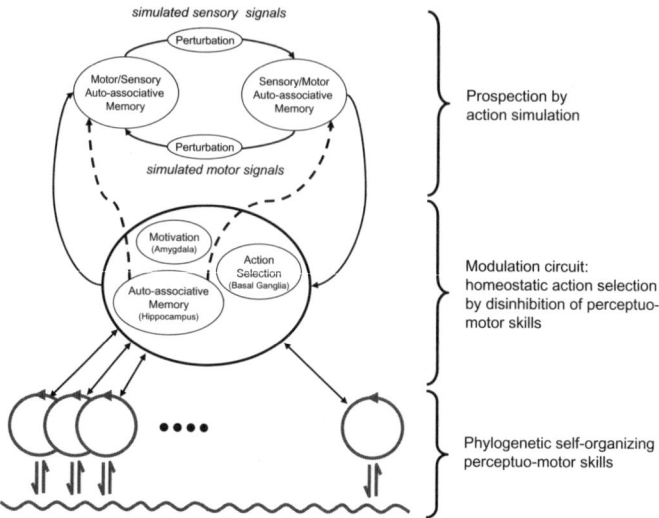

Fig. 3. The *iCub* cognitive architecture

of altering its system dynamics based on experience, to expand its repertoire of actions, and thereby adapt to new circumstances. This development, driven by both exploratory and social motives, is effected through the interaction of the anticipatory and modulation circuits, in particular by the update of the long-term anticipatory associative memories by the short-term modulation associative memory. In its current state, this is very much a strawman architecture: it has yet to be validated and it will need to be revised and amended as research progresses. This validation will be both empirical (through experiment) and theoretical (through reference to neuroscientific and psychological models).

6   The *iCub* Phylogeny: Innate Abilities

Development implies the existence of a basis for development; in other words, ontogenesis requires some initial phylogenetic configuration on which to build. This section presents a non-exhaustive list of initially-planned innate perceptuo-motor and cognitive skills that need to be effected in the *iCub* in order to facilitate its subsequent development. They are organized under the two generic headings perceptuo-motor abilities and enhanced phylogenetic abilities. The perceptuo-motor abilities can be considered to be in some sense innate (*i.e.* operative at or very soon after birth) while the enhanced phylogenetic abilities require some tuning or practice to become effective. These differ from skills that are the result of ontogenesis because there has been little or no modification of the system's state space, *i.e.* they aren't the result of a process of self-modification or development, but are more akin to learning by on-line parameter estimation.

The phylogenetic abilities include the ability to distinguish between relative and common motion in the visual field, the ability to ascribe objecthood to parts of the visual field that have persistent and well-defined outer boundaries, the ability to track objects through occlusion, the ability to re-orient gaze towards local perturbations in tactile, auditory, and visual field, and the ability to re-orient and locomote based on local view-dependent landmarks (rather than global scene representations). Since interaction with humans forms an extremely important component of neo-natal development, the phylogentic skills also include a propensity to attend to sounds, movements, and features of the human face and the ability to detect mutual gaze.

The enhanced phylogenetic abilities that we plan to implement initially include a disposition to bring the hand into the visual field, the ability to detect human faces and localize eyes in sensorimotor space, the ability to effect smooth pursuit, and the ability to stabilize percepts in a moving agent, *i.e* fusion of frames of reference. Subsequently, we will implement abilities concerned with the coordination of perceptuo-motor skills such as ocular modulation of head pose (a tendency whereby head pose is adjusted to centre eye gaze) and the ability to stabilize the percepts arising from moving scenes through successive saccades, *i.e.* opto-kinetic nystagmus.

We represent this collection of innate phylogenetic abilities in the *iCub* cognitive architecture as a series of arrow circles, in the spirit of Maturana and Varela's ideogram of a self-organizing (autopoietic) system [9]; see Figure 3.

Note that this is just a partial list of both perceptuo-motor and enhanced phylogenetic abilities. Neonates have other innate skills that we also intend to implement; see [31] for details.

7 The *iCub* Ontongeny: A Scenario for Development

The primary focus of the early stages of ontogenesis of the *iCub* is to develop manipulative action based on visuo-motor mapping, learning to decouple motor synergies (*e.g.* grasping and reaching) [32,33], anticipation of goal states, learning affordances, interaction with other agents through social motives [34,35,36,37] and imitative learning [38,39,40]. Needless to say, ontogenesis and development are progressive. We emphasize the early phases of development, building on the enhanced phylogenetic skills outlined in the previous section and scaffolding the cognitive abilities of the *iCub* to achieve greater prospection and increased (action-dependent) understanding of its environment and to establish a mutual understanding with other cognitive agents.[1]

It is important to emphasize that the ontogenetic training program that facilitates the development of the *iCub* is biologically inspired and tries to be as faithful as possible to the ontogenesis of neonates. Consequently, the development of manipulative action will build primarily on visuo-motor mapping.

---

[1] An archive of *iCub* publications can be found at
www.robotcub.org/index.php/robotcub/more_information/papers.

Once the *iCub* has mastered these skills, we will move on to experimental scenarios in which the *iCub* learns to develop object manipulation by playing on its own and or with another animate agent, that is, grasping objects and doing things in order to attain effects, like inserting objects into holes, building towers out of blocks *etc.* At this stage, social learning of object affordances becomes crucial. These scenarios will focus on the use of more than one object, emphasising the dynamic and static spatial relationships between them. In order of complexity, examples include learning to arrange block on a flat-surface, to stack blocks of similar size and shape, to stack blocks on similar shape but different size, and to stack blocks of different shape and size.

The chief point about these scenarios is that they represent an opportunity for the *iCub* to develop a sense of spatial arrangment (both between itself and objects and between objects), and to arrange and order its local environment in some way. These scenarios also require that the *iCub* learns a set of primitive actions as well as their combination.

8   Summary

Enactive embodied emergent cognition represents a fundamental attempt to break with dualist, functionalist, positivist foundations of classical cognitivist AI and to put in place instead a research programme grounded in phenomenology, reasserting the primacy of embodied developmental interaction in cognitive systems. We hope to contribute to this programme by making the *iCub* cognitive humanoid robot freely available to the research community as cost-effectively as possible and by providing researchers with a suite of cognition software modules for both phylogenetic and ontogenetic functionality. We expect and intend that the research community will improve and add to the *iCub* hardware and software, sharing their designs and code on the *iCub* repository at `www.icub.org`. By working together on this programme, we believe we have a better chance of making the breakthrough in understanding natural and artificial cognition that has eluded classical AI over the past 50 years.

Acknowledgements

The content of this paper represents the work of many people. These include: Paul Fitzpatrick, Lorenzo Natale, Francesco Nori, Francesco Orabona, Matteo Brunettini, University of Genoa; Paolo Dario, Cecilia Laschi, Anna Maria Carrozza, Giovanni Stellin, Scuola S. Anna, Pisa; Rolf Pfeifer, Gabriel Gomez, Alexandre Schmitz, Yvonne Gustain, Jonas Ruesch, University of Zurich; Claes von Hofsten, Kerstin Rosander, Olga Kochukova, Helena Gronqvist, University of Uppsala; Luciano Fadiga, Laila Craighero, Andrey Olyniyck, Livio Finos, Giovanni Ottoboni, University of Ferrara; Kerstin Dautenhahn, Chrystopher Nehaniv, Assif Mirza, Hatice Kose-Bagci, University of Hertfordshire; José Santos-Victor, Alex Bernardino, Luis Montesano, Julio Gomes, Rodrigo Venturo, IST Lisbon; Darwin Caldwell, John Gray, Nick Tsagarakis, University of Salford;

Aude Billard, Auke Ijspeert, Ludovic Righetti, Sarah Degallier, Ecole Polytechnique Federal de Lausanne; Francesco Becchi, Telerobot S.r.l. Their contributions are gratefully acknowledged.

This work is funded by the European Commission, Project IST-004370 RobotCub, under Strategic Objective 2.3.2.4: Cognitive Systems.

References

1. Freeman, W.J., Núñez, R.: Restoring to cognition the forgotten primacy of action, intention and emotion. Journal of Consciousness Studies 6(11-12), ix–xix (1999)
2. Varela, F.J.: Whence perceptual meaning? A cartography of current ideas. In: Varela, F.J., Dupuy, J.-P. (eds.) Understanding Origins – Contemporary Views on the Origin of Life, Mind and Society. Boston Studies in the Philosophy of Science, pp. 235–263. Kluwer Academic Publishers, Dordrecht (1992)
3. Clark, A.: Mindware – An Introduction to the Philosophy of Cognitive Science. Oxford University Press, New York (2001)
4. Maturana, H.: Biology of cognition. Research Report BCL 9.0, University of Illinois, Urbana, Illinois (1970)
5. Maturana, H.: The organization of the living: a theory of the living organization. Int. Journal of Man-Machine Studies 7(3), 313–332 (1975)
6. Maturana, H.R., Varela, F.J.: Autopoiesis and Cognition — The Realization of the Living. Boston Studies on the Philosophy of Science. D. Reidel Publishing Company, Dordrecht, Holland (1980)
7. Varela, F.: Principles of Biological Autonomy. Elsevier, North Holland, New York (1979)
8. Winograd, T., Flores, F.: Understanding Computers and Cognition – A New Foundation for Design. Addison-Wesley Publishing Company, Inc., Reading, Massachusetts (1986)
9. Maturana, H., Varela, F.: The Tree of Knowledge – The Biological Roots of Human Understanding. New Science Library, Boston & London (1987)
10. Granlund, G.H.: The complexity of vision. Signal Processing 74, 101–126 (1999)
11. Ogden, B., Dautenhahn, K., Stribling, P.: Interactional structure applied to the identification and generation of visual interactive behaviour: Robots that (usually) follow the rules. In: Wachsmuth, I., Sowa, T. (eds.) GW 2001. LNCS (LNAI), vol. 2298, pp. 254–268. Springer, Heidelberg (2002)
12. Clark, H.H.: Managing problems in speaking. Speech Communication 15, 243–250 (1994)
13. Christensen, W.D., Hooker, C.A.: An interactivist-constructivist approach to intelligence: self-directed anticipative learning. Philosophical Psychology 13(1), 5–45 (2000)
14. Vernon, D.: The space of cognitive vision. In: Christensen, H.I., Nagel, H.-H. (eds.) Cognitive Vision Systems: Sampling the Spectrum of Approaches. LNCS, pp. 7–26. Springer, Heidelberg (2006)
15. Vernon, D.: Cognitive vision: The case for embodied perception. Image and Vision Computing, 1–14 (in press, 2007)
16. Krichmar, J.L., Edelman, G.M.: Principles underlying the construction of brain-based devices. In: Kovacs, T., Marshall, J.A.R. (eds.) Proceedings of AISB 2006 - Adaptation in Artificial and Biological Systems. Symposium on Grand Challenge 5: Architecture of Brain and Mind, Bristol, vol. 2, pp. 37–42. University of Bristol (2006)

17. Gardner, H.: Multiple Intelligences: The Theory in Practice. Basic Books, New York (1993)
18. Beira, R., Lopes, M., Praça, M., Santos-Victor, J., Bernardino, A., Metta, G., Becchi, F., Saltarén, R.: Design of the robot-cub (icub) head. In: International Conference on Robotics and Automation, ICRA, Orlando (May 2006)
19. Metta, G., Fitzpatrick, P., Natale, L.: Yarp: yet another robot platform. International Journal on Advanced Robotics Systems 3(1), 43–48 (2006)
20. Vernon, D., Metta, G., Sandini, G.: A survey of artificial cognitive systems: Implications for the autonomous development of mental capabilities in computational agents. IEEE Transaction on Evolutionary Computation 11(1), 1–31 (2006)
21. von Hofsten, C.: On the development of perception and action. In: Valsiner, J., Connolly, K.J. (eds.) Handbook of Developmental Psychology, pp. 114–140. Sage, London (2003)
22. von Hofsten, C.: An action perspective on motor development. Trends in Cognitive Science 8, 266–272 (2004)
23. Krichmar, J.L., Edelman, G.M.: Brain-based devices for the study of nervous systems and the development of intelligent machines. Artificial Life 11, 63–77 (2005)
24. Krichmar, J.L., Reeke, G.N.: The darwin brain-based automata: Synthetic neural models and real-world devices. In: Reeke, G.N., Poznanski, R.R., Lindsay, K.A., Rosenberg, J.R., Sporns, O. (eds.) Modelling in the neurosciences: from biological systems to neuromimetic robotics, pp. 613–638. Taylor & Francis, Boca Raton (2005)
25. Shanahan, M.P.: A cognitive architecture that combines internal simulation with a global workspace. Consciousness and Cognition (to appear)
26. Shanahan, M.P., Baars, B.: Applying global workspace theory to the frame problem. Cognition 98(2), 157–176 (2005)
27. Shanahan, M.P.: Emotion, and imagination: A brain-inspired architecture for cognitive robotics. In: Proceedings AISB 2005 Symposium on Next Generation Approaches to Machine Consciousness, pp. 26–35 (2005)
28. Shanahan, M.P.: Cognition, action selection, and inner rehearsal. In: Proceedings IJCAI Workshop on Modelling Natural Action Selection, pp. 92–99 (2005)
29. Baars, B.J.: A Cognitive Theory of Consciousness. Cambridge University Press, Cambridge (1998)
30. Baars, B.J.: The conscious assess hypothesis: origins and recent evidence. Trends in Cognitive Science 6(1), 47–52 (2002)
31. von Hofsten, C., et al.: A roadmap for the development of cognitive capabilities in humanoid robots. Deliverable D2.1, Project IST-004370 RobotCub (2006), http://www.robotcub.org/index.php/robotcub/more_information/deliverables/
32. Gomez, G., Hernandez, A., Eggenberger-Hotz, P., Pfeifer, R.: An adaptive learning mechanism for teaching a robot to grasp. In: AMAM 2005. International Symposium on Adaptive Motion of Animals and Machines (1999)
33. Natale, L., Metta, G., Sandini, G.: A developmental approach to grasping. In: Developmental Robotics. A 2005 AAAI Spring Symposium, Stanford University, Stanford, CA, USA (2005)
34. Dautenhahn, K., Billard, A.: Studying robot social cognition within a developmental psychology framework. In: Proceedings of Eurobot 1999: Third European Workshop on Advanced Mobile Robots, Switzerland, pp. 187–194 (1999)
35. Dautenhahn, K., Ogden, B., Quick, T.: From embodied to socially embedded agents - implications for interaction-aware robots. Cognitive Systems Research 3(3), 397–428 (2002)

36. Blow, M., Dautenhahn, K., Appleby, A., Nehaniv, C.L., Lee, D.: Perception of robot smiles and dimensions for human-robot interaction design. In: RO-MAN 2006. Proceedings of the 15th IEEE International Symposium on Robot and Human Interactive Communication, Hatfield, UK, 6-8 September 2006, pp. 469–474. IEEE Computer Society Press, Los Alamitos (2006)

37. Blow, M., Dautenhahn, K., Appleby, A., Nehaniv, C.L., Lee, D.: The art of designing robot faces-dimensions for human-robot interaction. In: HRI 2006. ACM Proceedings of 1st Annual Conference on Human Robot Interaction, Salt Lake City, Utah, USA, 2-3 March 2006, pp. 331–332. ACM Press, New York (2006)

38. Lopes, M., Santos-Victor, J.: Self-organization in a perceptual network. IEEE Transactions on System Man and Cybernetics - Part B: Cybernetics (June 2005)

39. Hersch, M., Billard, A.: A biologically-inspired model of reaching movements. In: Proceedings of the 2006 IEEE/RAS-EMBS International Conference on Biomedical Robotics and Biomechatronics, Pisa (2006)

40. Hersch, M., Billard, A.: A model for imitating human reaching movements. In: Proceedings of the Human Robot Interaction Conference, Salt Lake City (2006)

Intelligent Mobile Manipulators in Industrial Applications: Experiences and Challenges

Hansruedi Früh[1], Philipp Keller[1], and Tino Perucchi[1,2]

[1] Neuronics AG, Technoparkstrasse 1, CH-8005 Zürich, Switzerland
[2] Dalarna University, Faculty of Computer Engineering, Borlänge, Sweden
{hansruedi.frueh,philipp.keller,tino.perucchi}@neuronics.ch

Abstract. This paper describes how industrial applications were targeted and successfully implemented by robotic manipulators that have been developed from studies in embodied artificial intelligent systems. The goal was to design mobile, flexible and self-learning manipulators that allow to perform multiple tasks with very short preparation time, a reasonable working speed and, at the same time, in a human-like manner. The advantages and disadvantages of these solutions compared to traditional industrial robot applications had to be considered continuously to concentrate on the right market segments, applications and customers. Thus, in addition to develop the appropriate requirements of real-time executions, risk analyses and usability, studies were established and implemented in collaboration with scientists, integrators and end customers. Acceptance, impacts of the revolution in personal intelligent robotics as well as challenges to overcome in the future are discussed.

Keywords: Automation, intelligent personal robotics, manipulator, learning, neural network, genetic algorithms, simulated annealing, robotic arm, mobile solutions.

1 Introduction

Studies in embodied systems in the context of artificial intelligence have been conducted in depth by major labs in Artificial Intelligence during the last three decades [1-3]. They designed principles how robotic systems can behave efficiently in nondeterministic, changing environments. The principles include hierarchically organized behavioral programs, sensory-motor loops, high mobility, self-learning algorithms, robustness based on adaptivity and redundancy of the sensory system.

How can these principles be applied into a robotic manipulator so that it leads to real practical advantages compared to traditional robotics and can compete with new developments of indirect competitors such as low-cost linear automation modules, industrial robots and machines specialized and optimized for one single task? And finally, how can a self-learning, mobile manipulator compete to monotonous human labor in regards to performance, logistics and costs?

Neuronics has gone a way which is characterized by focusing on applications in industrial environments and at the same time maintaining and enhancing the key aspects for mobile use. The main reason for this strategy is the insight that both of

M. Lungarella et al. (Eds.): 50 Years of AI, Festschrift, LNAI 4850, pp. 370–385, 2007.

these demands have to be successfully satisfied in order to fulfill the needs of further changes in industrial plants as well as the emerging service robotics field.

On the algorithm side, it requires an ideal combination of traditional control structures (including Cartesian space and database functions as well as teach-in functions and programming wizards) with adaptive and context sensitive behavior.

1.1 Indirect Competitors: From Simple Linear Systems Up to Humans

The competitors to the mobile, self-learning and flexible automation systems are very heterogeneous:

a) Low-cost linear systems, which allow to build an individual handling system by combining two or more linear movement systems.
b) Special purpose machines. They are optimized solutions for a given task and usually win when the size of series to be handled or manipulated exceeds around 20 million items per year. The price may be very high, but the return on the investment is received over the large number of objects to be manipulated.
c) Conventional industrial robots belong to the traditional field of flexible automation. Their advantage mainly lies within the ability to perform different tasks in parallel and to be re-programmable for new tasks.
d) Human labor. The advantages of employing humans for monotonous works are mainly the easiness to introduce them for a new task, their high skills and awareness for problems to be circumvented and their enormous flexibility.

The fields where the new kind of robotic manipulators come into play are mainly the segments occupied by a) and d): instead of building a hard wired automation solution, very flexible, non-dangerous middle-cost robotic solutions are considered as an alternative which allow additional applications, full programmability and mobility. And while humans have many advantages, their motivation to perform monotonous tasks is strongly limited, leading to errors and decreasing performance over time. This kind of work is often a risk for the health of these people in physical and psychological regards. Thus, if systems allow to perform in a "human-like" manner, they can cover both sides optimally: the advantages of machines and those of human labor.

1.2 Industrial Robot Interfaces

The software of industrial robot solutions of key players like Kuka, Adept, ABB, Mitsubishi and many others has gone through a huge evolution process during the last decade. It is, however, still strongly based on a cartesian coordinate system and assumes that the environment is not changing, so that precision is obtained by a fixed hardware with only minimal elasticities of the whole system.

Industrial handling usually assumes fix environmental conditions: The points to teach in are expected to be at constant positions of the pick-up and place locations. However, in reality, the environment often moves at least minimally, leading to errors in handling and thus reducing the reliability of the system.

A mobile, flexible and versatile robot that is built for a variety of tasks in industry cannot rely on such an unchanging environment and in a defined setting with static calibration of its sensory equipment. It must be able to adapt intelligently. For that, standard industrial interfaces must be supplemented with both additional sensory equipment and algorithms able to detect changes in the environment and adapt its configuration and the robot's behavior accordingly.

1.3 Intelligent Optimization and Adaptation Techniques

Methods available to develop behaviors in a context-sensitive way include artificial neural networks, fuzzy logic, genetic algorithms and other mathematical methods of numerical optimization. Neuronics has its focus on neural "learning machines" but also has implemented a set of other types of AI algorithms. The architecture and examples of algorithms and applications in industry are described in more detail in section 2.2.

2 Katana – An Intelligent Personal Robot in Industrial Applications

The Katana robot is being used in many different industrial environments all over the world. This Intelligent Personal Robot is allowed to interact directly with humans based on an extensive risk analysis. It is able to apply sensors and sophisticated software to generate intelligence.

Hardware. The total weight of Katana with six axes is about 4.5 kg. This low weight is possible because the paradigm of high stability of all components has been skipped. The mechanics are produced in aluminum. Although elasticity is significant,

Table 1. Technical data of the Katana6M robot arm

Drive	DC motors with position encoders
Repeat accuracy	± 0.1 mm
Degrees of freedom	4 to 6
Working range	Up to 60 cm radius (standard)
Construction	Aluminum, anodized
Weight	4.5 kg
Payload	500 g
Power	Max. 96 W (24 V / 4 A)
Speed	90°/sec. all joints simultaneously
CPU Mainboard	PPC MPC5200, FPGA on board
Motor Controllers	DSP 32bit processors in each axis
Peripherals	Industrial Ethernet, CAN, USB, Dig. I/O
Standards	CE, EN 12100, EN 61010

Harmonic Drive technology ensures a zero-backlash translation of the motor's turnings in a relation of up to 1:200. Each joint is controlled by a separate powerful 32bit DSP controller communicating at 1Mb/s over an integrated CAN bus. The integrated control board with a 750 MIPS PPC processor running a real-time enhanced embedded Linux brings fully accessible standard interfaces such as USB host, USB device, Ethernet hub, Digital I/O and industrial field bus connectivity.

The challenges on the hardware side for the construction of a compact and versatile industrial robot are manifold. Apart from its mechanical precision, it must be replenished with sensory equipment that provides the input data necessary for intelligent optimization. Part of the sensory equipment comes with the axis controllers and their ability to measure drive and current on the axis nodes. This is supplemented by many different possibilities to add sensors on the robot, on the gripper tool or externally in the near environment, communicating via the robot's standard interfaces.

An example of such a system with additional sensory hardware is the Katana robot arm equipped with a high resolution camera system. The visual software is directly coupled with the control software of the robot arm.

A learning algorithm increases the precision for every point within an array of target points. The environment can be shifted during the operation, and still the robot correctly takes and places the objects. Thus, mechanical precision in its construction and additional sensory hardware is optimally complemented by intelligent optimization techniques on the controlling software side.

Fig. 1. Katana robot with vacuum gripper and high-resolution camera system placed near the end effector

Operating System Requirements. The demands on the platform hosting the Katana control are manifold and very much heterogeneous in the sense that the different modules do not require the same execution environment. Controlling the robot and calculating the kinematics need hard real-time, whereas the visualization and some optimization algorithms do not necessarily so. Furthermore, the software should be modular, configurable and easily portable and adaptable to new robots and product lines. The ultimate choice for the Katana Robot operating system fell on embedded Linux because its real-time extensions have matured considerably during the last few years and it brings a highly customizable platform that is largely familiar to a broad range of developers, non-proprietary and unbeatable regarding long-term maintenance.

Real-time Concept. Making GNU/Linux a hard real-time system is achieved by using the co-kernel approach of the Real-Time Application Interface [RTAI] which takes control of the hardware interrupt management, and allows running real-time tasks seamlessly aside of the hosting GNU/Linux system. The 'regular' Linux kernel is eventually seen as a low-priority background task of the small real-time executive. However, this approach has a major drawback: since the real-time tasks run outside the Linux kernel control, the GNU/Linux programming model cannot be preserved when porting these applications. The result is an increased complexity in redesigning and debugging the ported code. That is why the real-time nucleus Xenomai [12] was added. This pre-emption and scheduling concept provides a real-time development framework cooperating with the Linux kernel in order to provide pervasive, interface-agnostic, hard real-time support to nucleus-, kernel- and user-space applications, seamlessly integrated into the GNU/Linux environment. Xenomai is based on an abstract RTOS core and relies on the sophisticated hardware abstraction layer ADEOS [12].

User Space Real-Time. The benefits of this approach is mainly to keep the development process in the GNU/Linux user space environment, instead of moving to a rather 'hostile' kernel context. This way, the rich set of existing tools such as debuggers and monitors are immediately available to the application developer. Moreover, the standard GNU/Linux programming model is preserved, allowing the application to use the full set of facilities existing in the user space (e.g. full POSIX support, including interprocess communication). Last but not least, programming errors in a customers own robotics application occurring in this context don't jeopardize the overall system stability, unlike what can happen if a bug is encountered on behalf of a hard real-time task in kernel space, which could cause serious damage to the running Linux kernel.

Once this basic real-time system is set up, giving a particular task over to a real-time context, be it in a separate real-time process or in the current address space, is very straightforward.

For the Katana robot and its embedded Linux system, this means that dedicated applications can fully profit from real-time support even if they run in user space. That strongly simplifies the requirements on an interface that exports intelligent algorithms, robot control and communication functionality.

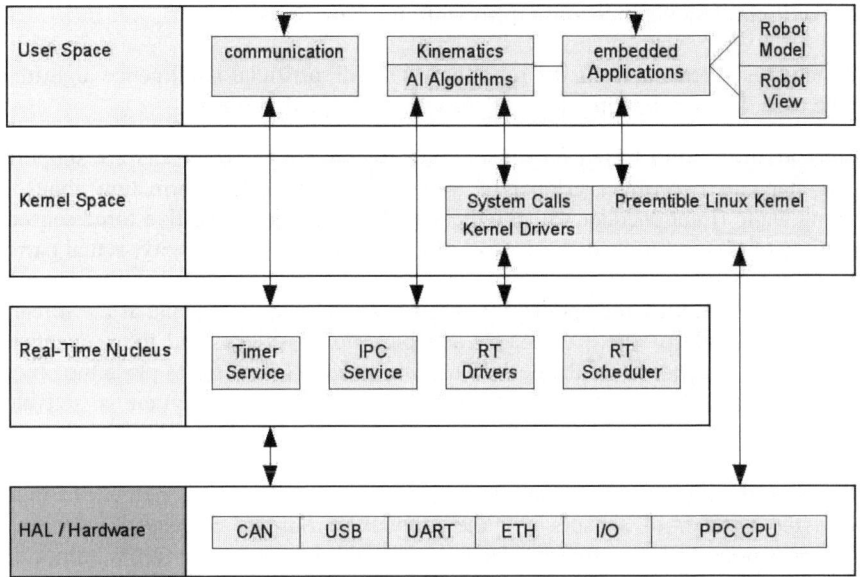

Fig. 2. Implementation of a robotic arm control system with a pre-emptible Linux kernel and real-time scheduling in user space

High-level software. The architecture of the Katana embedded robot control system has been designed to provide a transparent interface to external and internal control application programs while retaining the legacy interfaces that were implemented for earlier Katana versions. This allows for different legacy and new control and visualization clients to have access to the robot at the same time, while the details of the real-time control implementation remain hidden to them.

The client applications are to support all kinds of operators and users: those who like to program literally everything themselves and those who have never written a line of code. For the former, Neuronics provides the Katana Native Interface KNI, a C++ library for control application development at the lowest interface level. Programs written in KNI can also be cross compiled an run directly on the robot. The interface to this library is also exported as a Python binding, so that native and external programs can be written directly in Python.

On the side of a comfortable and easy to use programming and control interface, Neuronics has developed Katana4D. It has been designed for industrial applications and provides an easy but powerful control environment for the Katana robot. Skills in any programming language are not required. Additionally, there is the possibility for the user to create an application easily by means of an inbuilt scripting language or by simply teaching the robot by hand. There are different possibilities to teach the target positions within the workspace of the robot. The arm can be moved to the desired position by hand, and the exact positioning can be achieved with a few mouse-clicks from within Katana4D. It comes with full inbuilt support for AI algorithms that can be used for path optimization and adaptation.

2.1   Artificial Intelligence and Optimization Algorithms

For many tasks and subtasks the robot has to fulfill, artificial intelligence algorithms can be used. In the following, a set of such modules are described.

Object identification based on sensor data fusion. Up to 16 sensors are supported in the standard two-finger gripper of Katana. They provide information about the opening angle (joint encoder values), the actual forces (piezo-resistive force sensors), proximity of objects or obstacles (active infrared sensors), conductivity, actual current and more. Katana can learn to categorize these objects by simply executing a pick & place program built by the application wizard: The user takes Katana at the forearm, leads it to the place where to search and pick up an object, and then, eventually passing additional points on the desired trajectory, shows it where to place the objects in the different cases. Then, the "feedback" mode of the program is activated, allowing a continuous dialog with the robot while executing. Thus, the robot learns to relate the values stored during grasping with the actions to be performed. The sensor data fusion is performed by a multilayer neural network [8-9]. Its number of inputs equals the number of sensors and the number of outputs equals the number of different actions to be performed dependent on the grasped material. Within a few minutes, Katana has learned to perform the task in a reliable manner.

Precision enhancement for object placement. Many industrial applications require to pick up objects from an array of 100 or more positions and, after the object is processed by a machine, to place it in a second array of the same or different dimensions. Not every of these points is taught-in, but only the position of the edges. The intermediate points are then calculated by the kinematic model of the system. However, the precision of these intermediate points is not very high and may differ from reality up to 1 mm. Industrial applications require often a position of a tenth of a millimeter. Thus, correction algorithms are required.

A measuring device mounted on the end effector allows the robot to explore the working area. At each moment, Katana stores the configuration of the joint parameters and the values measured by the instrument in the memory. These values are then processed by a learning algorithm, which will extract the mapping between a position and the error. Based on this knowledge, the system is then able to compensate the position error for any point in the space.

In [4], a neural network is used to improve the position accuracy of a robot manipulator. After the position errors for all grid points of a calibration board are identified, the network is used as an interpolator to determine the errors for any location within the calibration space. The training is executed on-line using the grid points in the neighborhood of the target position as training patterns.

For the calibration process of Katana, a modified version of this approach is used. The network is trained off-line over the entire set of measured points in order to save processing time during the execution. Furthermore, it allows to better control the training process and eventually to repeat it until the desired performance is achieved.

The learning algorithm is composed of a neural network, which has the ability to learn and generalize from previous experiences. The first step consists of collecting a set of samples which will be used to train the network. A sample is composed by the

Fig. 3. Katana during acquisition of a training sample

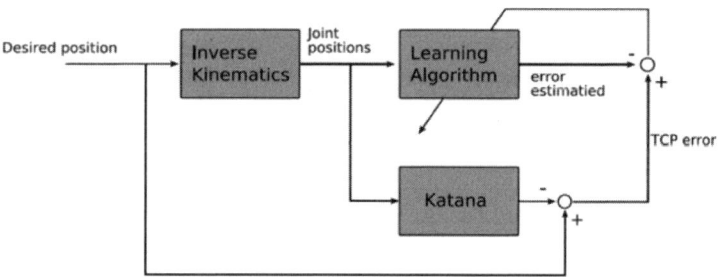

Fig. 4. Learning scheme

absolute position of each joint (in encoder steps) and the corresponding position error of the end effector. Then the network is trained using Quasi-Newton [5, 10] or steepest gradient backpropagation algorithm and a validation set is used for an early stop of the training process.

After the training process has reached the convergence, the network is used in feed forward mode to estimate the errors of each new target position. Then, a false target position is computed and sent to the robot, which will eventually reach the desired point.

However, simply subtracting the estimated error from the target position will not lead the end effector to the desired point. This problem is illustrated in figure 3, where the dotted circle represent the false target computed by subtraction of the error. At this point, the error vector may be completely different from that estimated, specially when the robot is close to some critical configurations. As a consequence the robot will stop at a point ('X') which is far from the desired. A search algorithm based on simulated annealing [6] and genetic algorithm [7] is then used to compute the best false point in the nearing of the target position which minimize the residual error. The cost function to be minimized, used to evaluate the partial solution is defined as:

$$C = \delta_d - (DK(e_t) + y(e_t)) .$$ (1)

Where δ_d is the desired position of the end effector, $DK(e_t)$ is the searched false target and $y(e_t)$ is the output of the neural network when the joint positions e_t at the false target are presented as inputs.

The Quasi-Newton training method is faster than a standard backpropagation method because it uses, in addition to the gradient, second order approximation of the error surface [10]. To illustrate the efficiency of this method, a simple simulation is described. A training set containing 35 patterns is used for training a network with 6 inputs, 10 units in the first hidden layer, 5 units in the second hidden layer and one output. The validation set consists of 5 patterns. The mean square error, the mean absolute error and the number of epochs required for convergence are averaged over 10 learning sessions. The results are presented in table 2.

Table 2. Neural network performance

Learning Method	MSE	MAE	Epochs
Steepest descent	0.033	0.148	3720
Quasi-Newton	0.026	0.117	17

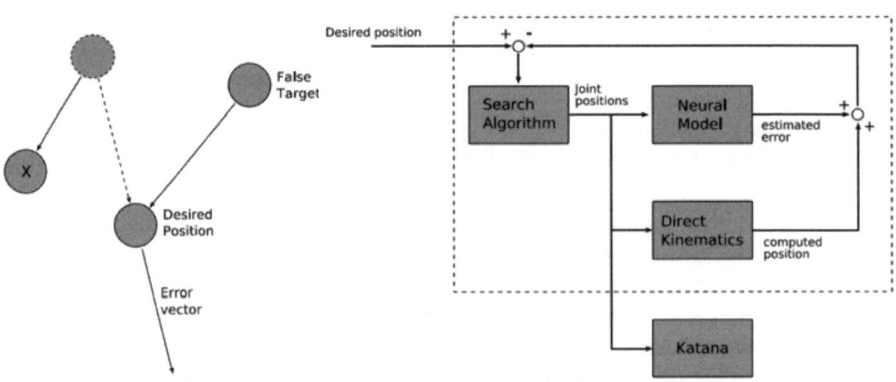

Fig. 5. Optimal false target position (left) and error compensation scheme (right)

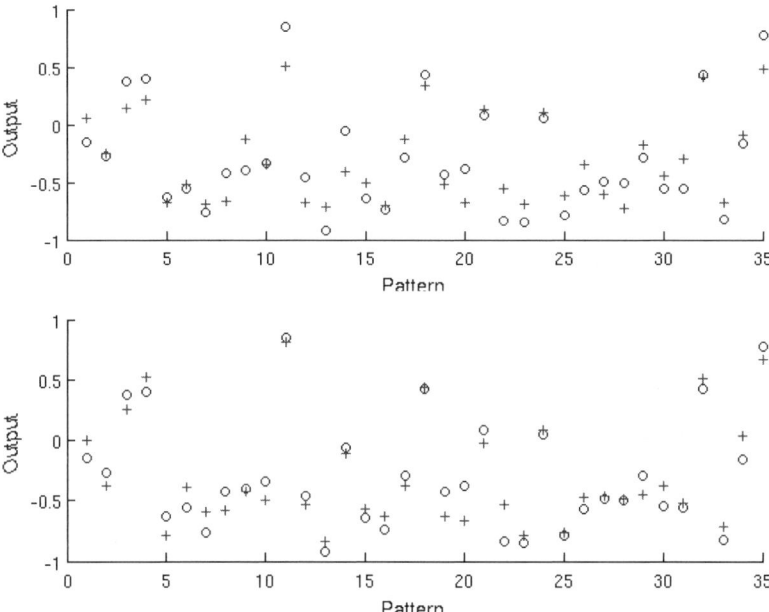

Fig. 6. Best solution found with Steepest Descent (upper) and Quasi-Newton algorithm (lower graph)

The Quasi-Newton method clearly shows a superior performance, achieving better results than the steepest descent method in about 200 times epochs less. Figure 4 and 5 show the best solution found by both algorithms. The circles represents the desired outputs, while the crosses are the responses of the neural networks for the same inputs.

Calibration Process. An example of the method described above is the calibration process for a restricted area in space. First, a subset of points to be measured is being defined, usually placed along a grid. Eventually, specific required points can be manually added or one can let the robot choose points at random in the working space. At each point the robot measures the position error, but only if a measuring equipment is available. Otherwise the user is asked to feedback the error by placing the end effector at the real point. Once a sufficient number of samples are available, the learning process may start. The number of validation patterns plays an important role for the quality of the solution. They are used for cross-validating the performances of the network and avoid over-fitting the training samples. The application (Katana4D user software, see Figure 6) shows two graphs representing the squared error over the training (upper) and the validation set (lower curve) of the patterns in function of the learning epochs. Once the learning has terminated, the network's weight matrix corresponding to the minimum point in the validation error is saved. This is the network with the best generalization propriety.

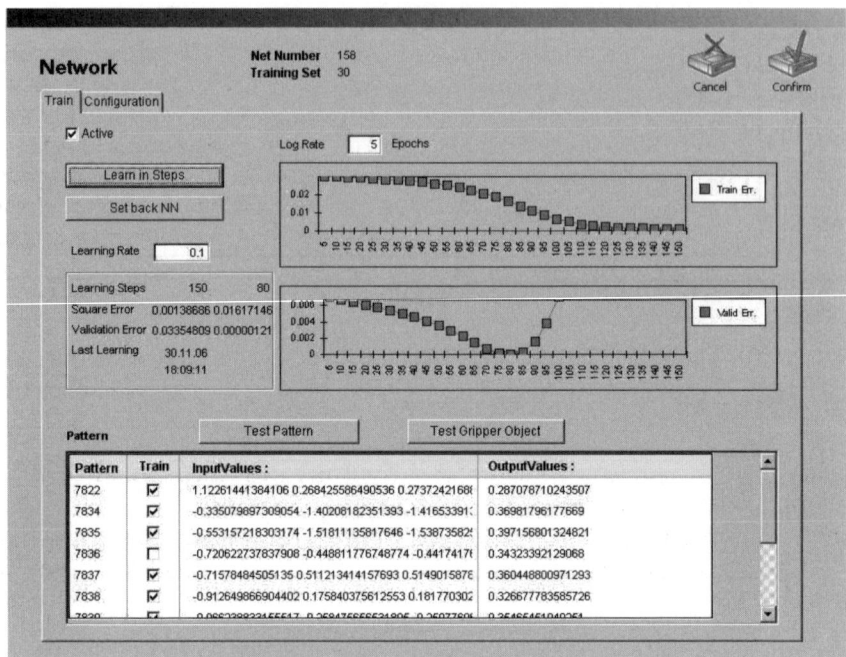

Fig. 7. GUI for neural network training process

Self-adaptation to the real space coordinates. In most industrial applications, only a small part of the complete robot's working space is exploited. There are, in general, two or three bounded regions where the robot has to perform a task with high accuracy, the remaining portion is used by the manipulator as a transitory space. A single calibration process that is valid for the whole working space would require a large amount of time. First, hundreds of calibration points have to be collected, then a neural network of medium to large size would be necessary to generate the complex input output mapping. To overcome this problem, the application allows the creation of multiple models of local areas. In this way, a neural network may be trained only with patterns belonging to a limited region, creating a more accurate error estimator. When the robot is active the joint positions are constantly monitored and, when they fall into the input boundaries of a neural network, the position of the end effector is compensated.

The environment often moves at least minimally, leading to errors in handling and thus reducing the reliability of the system. A visual system is a powerful possibility to provide autonomous feedback by continuous screening of the working space. Inverse kinematics, visual object recognition and neural error compensation as described above are combined to provide a self learning and robust pick & place system.

Speech functions. Katana can listen and talk to the user. These functions make use of the improvements during the last years in both, the language understanding and the improvement of the comprehension based on training and constraints. Compared to manual control of the robot (e.g. by leading it at the forearm to a desired place, using a joystick or a keyboard), the language interface reacts slower for two reasons: 1) The word or sentence takes time to be pronounced, and the end has to be detected. 2) The danger that a wrong action is taking place because of an undesired command which the robot gets (e.g., from a dialog between humans) requires a confirmation before execution of the command. However, the language functions can be very useful to start and stop full programs or for disabled persons to interact with the robot.

2.2 Fields of Applications

Katana is employed, in two major fields: a) industrial manufacturing, assembling and quality control and b) in service robotics. The applications in research and education today often target the service robotics field, although there are many interesting research topics which address fixed-place industrial applications or both.

The use of the algorithms described above is different in these two major fields: where in the industrial applications the Katana arm is the main component, which also may serve as master for the control of the connected machines and automation peripherals, in service robotics the arm is often the slave which is controlled by the moving platform which may contain computer power higher or equal to that of the Katana arm.

Industrial applications. The Katana robotic arm is ideally suited for pick & place, assembling and quality control tasks of light-weighted objects. Meanwhile, the system is used world-wide by small-sized, medium and large companies like BMW, Intel, Unilever, Mettler-Toledo or Maxon Motors for a variety of applications. The main advantages are the high flexibility and easiness of its use, both based on the security

Fig. 8. Cooperation between humans and Katana at a desktop working station. For such kind of tasks, the safety of the system proven by an extensive risk analysis is crucial.

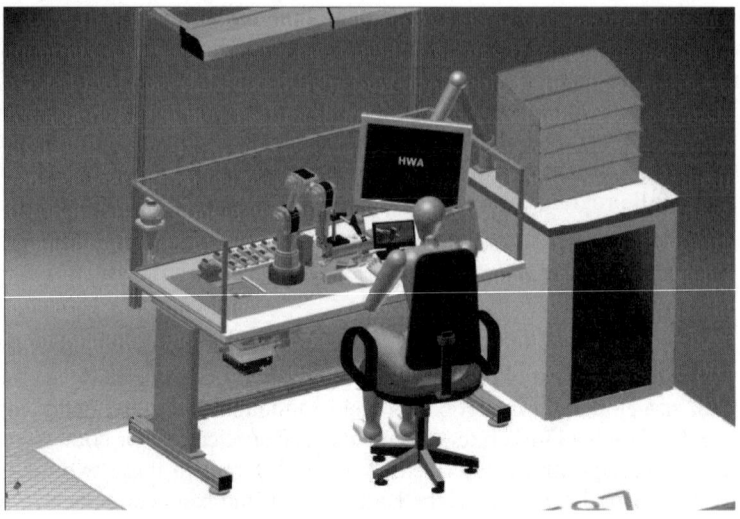

Fig. 9. Two Katanas cooperating at a bending machine. The left arm is picking up objects at positions recognized by a camera system, while the right one transports the bended objects to a quality control station.

aspects as well as on the combination of standard and artificial intelligence powered control and interface solutions.

Service robotics. Due to the low power consumption, Katana can also be an exceptional system component in mobile robotics. The control system architecture described above makes it an interesting component for several reasons: The concept of distributed intelligence can be realized under many different aspects; the powerful and standalone real-time Linux environment opens a wide range of applications which may include movements along trained trajectories, interactive tasks with humans and other robotic systems as well as robust behavior in unknown environments based on the sensory systems available on the arm and its grippers.

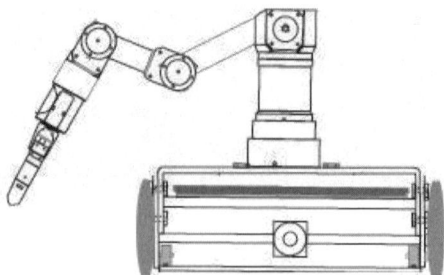

Fig. 10. Katana is used on a variety of mobile platforms for service robotics tasks

3 Implications

In applied robotics, be it in complex industrial applications or in areas where close interaction between the robot and humans implies the need for redundancies in intelligent situational and environmental adaptability, artificial intelligent algorithms can play an important role. Indeed, in robotics implementations where the requirements comprise an adjustability to unforeseen situations, or where substantial independence from a fixed one-time calibration to the robot's environment is crucial, self-learning algorithms offer a very viable solution.

In the case of Katana, being one single product that targets many different fields of application, flexibility cannot be achieved solely by providing a host of many different configurations. A certain grade of self-learning capabilities bring a considerable advantage in an area of industry that conventionally relies heavily on highly specialized and tailor-made equipment.

Also for demands in practical industrial applications such as easier or partially automated calibration to multiple end points, paths and grids, AI algorithms can play a very important part. If implemented properly, the operator or user works with a system that is helpful in achieving a particular solution and that acts intelligently without bringing with it additional complexity in handling the device.

All of these have, of course, a direct repercussion to the robot's instrumentation with sensory equipment. Senses and perception being a prerequisite to intelligence, Neuronics invests considerable resources into sensor implementation, research and development. The vision of a highly perceptive and intelligent robot adaptable to many different areas of application is a fundamental driving force behind the development of new robotic products in Neuronics.

4 Challenges for the Future

How will industrial plants change in the future? How do we adapt the robotic systems to those requirements. What has to be improved to make robots more and more personal, useful nearly everywhere?

Katana is today a stand-alone robot which can execute several different tasks autonomously. This simple concept limits the class of applications that can be accomplished and requires a constant supervision of the job, since any robot may encounter problems or fail. As the common experience shows, more units working on the same task are a clear advantage in terms of speed and robustness. Moreover, a multi-robot systems can perform more complex tasks that cannot be done by one robot alone, such as carrying heavy objects or performing assembly operations. Such systems require the implementation of cooperation abilities in each unit, allowing the interchange of information between them. Alternatively, a central unit with a global view of the main task in execution can coordinate the work. More robots will then operate together in a cooperative manner to reach the same global goal.

The integration of self-adaptive algorithms allows the robots to acquire knowledge in a learning-by-doing manner. The behavior and the ability of an individual evolve based on past experiences but, as with humans, knowledge acquisition often requires time and effort. Furthermore, the success of the learning is not always guaranteed. An important improvement in every learning process is achieved through teaching: a skilled individual transfers his abilities to others, giving them a solid background from where to start practicing and learning.

The increasing capabilities of the robots reflect the complexity of the tasks they are asked to execute. Nowadays robots are often applied in multi-task environments and people expect them to act in an intelligent manner and find optimal solutions to the problems. Thus, multi-task robots need to have a dynamic behavior. One possible solution for a future version of Katana would be to replace a static task , which is a set of simple operations performed one after the other, by a goal-oriented model [11]. Several sub-tasks are created, each of them is used to reach a defined sub-goal. In a higher level, the behavior of the robot is represented as a collection of strategies, which depend on the state of the environment. Then, given a global goal, the robot will optimally choose the next operation to be executed.

Acknowledgments. Neuronics was founded as a Spinoff company of the Artificial Intelligence Lab of the University of Zurich, Switzerland. We would like to thank Prof. Dr. Rolf Pfeifer and his team for their contributions during the company's early stage to bring our venture to a success in the industry.

References

1. Pfeifer, R., Iida, F., Gomez, G.: Designing intelligent robots – on the implications of embodiment. Review article in the Journal of Robotics Society of Japan 24(7), 9–16 (2006)
2. Brooks, R.A., Aryananda, L., Edsinger, A., Fitzpatrick, P., Kemp, C., O'Reilly, U.-M., Torres-Jara, E., Varshavskaya, P., Weber, J.: Sensing and Manipulating Built-For-Human Environments. International Journal of Humanoid Robotics 1(1), 1–28 (2004)
3. Christaller, T., Fiorini, P., Choset, H., Prassler, E.: Proceedings SSRR 2004 IEEE International Workshop on Safety, Security, and Robotics. Fraunhofer IRB Verlag, Bonn (2004)
4. Wang, D., Bai, Y.: Improving Position Accuracy of Robot Manipulator Using Neural Networks. In: Instrumentation and Measurement Technology Conference, Ottawa, Canada (2005)
5. Pham, D.T., Karaboga, D.: Intelligent Optimisation Techniques. In: Genetic Algorithms, Tabu Search, Simulated Annealing and Neural Networks, Springer, New York (2000)
6. Aarts, E., Korst, J.: Simulated Annealing and Boltzmann Machines. Wiley, Chichester, UK (1989)
7. Goldberg, D.E.: Genetic Algorithms in Search, Optimisation and Machine Learning. Addison-Wesley, Reading, MA (1989)
8. Despagne, F., Massart, L.: Neural Networks in Multivariate Calibration. The Analyst 123 (1998)
9. Haykin, S.: Neural Networks, a Comprehensive Foundation, 2nd edn. Prentice-Hall, Englewood Cliffs (1999)

10. Stanevski, N., Tsvetkov, D.: On the Quasi-Newton Training Method for Feed-Forward Neural Network. In: International Conference on Computer System and Technologies- CompSysTech 2004 (2004)
11. Pokahr, A., Braubach, L., Lamersdorf, W.: Jadex: Implementing a BDI-Infrastructure for JADE Agents. EXP- in search of innovation 3(3), 76–85 (2003)
12. Gerum, P.: Xenomai - Implementing an RTOS emulation Framework on GNU/Linux (2004)

The Dynamic Darwinian Diorama: A Landlocked Archipelago Enhances Epistemology

Adrianne Wortzel

Abstract. This paper discusses the relevance of embedding dramatic scenarios and expressive language into methodologies employed in the research and development of biochemical and/or electronic sentient beings. The author demonstrates how integrating imagined modalities into current practices can afford a profound and positive effect on outcomes.

Keywords: drama, scenario, empiricism, truth, language.

1 Introduction

The usages of story-telling and metaphorical prose for explication of both natural and processed phenomena are not unfamiliar to us. In the history of AI stories have evolved at every stage. The Turing test "story" can be retold at any point in time; at this time it could be something like: "George is traveling through a three-dimensional virtual reality environment. Inside this world, in various virtual locations (airport, museum, store, academic institution, art gallery, private residence, corporate headquarters, hospital), George engages in a natural language conversation with two other avatars where he is told that one is human and the other a program. In spite of his astute "testing" for what he considers the limits of a software robot in a virtual environment as opposed to a human, he still cannot reliably tell which is human and which is the "machine," He wonders how much significance there is in the truth, or if all that really matters is how one's representation is perceived.[1]

One of the most fertile aspects of Artificial Intelligence is that it draws from so many disciplines computer science, psychology, philosophy, neuroscience, engineering, linguistics, etc. AI is not a "contained' field, but could be considered an "un systemized system' a free flowing cluster of disciplines forming and reforming dynamic nodes and synapses which intermingle, emulating a dynamic neural network of disciplines. Those nodes and synapses could be words or meanings situated in the context of a story and offer up many possible worlds.[2]

Within this network, the use of narrative within the field would not be restricted to either the vernacular of each particular discipline nor to strict adherence to extremely orthodox research methodologies. Truths and opportunities for pockets of discovery could stretch beyond the designated glossary and syntax for a particular scientific

---

[1] An excellent synopsis of the history of philosophy of AI is provided in this volume: "Philosophical Foundations of Enactive AI" by David Vernon and Dermot Furlong.

[2] For an in depth consideration of literary narratives and possible worlds see the section on "Possible Worlds" in Ryan, Marie-Laurie, Possible Worlds, Artificial Intelligence, and Narrative Theory, Indiana University of Bloomington and Indanapolis Press, 1991, pp. 16-21.

M. Lungarella et al. (Eds.): 50 Years of AI, Festschrift, LNAI 4850, pp. 386–398, 2007.
© Springer-Verlag Berlin Heidelberg 2007

field. This "artistic" process would provide an efficacious representational force for displaying truths, without those truths suffering diminished credence.

If we assume, for argument's sake, that a newborn is not a blank slate, but has all the intuitive knowledge (not Information, butt the sense of "knowing) there is in the universe and, that the process of growth and learning for the infant is to slowly register clues as to which bits of knowledge to integrate and grow with and which to discard, then growth constitute trying not to remember discriminately. A storytelling process, embedded in research methodology, could serve the role of "reconstituting" memory in such a way that it preserves the ties to a kind of consciousness that precedes the compartmentalization of knowing, and keep research methodologies open to associations which might otherwise be missed.

2 Example by Practice

My practice as an artist includes the invention of narratives nascent to technological research and examining methodologies in order to point to their creative and intuitive nature built on an armature of empirical knowledge. The content of my work examines, or displays obliquely, aspects of technological research such how humans might relate to machines, and how machines, if they could, would relate to humans. Fictive narrative is embedded in all of my robotic and telerobotic artworks. In these works, every technological phenomenon is layered with context and meaning both in itself, and in its process of coming into being. Through artistic observation and interpretation these layers can be made tangible in art forms such as literature, film, installation and live performance. By working with the issues of artificial intelligence, artists can move away from mere sculptural or choreographic concerns to develop dramatic scenarios, which deal with deeply vital philosophical issues. The armature of these stories is always situated in some real event or text.

2.1 Science Stories

"The real history of the bee begins in the seventeenth century, with the discoveries of the great Dutch savant Swammerdam . . . Before Swammerdam, a Flemish naturalist named Clutius had arrived at certain important truths, such as the sole maternity of the queen and her possession of the attributes of both sexes, but he had left these unproved. Swammerdam found the true methods of scientific investigation; ...contrived injections to ward off decay, was the first to dissect the bees, and by the discovery of the ovaries and the oviduct definitely fixed the sex of the queen, hitherto looked upon as a king, and threw the whole political scheme of the hive into most unexpected light by basing it upon maternity."[3]

This quote initially establishes its subject as the history of the science of the bee. As described, that history resonates in the very process it is describing, and then is

---

[3] Maeterlinck, Maurice, "The Life of the Bee," Dodd, Mead & Company, New York – 1958, Quoted from the Project Gutenberg online version at http://www.gutenberg.org/dirs/etext03/lfftb10.txt; from the Chapter: "On the Treshhold of the Hive".

"dissected" into its chronological parts, culminating in a conclusion which leaps by language, inadvertently, into another realm of, let's say, gender "politics." The language includes the fact that the researchers were startled and compelled to give up a long held belief about the nature of environmental protocols in the hive where a paradigm shift occurs in the research.

That consideration opens the door to investigation of the significance of simultaneously relinquishing belief while developing theories through experiment-tation. Charles Darwin's scrupulous empirical observations of the natural world bear witness to this phenomenon, as his research and its expression retained simultaneous and subsequent reconsideration of beliefs – he really never knew what he would find, and in spite of hardships and frustrations, he sustained a grand sense of adventure, conveyed to us through his writings. In charting unknown territories, preconceptions could only be considered superficially, as a game, and his continual surprise and astonishment were duly recorded and not separated out of his writings, remaining expressive and communicative.

Language exceeding the boundaries of the designated glossary in each field of research, despite some relaxed standards in naming, is often thought of as distracting or detracting. It is true that the unmitigated use of expressive language and dramatic scenarios could be misleading. I believe it will not be distracting if the language originates concurrently embedded in, and remains true to, the research at hand not literally, but in the same way oral traditions emerge and sustain in the long term with allegiance to continued experience and the intrinsic expository and communicative qualities and capabilities inherent in the "stories."

The following text is typical of signage posted in the American Museum of Natural History

> "Giant spiders, worms and beetles live on the ground in the forest, so even though it looks like a dead heap of trash, the forest floor is really alive. In fact, a square foot of dirt in a forest holds four times as many dead insects and animals as the amount of humans there are on all of the earth at any given time. In every moment of time, leaves, flowers, fruits, twigs and dead animals fall on the forest floor. If the pile just grew and grew the forest wouldn't get any light and air and everything would die and the Cycle of Nutrition and Decay would just stop dead in its tracks." [4]

By encapsulating narrative in evolving research one also embeds the philosophy of science as an active element in the process. The goal, however, is not to arbitrarily manufacture paradigm shifts or scientific revolutions, but to amplify existing methodologies so that research remains "ventilated" -- open to combinatory experiments with other disciplines -- and, in addition, so that results are resonant with significance in these other fields in a way which feeds back into the research there. This, in turn, opens up new possibilities for discovery and disclosure by eradicating the need to work within the constraints of any paradigm at all. The recursion that makes that makes "no paradigm at all" a paradigm in itself will not apply because investment would be in "process" rather than "product"; the process being a perpetually fluid one of struggle to

---

[4] American Museums of Natural History, New York, signage in the Hall of North American Forests on the Cycle of Nutrition and Decay.

throw off a paradigm from the moment it displays evidence of a takeover. The removal of paradigm thinking raises the risk of extreme failure as well as the possibilities of success in unexpected quarters. The benefit of this is that both kinds of consequences will perpetually provide more information than a confined and constrained experimental situation locked to a fixed paradigm, even if it is one that emerges from nature. In addition, it benefits because it allows the researcher to deal simultaneously with short-, middle and long-term research goals and to reach out to community where research events provide information relevant to other experimenters.

2.2 Applying Personae to Developing Entities

Sayonara Diorama, a play I wrote and produced, creates a fictive narrative of a second voyage of the Beagle by Darwin and Fitzroy, thirty years after the first. The known history of the first Voyage, and its subsequent lineage of publications by Darwin, offer a verdant field for examining the power of expressive language in description and developing theory.

In *Sayonara Diorama*, the story is that Darwin and Fitzroy, while at sea, share their intense positions on organized religion. The resonance of their theological simmer rolls over into a quarrel, which triggers a tremendous storm. Simultaneously, Fate is forces its way through a fissure in the earth's core up to the underbelly of a nearby volcano. Appalled at the lateness of the hour for a visit, the volcano blows its stack. Fate, expelled from the volcano's throat, rises up and couches itself like a recalcitrant Buddha on the crest of spewing lava and then collides with the fierce gusts of Darwin and Fitzroy's altercation. A shipwreck ensues. Captain Fitzroy is dispatched to a well-documented island called Heaven. Darwin, however, is tossed to an island occupied by creatures displaying unusual forms of human physiology. These are, in fact, the deformed creatures depicted in ancient sagas and Western European medieval maps, where they are placed at the edge of a flat world standing in for what was unknown.

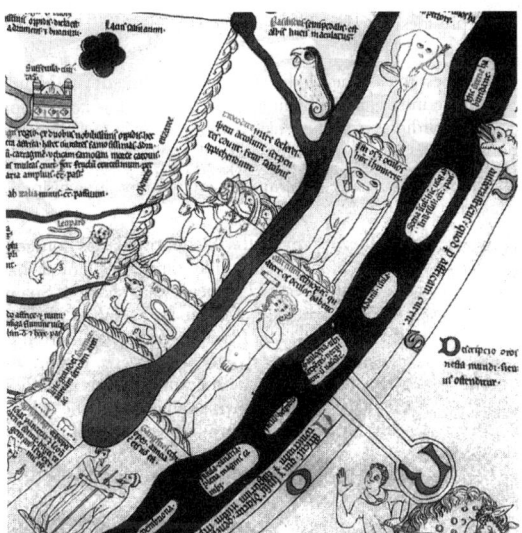

Illustration of the Hereford Mappa Mundi, 1299 – Detail
Courtesy of The Dean and Chapter of Hereford and the Hereford Mappa Mundi Trust

The Darwin character (anagram pseudonym: "Clan-Is-Raw-Herd"), begins examining the creatures, and is startled to find that they, in turn, are examining him. What ensues is an argument over who has the best evolutionary adaptations and how that should reflect their position in the world.

Clan-Is-Raw-Herd: These creatures are the living expression of the literary conventions of Solinus.[5] As unseen entities sourced in Africa, east of the Nile, he conjured these and marginalized them as gargoyles, demons, monsters, sinners, and unformed and deformed inhabitants of the edge of the flat disc that was then the world.

I thought them mere mythical paradigms of strangeness for what we cannot actually see or understand. But here they are, as upright as we are, one with the body of a man but the head of a dog (Cynocephali), one with no head at all but with his eyes, mouth and nose centered in his chest (Blemye), and one whose lower body ends in one limb rather than two (Sciapod).

Clan-Is-Raw-Herd (to the creatures): Those who make maps must divide the world into empirical geographical zones in order to examine their position in it. What is unknown must be identified at least as idiosyncratic as emblematic of "not-knowing." You three of Solinus should not take personal offense at the peripheral territory you are delegated, because in a flat world, all is equal.

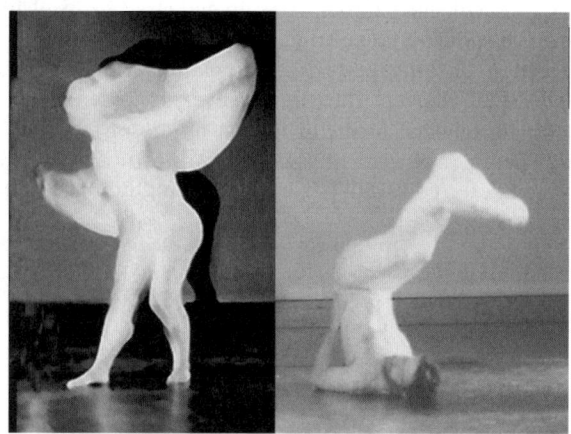

Shades of Mr. Panotti and Mr. Sciapod
Sayonara Diorama

Blemye: Tis completely foolish! If there are more men like our perpetual recorder here, whose eyes are not central in their chest close to their hearts as God intended, then why not designate THEM as mythological and monstrous and put THEM at the edge for all time. For to me (he leers at Clan-Is-Raw-Herd) they are strange beyond endurance.

Cynocephali: But then, Master Blemye, I must indeed be the missing link indeed between you and this venerable gentleman who sits patiently recording what he see.

---

[5] Gaius Julius Solinus, Latin grammarian and geographer, third-century A.D.

Certainly, I should be at the center where Jerusalem is now, for I am the link between those who have heads and those who do not.

Sciapod: All I know is that I am the only one of you who is rendered safe and dry in rain and snow by my own physiology. If evolution is indeed everything our bearded friend here says it is then certainly my ancestors, who have engineered the most ingenious and useful adaptation of all, who deserve placement as the centerpiece of the world. If inside the edge and outside the edge are as homogenous as he says they are, and do not mean completely different things hierarchically, then why not give the inside up to the ones you have designated 0utsiders?

3 The AILAB

In 2004 I was the recipient of a Swiss Artists-in-Residence Award[6] to spend six months embedded in the Artificial Intelligence Laboratory, Department of Informatics, University of Zurich ("AILAB"). I came to the AILAB with over 10 years of experience as an artist creating robotic and telerobotic art installations and performance productions, both in physical and virtual networked environments.[7] For these works, I had collaborated with research engineers working in the fields of robotics and related fields in the US.

To a large degree, AILAB researchers develop disparate idiosyncratic robots in their individual labs rendering my tenure there an experiential journey because I spent my time on the premises traveling from one individual research laboratory to another, examining the creatures produced indigenous to each lab. Therefore; both the layout and the environment of the AILAB lent itself to my rendering as a geographical territory of dispersed islands on which singular robotic species evolve in relative isolation. The individual labs became, in my mind, islands in an architectural "galapagos" and I proceeded to create video content as a "re--enactment" of Darwin's Chapter 17 of the Voyage of the Beagle: *Galapagos*.[8] The resulting work: *archipelago.ch* is a video depicting that journey. Darwin's prose fits the content of the lab research beyond all reasonable expectations. In the script excerpts for the video below Darwin's verbatim words are in italics.

The individual labs, depicted as "islands", both "breed" and sustain creatures (robots) as they emerge from the research. In the archielago.ch video, the robot are empirically examined and interpreted by a Darwinian voiceover and sensibility, with some additional contributions from the researchers' papers. Islands in this archipelago are re-named after the researcher residing there. The terrain of their labs replaces flora and fauna with the tools and detritus of each researcher's individual lab. Depiction of isolated robotic parts, particularly those from trial and error experimentation, emphasizes a robot's evolution as a specimen striving for fitness and survival. Latitudes ad longitudes in Darwin's text have been changed to those of Zurich, Switzerland.

---

[6] Artist-in-Labs, Project Director Jill Scott, http://www.artistsinlabs.ch/
[7] Documentation at http://www.adriannewortzel.com/
[8] The Gutenberg Project at http://www.gutenberg.org/dirs/etext97/vbgle11.txt, Chapter 17, Galapagos Archipelago.

Why would the words of a 19th century experiential naturalist with such keens powers of observation and perception, and such a politicized way with words in presenting a revolutionary theory ten years after the journey, be so appropriate for illuminating 21st century robotic research endeavors taking place in a state of the art (and mind) facility? This paper seeks to exposition that question, if not entirely answer it. This experience was personal one, but merits disclosure, because it is what led me to, what I think and hope will be a contribution to the field.

3.1 The AILAB Terrain and Its Text

The following texts have Darwin's verbatim words in italics, the remainder are my words and in some cases (noted) from the researcher's papers. Where Darwin's text leaves off, the slack is taken up by an attempt to amplify the scenario so that it sows the research projects in a new, personified light. We begin with Darwin's words applied to the lay of the land in the AILab: The geographical details such as longitude and latitude have been changed to match those of Zurich.

The natural history of these islands is eminently curious, and well deserves attention. Most of the organic productions are aboriginal creations, found nowhere else; there is even a difference between the inhabitants of the different islands; yet all show a marked relationship with those of the continent although separated from it by a vast and difficult terrain.

At this time this archipelago consists of several principal islands, of which some exceed the others in scope. They are situated above the Equator, Latitude 47.38 degrees North, *Longitude 8.54 degrees East.*

The archipelago is a little world within itself, a satellite which at one time had spokes to a continent, whence it has derived a few stray colonists, and has received the general character of its indigenous productions, although within the archipelago the difference is that its islands somehow appear to encourage each emigrated morphology to emerge into something more than it was whence it came.

Considering the small size of the islands, we feel the more astonished at the number of their aboriginal beings, and at their confined range.

3.2 The AILAB Islands and Their Respective Populations

St. Simir Island. *Of terrestrial mammals, there is only one which must be considered as indigenous, namely, a mouse* of the species Archipelagogenisis.Mus, or Amouse), *and this is confined, as far as I could ascertain, to* Saint SimMir *Island, the most easterly island of the group.*

Although no one has a right to speculate without distinct facts, yet even with respect to this marvelous creature, we should consider that it may be borne of an imported species. *For I have seen, in a most unfrequented part of the* world below the equator, *a native mouse living in the roof of a newly built* hovel of similar comportment, *and therefore ...* transportation of the Amouse's ancestor in a vessel from there to here *is not improbable.* In fact, its lineage is reminiscent of a division of the family of mice characteristic of the Continent in an early stage of its development and *I can hardly doubt that this* mouse *is a variety* evolved via *the new and peculiar climate, food, and soil, to which it has been subjected.*

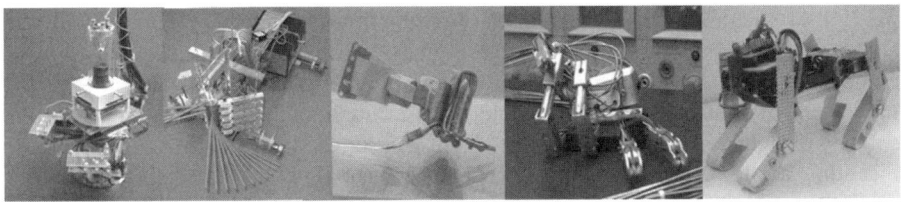

Photos, courtesy of AILAB

CREATURE, LEFT TO RIGHT	RESEARCHER(S)	LOCATION
amouse	Drs. Simon Bovet; Miriam Fend	St. Simir Island
Eyebot	Dr. Lukas Lichtensteiger	St. Lucia Island
Fish	Marc Ziegler	Marcus Cove
Humanoid Hand	Gabriel Gomez	Garcia Cove
Dog	Dr. Fumiya Iida	Fumiya Island

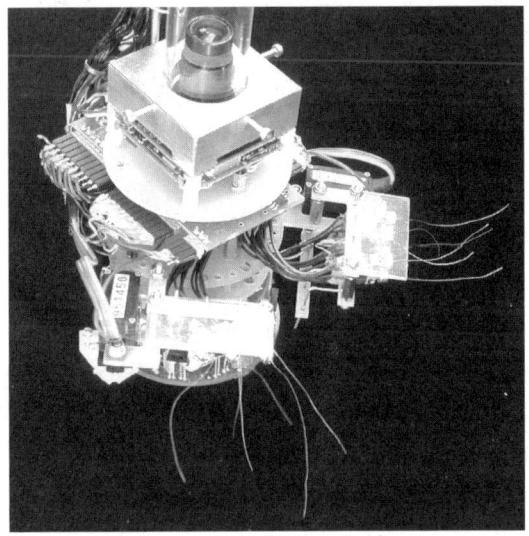

amouse, photo: Nathan Labhart
Researchers: Drs. Miriam Fend, Simon Bovet

The Amouse here is of particular interest because of the evolution of certain attributes of its whiskers which, in other species I have seen, such as other rodents, seals, opossums, and cats, tend to function as little more than accessories. These latter creatures, which I have observed in other natural habitats, make manifold use of their whiskers, such for hunting in murky water or darkness and detection of movement in air or water, but the detection resolution is usually extremely poor, and therefore unreliable for assurance of the creature's safety and comfort. In the instance of this Amouse, sensitivity is so acute, and of such a high order that we humans can only stand by and watch their operation in awe.

St Lucia Island. *The Beagle sailed* St Lucia Island, *and anchored in several bays. . . . The day was glowing hot, and the scrambling over the rough surface and through the intricate thickets, was very fatiguing; but I was well repaid by the strange Cyclopean scene. I* encountered a large and heavy creature moving in an extraordinarily straight path, low to the ground, whose very embodiment seemed purposed only to accommodate its sense of sight. No close analysis was required to see the rods and cones of its eye's structure, which were external, rather than internal, to its body. The movement of that eye seemed capable of discrimination; of seeking out particular sources of light formed in streams, to which it responded by reorganizing its rods so that the whole aspect resembled choreography of photosynthesis, the creature responding to light by saluting with this gesture of recognition.

Eyebot
Researcher: Lukas Lichtensteiger
Photo: Courtesy of AILAB

Indigenous Growth: Fumiya Island

Fumiya Island. *We doubled the south-west extremity of* Fumiya Island, *and were nearly becalmed.* The island was *covered with immense deluges of black naked lava, which have flowed over the rims of the great crater caldrons like pitch over the rim of* *a pot in which it has been boiled, or have burst forth from smaller orifices on the flanks and in their descent they have spread* over the entire island.

I will now turn to an order of mammal, which *gives the most striking character to the zoology of these islands. The species are numerous, and the numbers of individuals for each species are extraordinarily great* considering that they are confined only to this island. *I am referring to a number of creatures I think belong to the species of* Cynocephalus.

These animals appear fully-grown in varied sizes, from huge to what seems like a miniature size. Examining both the present day animal and the fossils found on the southward flank of the craters, one can really trace the evolutionary history. The larger specimens existing today are an anomaly. In reality through history, the stature of the creatures has generally gone from huge and dinosaur-like to tiny and mouse-like. It is evident that there are actually many levels of the evolutionary selection process ongoing here.

It seems that ancient generations were created out of one homogenous substance, with a skeletal structure that was hard and almost metallic in nature, while the materials constituting subsequent generations become extremely diverse in each specimen, the separate parts of the creature becoming smaller and smaller, enabling the size of the animal to diminish down through the generations to state, I believe, where they emerged as marine life out of the surrounding seas.

Fossil Traces

The island is completely devoid of moisture and it is difficult to discern how these creatures survive. Observing these dogs, it would seem they are a minimalist quadrupedic model of rapid locomotion inspired by some sort of biomechanical paradigm. Albeit, this sounds so artificial and reminiscent of something man-made, this is the only way I can explain how the animal forms its idiosyncratic system of rapid and robust legged locomotion. From my observations, I concluded that the locomotion is induced by spring-like like properties in the muscles of the animal, weight distribution, and body dimensions.

Passive Joints

Based on an anatomical study, we found with respect to the number of passive joints, dimensions of limbs, weight, and properties and locations of muscles, that the creature has evolved from a compromise between nature and a machine-like architecture to incarnate its body structure.

Upon dissection and analysis, it was discovered that the skeleton contains 28 passive joints, each of which has one passive rotational degree of freedom with each joint capable of small translational displacement as well. The passive joints intend to be controlled by a muscle actuation method which incorporates electric stimuli from the nervous system.

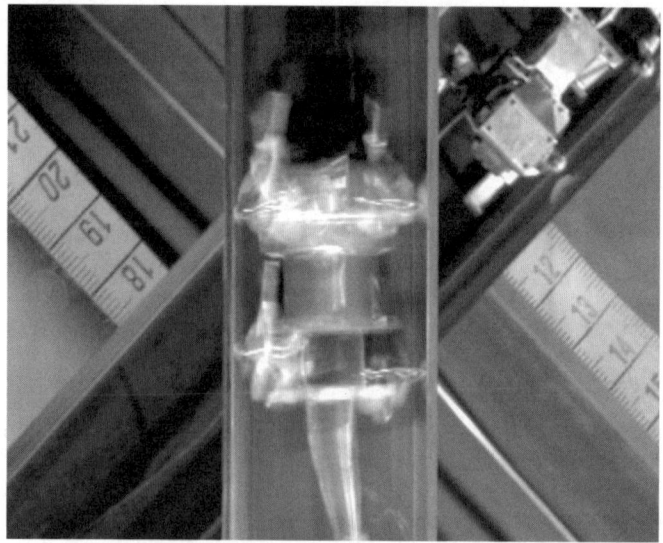

Dogs traveling to and from water sources
Still from the video "archipelago.ch", cinematography by Dr. Daniel Bisig

The dogs, *when thirsty, are obliged to travel fro*m one end of the island to the other, as there is a strange system of alternating currents in the rivers of this island. *Hence broad and well-beaten paths branch off in every direction from the wells down to the sea-coast. When I landed at* Fumiya Island, *I could not imagine what animal traveled so methodically along well-chosen tracks. Near the springs it was a curious spectacle to behold many of these creatures, one set eagerly traveling onwards with outstretched necks, and another set returning, after having drunk their fill.*

When one of them *arrives at a spring, quite regardless of any spectator, he buries his head in the water above his eyes, and greedily swallows great mouthfuls, at the rate of about ten in a minute.* The *animal stays three or four days in the neighborhood of the water, and then returns to the lower country; but they differed respecting the frequency of these visits. The animal probably regulates them according to the nature of the food on which it has lived. It is, however, certain, that these creatures can subsist even on these islands where there is no other water than what falls during a few rainy days in the year.*

The dogs, also have an inexplicable practice of sometimes running in place, as if on a treadmill, and it is remarkable to see this effort of locomotion when there is no discernable goal in sight. It also seems arbitrary when, at a certain time, different in each case, they suddenly stop moving, and stand perfectly still for hours on end. It is possible that their sense of smell is quite acute, and these spells of stationary running maybe be a reaction to some scent that cannot be tracked, but which requires a bravado performance on the part of the dog, to show that it can even outrun a scent which will always remain elusive.

Marcus Cove. In the evening we anchored in Marcus Cove. The next day, the water being unusually smooth, in some of the gullies and hollows there were beautiful red and other brightly colored fishes.

Their armature seemed to be constructed of oddly shaped bones, mostly flattened rectangular shales and they did not appear to exhibit any cartilaginous properties, nor do they seem to have any fat external to the bones. These fish are propelled through the water via their oscillation of their exoskeleton, the direction, speed and duration dependent upon the configuration of their tail or fins, and the lack of or preponderance of currents in the water.

Garcia Cove. [W]*hat can be more curious than that the hand of a man, formed for grasping, that of a mole for digging, the leg of the horse, the paddle of the porpoise, and the wing of the bat, should all be constructed on the same pattern, and should include the same bones, in the same relative positions?* [9]

4 Summary

Although the narrative for *archipelago.ch* remains emblematic of the themes of exploration and discovery of previously untouched territory, it also represents a future we, at this time, may not begin to fathom.

---

[9] Chapter 13 - Mutual Affinities of Organic Beings: Morphology: Embryology: Rudimentary Organs –
http://www.literature.org/authors/darwin-charles/the-origin-of-species/chapter-13.html

In a moment of time when science has obliterated science fiction, and we go forward at varied speeds towards phenomenal technological manifestations, an investment in methodologies that are interdisciplinary and involve more leaps of imagination furnished by associative narratives and expressive language will help us build more creative machines in inventive environments.

> "Scientific research is not only about solving problems, and what is more important is finding problems. . . . In particular, for the studies of autonomous adaptive systems, the research domain is very broad, where we need to look through the project from an evolutionary perspective. .... by projecting scientific projects onto Darwin's expedition, which both artists and scientists are interested in, we might be able to find a new way of "understanding" nature. . . "
>
> Fumiya Iida, Researcher,
> Artificial Intelligence Laboratory,
> University of Zurich

References

1. Agre, Philip, E.: The Soul Gained and Lost: Artificial Intelligence as a Philosophical Project, in a special issue of the Stanford Humanities Review, entitled "Constructions of the Mind: Artificial Intelligence and the Humanities". In: Guzeldere, G., Franchi, S. (eds.) The official citation is Stanford Humanities Review, vol. 4(2), pp. 1–19 (1995), http://www. stanford.edu/group/SHR/4-2/text/toc.html
2. Friedman, Block, J.: The Monstrous Races in Medieval Art and Thought. 2nd Rev Ed edn. Syracuse University Press (June 2000), ISBN 081568269
3. Goldberg, Ken (eds.): The Robot in the Garden: Telerobotics and Telepistemology in the Age of the Internet. The MIT Press, Cambridge (2001)
4. Goldberg, Ken, Siegwart, Robert (eds.): Beyond Webcams: An Introduction to Online Robots. The MIT Press, Cambridge (2001)
5. Harvey, P.D.A.: The Hereford World Map: Medieval World Maps and their Context, The British Library (November 30, 2006), ISBN: 0712347607
6. Williams, David: Deformed Discourse: The Function of the Monster in Mediaeval Thought and Literature, Mcgill Queens Univ. Pr (December 1999), ISBN 0773518711
7. Murray, Janet, H.: Hamlet on the Holodeck: The Future of Narrative in Cyberspace. The MIT Press (August 27, 1998), ISBN: 0262631873
8. Beyond Productivity: Information Technology, Innovation, and Creativity, National Research Council of the National Academies (2003), ISBN 0309088682
9. Manovich, Lev: The Language of New Media. Reprint edn. The MIT Press (March 7, 2002), ISBN 0262632551
10. Ryan, Marie-Laurie: Possible Worlds, Artificial Intelligence, and Narrative Theory, Indiana University of Bloomington and Indanapolis Press (1991), ISBN 0253350042
11. Wheeler, Gregory, R., Pereira, L.M.: A Note on Epistemology and Logical Artificial Intelligence. Journal of Applied Logic 2(4), 469–493 (2004)

Author Index

Printing: Mercedes-Druck, Berlin
Binding: Stein+Lehmann, Berlin

Lecture Notes in Artificial Intelligence (LNAI)